Design of Nonlinear Control Systems with the Highest Derivative in Feedback

SERIES ON STABILITY, VIBRATION AND CONTROL OF SYSTEMS

About the Series

Rapid developments in system dynamics and control, areas related to many other topics in applied mathematics, call for comprehensive presentations of current topics. This series contains textbooks, monographs, treatises, conference proceedings and a collection of thematically organized research or pedagogical articles addressing dynamical systems and control.

The material is ideal for a general scientific and engineering readership, and is also mathematically precise enough to be a useful reference for research specialists in mechanics and control, nonlinear dynamics, and in applied mathematics and physics.

Selected Volumes in Series A

Vol. 4 Asymptotic Methods in Buckling Theory of Elastic Shells
Authors: P. E. Tovstik and A. L. Smirinov

Vol. 5 Generalized Point Models in Structural Mechanics
Author: I. V. Andronov

Vol. 6 Mathematical Problems of Control Theory: An Introduction
Author: G. A. Leonov

Vol. 7 Analytical and Numerical Methods for Wave Propagation in Fluid Media
Author: K. Murawski

Vol. 8 Wave Processes in Solids with Microstructure
Author: V. I. Erofeyev

Vol. 9 Amplification of Nonlinear Strain Waves in Solids
Author: A. V. Porubov

Vol. 10 Spatial Control of Vibration: Theory and Experiments
Authors: S. O. Reza Moheimani, D. Halim, and A. J. Fleming

Vol. 11 Selected Topics in Vibrational Mechanics
Editor: I. Blekhman

Vol. 12 The Calculus of Variations and Functional Analysis: With Optimal Control and Applications in Mechanics
Authors: L. P. Lebedev and M. J. Cloud

Vol. 13 Multiparameter Stability Theory with Mechanical Applications
Authors: A. P. Seyranian and A. A. Mailybaev

Vol. 14 Stability of Stationary Sets in Control Systems with Discontinuous Nonlinearities
Authors: V. A. Yakubovich, G. A. Leonov and A. Kh. Gelig

Vol. 15 Process Modelling and Simulation with Finite Element Methods
Author: W. B. J. Zimmerman

SERIES ON STABILITY, VIBRATION AND CONTROL OF SYSTEMS

Series A **Volume 16**

Design of Nonlinear Control Systems with the Highest Derivative in Feedback

Valery D. Yurkevich

Concordia University, Canada

World Scientific

NEW JERSEY • LONDON • SINGAPORE • BEIJING • SHANGHAI • HONG KONG • TAIPEI • CHENNAI

Published by

World Scientific Publishing Co. Pte. Ltd.

5 Toh Tuck Link, Singapore 596224

USA office: 27 Warren Street, Suite 401-402, Hackensack, NJ 07601

UK office: 57 Shelton Street, Covent Garden, London WC2H 9HE

British Library Cataloguing-in-Publication Data
A catalogue record for this book is available from the British Library.

DESIGN OF NONLINEAR CONTROL SYSTEMS WITH THE HIGHEST DERIVATIVE IN FEEDBACK

ISBN 981-238-899-0

Printed by FuIsland Offset Printing (S) Pte Ltd, Singapore

To my wife,
Lyudmila

and to the memory of my sister and brother,
Natalya and *Peter*

Preface

Methods for the analysis and design of nonlinear control systems are growing rapidly. These developments are motivated by extensive applications, in particular, to such areas as mechatronic systems, robotics, and aircraft flight control systems. A number of new ideas, results, and approaches has appeared in this area during the past few decades.

This text was developed as a systematic explanation of one such new approach to control system design, which can provide effective control of nonlinear systems on the assumption of uncertainty. The approach is based on an application of a dynamical control law with the highest derivative of the output signal in the feedback loop. A distinctive feature of the control systems thus designed is that two-time-scale motions are forced in the closed-loop system. Stability conditions imposed on the fast and slow modes, and a sufficiently large mode separation rate, can ensure that the full-order closed-loop system achieves desired properties: the output transient performances are as desired, and they are insensitive to parameter variations and external disturbances.

A general design methodology for control systems with the highest derivative in feedback for continuous-time single-input single-output (SISO) or multi-input multi-output (MIMO) systems, as well as their discrete-time counterparts, is presented in this book. The method of singular perturbation is used to analyze the closed-loop system properties throughout.

The material is structured into thirteen chapters, the contents of which could be outlined as follows.

Chapter 1: Regularly and singularly perturbed systems. The main purpose of this chapter is a short explanation of some preliminary mathematical results concerning the properties and analysis of perturbed differential equations. The results constitute a background for an approximate analysis and design of nonlinear control systems under uncertainty.

Chapter 2: Design goal and reference model. The problem statement of output regulation for nonlinear time-varying control systems and the basic step response parameters are discussed. The model of the desired output behavior in the form of a desired differential equation is introduced; its parameters are selected based on the required output step response parameters (overshoot, settling time). Particularities of the reference model construction, in order to obtain the required system type, are also discussed.

Chapter 3: Methods of control system design under uncertainty. In this chapter a short overview of robust control synthesis techniques on the assumption of uncertainty is given. Main attention is devoted to discussion of nonadaptive approaches, in particular, to control systems with the highest derivative of the output signal and high gain in the feedback loop, control systems with state vector and high gain in the feedback loop, and control systems with sliding motions.

Chapter 4: Design of SISO continuous-time control systems. The problem of output regulation of SISO nonlinear time-varying control systems is discussed. The control system is designed to provide robust zero steady-state error of the reference input realization. Moreover, the controlled output transients should have some desired behavior. These transients should not depend on the external disturbances and varying parameters of the plant model. The insensitivity condition of the output transient behavior with respect to external disturbances and varying parameters of the system is introduced. The highest derivative in the feedback loop is used in proposed control law structures. The limit behavior of control systems with the highest derivative of the output signal in the feedback loop is discussed. Closed-loop system properties are investigated on the basis of the two-time-scale technique and, as a result, slow and fast motion subsystems are considered separately.

Chapter 5: Advanced design of SISO continuous-time control systems. Problems related to implementation of continuous-time control systems with the highest derivative in feedback are discussed. In particular, control accuracy and robustness of the control system, various design techniques for choosing controller parameters, the influence of high-frequency noisy measurements, and noise attenuation are considered.

Chapter 6: Influence of unmodeled dynamics. The peculiarities of SISO continuous-time control system design with the highest derivative in feedback are discussed on the assumption of uncertainty in the model description caused by unmodeled dynamics. These dynamics reflect errors on the system degree (or relative degree). Their influences, such as a pure time delay in the feedback loop and the unstructured uncertainties, lead to a plant model in the form of perturbed and/or singularly perturbed systems of differential equations. Some particulars of control design in the presence of a nonsmooth nonlinearity in the control loop are discussed as well.

Chapter 7: Realizability of desired output behavior. The conditions of realizability of the desired output behavior are discussed in this chapter. These are connected with invertibility conditions, nonlinear inverse dynamics solutions, and the problem of internal behavior analysis. Concepts such as invertibility index (relative degree), normal form of a nonlinear system, internal stability analysis, degenerated system on condition of output stabilization, and zero-dynamics are discussed. The design methodology for SISO control systems with the relative highest derivative in feedback is considered in the presence of internal dynamics. Finally, the problem of switching controller design is discussed.

Chapter 8: Design of MIMO continuous-time systems. The problem of output regulation of MIMO nonlinear time-varying control systems is discussed. Here the goals of control system design are to provide output decoupling and disturbance rejection, i.e., each output should be independently controlled by a single input, and to provide desired output transient performance indices on the assumption of incomplete information about varying parameters of the plant model and unknown external disturbances. The design methodology for SISO control systems with the highest derivative in feedback are extended to cover MIMO nonlinear time-varying control systems. The control law structure with the relative highest derivative in feedback is used in order to provide desired dynamical properties and decoupling of the output transients in a specified region of the system state space. The systematic design procedure for the control laws with the relative highest output derivatives is presented. The output regulation problem is discussed on the assumption that the previously presented realizability of the desired output behavior is satisfied.

Chapter 9: Stabilization of internal dynamics. This chapter is devoted to consideration of control system design where the dimension of the control vector is large, as the dimension of the output vector and redundant control variables are used in order to obtain internal dynamics stabilization. By this, the presented design methodology may be extended to more general system types. The discussed problem of internal dynamics stabilization for linear time-invariant systems corresponds to the displacement of zeroes of the transfer function in the left half of the complex plane.

Chapter 10: Digital controller design based on pseudo-continuous approach. The design of digital controllers for continuous nonlinear time-varying systems is discussed. The control task is formulated as a tracking problem for the output variables, where the desired decoupled output transients are attained on the assumption of incomplete information about varying parameters of the system and external disturbances. A distinguishing feature of the approach is that a pseudo-continuous-time model of the control loop with a pure time delay is used, where the delay is the result of a zero-order-hold transfer function approximation. The linear continuous-time controller with the relative highest output derivatives in feedback is designed, where the control law parameters are selected in accordance with the requirements placed on output control accuracy and damping of fast-motion transients. In particular, the selection of the sampling period is provided based on the requirement placed on the phase margin of the fast-motion subsystem. Then the Tustin transformation is applied to calculate the parameters of a digital controller. In order to increase the sampling period, a control law with compensation of the pure time delay is introduced.

Chapter 11: Design of discrete-time control systems. The method of discrete-time control systems design to provide the desired output transients is introduced, and is related with the purely discrete-time systems. In the case of continuous-time plants, the first step to be performed is discretization of the plant model. As a result, the discrete-time model of the plant in the form of a difference equation is used. A procedure to analyze the fast and slow motions in the discrete-time control system is given. It has been shown that if a sufficient time-scale separation between the fast and slow modes in the closed loop system and stability of the fast motions are provided, then after damping of the fast motions the output behavior in the closed loop system corresponds to the reference model and is insensitive to parameter variations of the plant and external disturbances. The design methodology is the discrete-time counterpart of the previously discussed approach to continuous-time control system design with the highest derivative in feedback.

Chapter 12: Design of sampled-data control systems. In this chapter, a design methodology for the discrete-time control system with two-time-scale motions is extended for the purpose of sampled-data control system design, by taking into account the particulars of the model of a series connection between a zero-order hold and a continuous-time system with high sampling rate. As a result, an approach to derive an approximate discrete-time model for nonlinear time-varying systems preceded by zero-order hold (ZOH) in the form of a difference equation with a small parameter is represented, where the small parameter depends on the sampling period. The design of SISO as well as MIMO sampled-data control systems is discussed.

Chapter 13: Design of control systems with distributed parameters. The main points of the extension of the previously presented methodology for control system design with the highest derivative in feedback for distributed parameter systems are highlighted, based on consideration of the parabolic-type system.

The book aims to disseminate new results in the area of control system design under uncertainty, and may be used as a course textbook. It contains numerous examples with simulation results, as well as assignments suitable for courses in nonlinear control system design. The core of the book is based on a translation of an earlier book [Yurkevich (2000a)] and lecture notes used by the author over the last ten years with students in the Automation and Computer Engineering Department at Novosibirsk State Technical University.

The design methodology may be useful for graduate and postgraduate students in the field of nonlinear control systems design. It will also be of interest to researchers, engineers, and university lecturers who are taking aim at real-time control system design in order to solve practical problems in the control of aircraft, robots, chemical reactors, and electrical and electromechanical systems.

Any comments about the book (including any errors noticed) can be sent to ⟨*yurkev@mail.ru*⟩ with the subject heading ⟨*book*⟩. They will be sincerely appreciated.

It is with great pleasure that I express gratitude to many colleagues who contributed to this book through useful discussions and helpful suggestions. My students and colleagues from the Automation Department of Novosibirsk State Technical University and, in particular, Professors G.A. Frantsuzova, O.Ya. Shpilevaya, and A.S. Vostrikov, have provided me with stimulating discussions of the subject. Professors A.L. Fradkov (Institute for Problems of Mechanical Engineering, Academy of Sciences of Russia), A.I. Rouban (Krasnoyarsk State Technical University), S.D. Zemlyakov (Institute of Control Sciences, Academy of Sciences of Russia), and N.D. Egupov (Kaluga Branch of Bauman Moscow State Technical University), offered many helpful suggestions and much moral support during my work. I would like to thank Professors M.J. Blachuta and K.W. Wojciechowski (Institute of Automatics, Silesian Technical University), with whom I have had the pleasure of working. Reviews of the book, along with many useful comments and pieces of advice were kindly provided by Professors J-P. Barbot (Equipe Commande des Systèmes) and L.M. Fridman (Universidad Nacional Avtonoma de Mexico) and were very much appreciated. I am grateful to Professor A. Guran (series Editor-in-Chief, Institute of Structronics of Canada) for the opportunity to publish this book, Professor M. Cloud (Lawrence Technological University) for editing the entire manuscript, and Mr. Yeow-Hwa Quek (World Scientific editorial staff) for assistance in the production of this book. Finally, I am grateful to Professors N. Esmail and K. Khorasani for many-sided and considerate support, and for accommodating me with the possibility of creative and fruitful work on the Faculty of Engineering & Computer Science at Concordia University.

Most of all, I would like to thank my wife, Lyudmila, for her love and moral support in my life and work.

Montreal, 2003 *Valery D. Yurkevich*

Contents

Chapter 1

Regularly and singularly perturbed systems

The main purpose of this chapter is to briefly explain some preliminary mathematical results concerning the properties and analysis of perturbed differential equations. These are used throughout the book as background for a technique of approximate analysis and design of nonlinear control systems. In particular, the main notions of two-time analysis, as well as the conditions for the stability of regularly and singularly perturbed differential equations, are introduced. Quantitative criteria for degree of time-scale separation between fast and slow motions are considered.

1.1 Regularly perturbed systems

1.1.1 *Nonlinear nominal system*

Let us consider an autonomous (time-invariant) dynamical system given by

$$\dot{X} = f(X) + \mu g(X), \tag{1.1}$$

where

X is the state of the system (1.1), $X \in \mathbb{R}^n$, $X = \{x_1, x_2, \ldots, x_n\}^T$;

f and g are continuous functions of X on Ω_X;

Ω_X is an open bounded subset of $\mathbb{R}^n$;

μ is a positive small parameter.

Taking $\mu = 0$ in (1.1) we obtain the system

$$\dot{X} = f(X), \tag{1.2}$$

which is called the nominal system. The system (1.1) is called a perturbation or perturbed system of the nominal system (1.2).

First, let us make some assumptions regarding the properties of the nominal system.

Let $0 \in \Omega_X \subset \mathbb{R}^n$ and let $X = 0$ be an equilibrium point of (1.2), i.e., $f(X)|_{X=0} = 0$. Let us assume that a Lyapunov function $V(X)$ exists such that the inequalities

$$c_1\|X\|^2 \leq V(X) \leq c_2\|X\|^2, \tag{1.3}$$

$$\dot{V}(X) = \frac{\partial V}{\partial X} f(X) \leq -c_3\|X\|^2, \tag{1.4}$$

$$\left\|\frac{\partial V}{\partial X}\right\| \leq c_4\|X\| \tag{1.5}$$

are satisfied for all $X \in \Omega_X$, where c_i are some positive constants and

$$\frac{\partial V}{\partial X} = \left\{\frac{\partial V}{\partial x_1}, \frac{\partial V}{\partial x_2}, \ldots, \frac{\partial V}{\partial x_n}\right\}$$

is a row vector.

From (1.3) the inequalities

$$\frac{V(X)}{c_2} \leq \|X\|^2 \leq \frac{V(X)}{c_1} \tag{1.6}$$

result. Then from (1.4) and (1.6) we have

$$\dot{V}(X) \leq -c_3\|X\|^2 \leq -\frac{c_3}{c_2} V(X). \tag{1.7}$$

Consequently

$$\int_0^t \frac{dV}{V} = \ln \frac{V(X(t))}{V(X(0))} \leq -\frac{c_3}{c_2} t$$

and

$$V(X(t)) \leq V(X(0)) \exp\left(-\frac{c_3}{c_2} t\right), \tag{1.8}$$

where t denotes the time variable.

In accordance with (1.3), (1.6), and (1.8), we have

$$\begin{aligned} \|X(t)\| &\leq \frac{V^{1/2}(X(t))}{c_1^{1/2}} \leq \frac{V^{1/2}(X(0))}{c_1^{1/2}} \exp\left(-\frac{c_3}{2c_2} t\right) \\ &\leq \left[\frac{c_2}{c_1}\right]^{1/2} \|X(0)\| \exp\left(-\frac{c_3}{2c_2} t\right). \end{aligned} \tag{1.9}$$

From (1.9) it follows that

$$\lim_{t \to \infty} X(t) = 0$$

and, moreover, that $X = 0$ is the exponentially stable equilibrium point of (1.2). The result may be formulated as a theorem.

Theorem 1.1 *Let $X = 0$ be an equilibrium point for system (1.2), and suppose the Lyapunov function $V(X)$ exists such that the conditions (1.3) and (1.4) are satisfied. Then the origin of the system (1.2) is exponentially stable.*

1.1.2 *Linear nominal system*

Let us consider a linear time-invariant (LTI) dynamical system of the form

$$\dot{X} = AX, \tag{1.10}$$

where

(i) A is an $n \times n$ real matrix;
(ii) $\det(A) \neq 0$ and so $X = 0$ is the isolated equilibrium point of (1.10);
(iii) $\operatorname{Re} \lambda_i(A) < 0, \ \forall\, i = 1, \ldots, n$ and so A is a stability matrix[1] (Hurwitz matrix.)

Let us consider a quadratic Lyapunov function

$$V(X) = X^T P X, \tag{1.11}$$

where P is a real symmetric positive definite matrix and P is the unique solution of the Lyapunov equation

$$PA + A^T P = -Q \tag{1.12}$$

for the given real symmetric positive definite matrix Q.

---

[1] $\operatorname{Re} \lambda_i(A)$ is the real part of the eigenvalue λ_i of A.

Then the inequalities (1.3), (1.4), (1.5) appropriate to (1.10), (1.11) may be rewritten in the following form:

$$\lambda_{\min}(P)\|X\|_2^2 \leq V(X) \leq \lambda_{\max}(P)\|X\|_2^2, \tag{1.13}$$

$$\begin{aligned} -\lambda_{\max}(Q)\|X\|_2^2 \leq \dot{V}(X) &= \frac{\partial V}{\partial X}AX \\ &= -X^TQX \leq -\lambda_{\min}(Q)\|X\|_2^2, \end{aligned} \tag{1.14}$$

$$\left\|\frac{\partial V}{\partial X}\right\|_2 = \|2X^TP\|_2 \leq 2\|P\|_2\|X\|_2 = 2\lambda_{\max}(P)\|X\|_2. \tag{1.15}$$

Consequently, the inequality (1.7) may be rewritten as

$$\dot{V}(X) \leq -\lambda_{\min}(Q)\|X\|_2^2 \leq -\frac{\lambda_{\min}(Q)}{\lambda_{\max}(P)}V(X). \tag{1.16}$$

Then from (1.16) an upper bound for the Lyapunov function $V(X)$ follows:

$$V(X(t)) \leq V(X(0)) \exp\left(-\frac{\lambda_{\min}(Q)}{\lambda_{\max}(P)}t\right).$$

Therefore, instead of (1.9), from the above we have

$$\|X(t)\|_2 \leq \left[\frac{\lambda_{\max}(P)}{\lambda_{\min}(P)}\right]^{1/2} \|X(0)\|_2 \exp\left(-\frac{\lambda_{\min}(Q)}{2\lambda_{\max}(P)}t\right) \tag{1.17}$$

as an upper bound for the norm of the function $X(t)$.

Note that the ratio $\lambda_{\min}(Q)/\lambda_{\max}(P)$ is maximized if $Q = I$ (see in [Patel and Toda (1980)], [Khalil (2002), p. 372]).

Similarly, from (1.13) and (1.14) it follows that

$$\dot{V}(X) \geq -\lambda_{\max}(Q)\|X\|_2^2 \geq -\frac{\lambda_{\max}(Q)}{\lambda_{\min}(P)}V(X)$$

is a lower bound for the derivative $\dot{V}(X)$ of the Lyapunov function $V(X)$ with respect to t, and hence

$$V(X(t)) \geq V(X(0)) \exp\left(-\frac{\lambda_{\max}(Q)}{\lambda_{\min}(P)}t\right) \tag{1.18}$$

is a lower bound for the Lyapunov function $V(X)$. Finally, from (1.13) and (1.18), we find that

$$\|X(t)\|_2 \geq \left[\frac{\lambda_{\min}(P)}{\lambda_{\max}(P)}\right]^{1/2} \|X(0)\|_2 \exp\left(-\frac{\lambda_{\max}(Q)}{2\lambda_{\min}(P)}t\right) \tag{1.19}$$

is a lower bound for the norm of the solution $X(t)$ of the linear system (1.10).

1.1.3 *Vanishing perturbation*

Let us consider the system (1.1), where it is assumed that the above assumptions regarding the function f are satisfied and, moreover, that g is an unknown continuous function of X on Ω_X and $c_5 > 0$ exists such that the condition

$$\|g(X)\| \le c_5 \|X\|, \quad \forall\, X \in \Omega_X \tag{1.20}$$

holds.

From (1.20) follows that $g(X)|_{X=0} = 0$, and so the perturbation vanishes completely at the equilibrium point.

Obviously, the time derivative of $V(X)$ along trajectories of (1.1) is given by

$$\dot{V}(X) = \frac{\partial V}{\partial X} f(X) + \mu \frac{\partial V}{\partial X} g(X). \tag{1.21}$$

In accordance with the above assumption, (1.4)–(1.5), and (1.20), it is easy to see that

$$\dot{V}(X) \le -c_3 \|X\|^2 + \mu \left\| \frac{\partial V}{\partial X} \right\| \|g(X)\| \le -(c_3 - \mu c_4 c_5)\|X\|^2. \tag{1.22}$$

As a result, if the inequality

$$0 < \mu < \frac{c_3}{c_4 c_5} \tag{1.23}$$

is satisfied, then

$$c_3 - \mu c_4 c_5 > 0$$

and, accordingly, we have

$$\dot{V}(X) < 0, \quad \forall\, X \neq 0, \quad \forall X \in \Omega_X. \tag{1.24}$$

From (1.3), (1.22), and (1.24) it follows that the origin is an exponentially stable equilibrium point of the perturbed system (1.1) if the parameter μ is small enough. So the result may be formulated as a theorem.

Theorem 1.2 *Let the origin of the nominal system (1.2) be an exponentially stable equilibrium point, and suppose the requirement (1.20) for the* *continuous function g holds. Then there exists $\mu^\star > 0$ such that for all $\mu \in (0, \mu^\star)$, the origin of perturbed system (1.1) is exponentially stable.*

Instead of (1.1), let us consider the system given by

$$\dot{X} = AX + \mu g(X), \tag{1.25}$$

which is the perturbation of the stable linear nominal system (1.10). Then the inequality (1.22) appropriate to (1.25) may be rewritten as

$$\dot{V}(X) \leq -\lambda_{\min}(Q)\|X\|_2^2 + 2\mu\lambda_{\max}(P)c_5\|X\|_2^2$$

and, from (1.23), the inequality

$$0 < \mu < \frac{\lambda_{\min}(Q)}{2\lambda_{\max}(P)c_5}$$

follows, where P is the solution of the Lyapunov equation (1.12).

1.1.4 *Nonvanishing perturbation*

Instead of (1.1), let us consider the perturbed system given by

$$\dot{X} = f(X) + \mu g(X, w), \tag{1.26}$$

where the above assumptions regarding the function f are satisfied and

(i) g is an unknown continuous function of X on Ω_X and of w on Ω_w;
(ii) w serves to represent a vector of external disturbances and varying parameters;
(iii) $w \in \Omega_w$, where Ω_w is a bounded subset of $\mathbb{R}^l$.

When we refer to a nonvanishing perturbation of the nominal system (1.2), we have in mind that

$$\exists\, \bar{w} \in \Omega_w \mid g(X = 0, \bar{w}) \neq 0 \tag{1.27}$$

and that a positive constant c_6 exists such that

$$\|g(X, w)\| \leq c_6, \quad \forall\, X \in \Omega_X \text{ and } \forall\, w \in \Omega_w.$$

Then, in accordance with (1.26), (1.4), and (1.5), the time derivative of $V(X)$ along trajectories of (1.26) can be found using the chain rule. It is given by

$$\dot{V}(X) = \frac{\partial V}{\partial X} f(X) + \mu \frac{\partial V}{\partial X} g(X, w) \le -c_3 \|X\|^2 + \mu c_4 c_6 \|X\|$$
$$= -(1-d)c_3 \|X\|^2 + (\mu c_4 c_6 - d c_3 \|X\|)\|X\|.$$

Let $0 < d < 1$. If the inequality

$$\|X\| \ge \frac{\mu c_4 c_6}{d c_3}$$

is satisfied, then

$$\dot{V}(X) \le -(1-d)c_3 \|X\|^2. \tag{1.28}$$

Hence, some finite time t_1 exists such that the condition (1.28) holds for all $t \in [0, t_1)$. Therefore, similar to (1.9), we have an upper bound for $\|X(t)\|$ on this finite time interval given by

$$\|X(t)\| \le \left[\frac{c_2}{c_1}\right]^{1/2} \|X(0)\| e^{-\frac{(1-d)c_3}{2c_2} t}, \quad \forall\, 0 \le t < t_1$$

and an upper bound for $\|X(t)\|$ on infinity defined by the inequality

$$\|X(t)\| \le \frac{\mu c_4 c_6}{d c_3}, \quad \forall\, t \ge t_1. \tag{1.29}$$

So in the presence of the nonvanishing bounded perturbation discussed above, the solutions of (1.26) are ultimately bounded with an ultimate bound (1.29) that approaches zero as $\mu \to 0$.

Let us reconsider the perturbed system of the form (1.25) with the linear nominal model (1.10) and in the presence of the nonvanishing bounded perturbation (1.27). Then from (1.15) and (1.16) it follows that the inequality (1.29) may be rewritten as

$$\|X(t)\|_2 \le \frac{2\mu \lambda_{\max}(P) c_6}{d \lambda_{\min}(Q)}, \quad \forall\, t \ge t_1.$$

1.2 Singularly perturbed systems

1.2.1 *Singular perturbation*

Let us consider the following set of differential equations:

$$\dot{X} = f(X, Z), \quad X(0) = X^0, \tag{1.30}$$
$$\mu \dot{Z} = g(X, Z), \quad Z(0) = Z^0, \tag{1.31}$$

where μ is a small positive parameter, $X \in \mathbb{R}^n$, $Z \in \mathbb{R}^m$, and f and g are continuously differentiable functions of X and Z.

The system (1.30)–(1.31) is called the standard singular perturbation model of a finite-dimensional dynamical system.

Let us release Z from the initial condition; then, with $\mu = 0$, the system (1.30)–(1.31) of dimension $n + m$ degenerates into

$$\dot{X} = f(X, Z), \quad X(0) = X^0, \tag{1.32}$$

$$0 = g(X, Z), \tag{1.33}$$

where the system (1.32)–(1.33) has dimension n.

In accordance with the implicit function theorem, assume that

$$\det\left\{\frac{\partial g(X, Z)}{\partial Z}\right\} \neq 0, \quad \forall\, Z \in \Omega_Z; \tag{1.34}$$

then a function

$$\bar{Z} = h(X) \tag{1.35}$$

exists such that the function (1.35) is an unique solution of the equation $g(X, \bar{Z}) = 0$. Accordingly, the equality

$$g(X, h(X)) = 0, \quad \forall\, X \in \Omega_X$$

holds.

Then the set

$$M = \{(X, Z) \mid g(X, Z) = 0\} \tag{1.36}$$

is an n dimensional manifold in the original $n + m$ dimensional state space and, in accordance with (1.32) and (1.33), the behavior of $X(t)$ on this manifold is described by the reduced system

$$\dot{X} = f(X, h(X)), \quad X(0) = X^0. \tag{1.37}$$

1.2.2 *Two-time-scale motions*

If a pair of functions $X(t), Z(t)$ is such that

$$g(X(t), Z(t)) = 0, \ \forall\, t \geq 0$$

then the equality

$$dg(X(t), Z(t))/dt = 0, \ \forall\, t \geq 0$$

also holds. Accordingly, we have

$$\frac{\partial g}{\partial X}\dot{X} + \frac{\partial g}{\partial Z}\dot{Z} = 0 \tag{1.38}$$

and from (1.34) and (1.38) it follows that the behavior of $Z(t)$ on the manifold (1.36) is described by the equation

$$\dot{Z} = -\left\{\frac{\partial g}{\partial Z}\right\}^{-1} \frac{\partial g}{\partial X}\dot{X}.$$

It follows that on the manifold M the ratio

$$\frac{\|\dot{Z}\|}{\|\dot{X}\|} \leq \left\| \left\{\frac{\partial g}{\partial Z}\right\}^{-1} \frac{\partial g}{\partial X} \right\|$$

is some regular numerical value that depends only on the functions f, g.

At the same time, in accordance with the system of equations (1.30) and (1.31), we find that at an arbitrary point $(X, Z) \notin M$ of the $n + m$ dimensional state space this ratio is given by

$$\frac{\|\dot{Z}\|}{\|\dot{X}\|} = \frac{1}{\mu}\frac{\|g(X,Z)\|}{\|f(X,Z)\|}$$

and depends on the small parameter μ. So if $\mu \to 0$, then beyond the manifold M two-time-scale motions appear in the solutions of the equations (1.30)–(1.31), where Z is a fast changing variable and X is a slow changing variable as shown in Fig. 1.1.

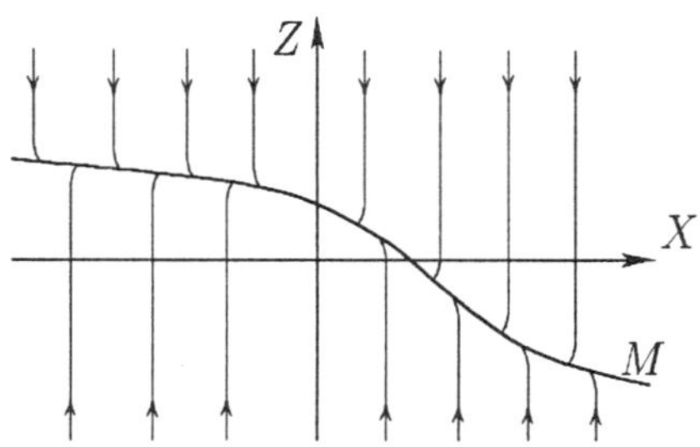

Fig. 1.1 Typical phase portrait in the case of a singularly perturbed system.

1.2.3 Boundary-layer system

Let us introduce a new variable $Y = Z - h(X)$, where Y is a deviation of Z from manifold (1.36). Then the equations (1.30)–(1.31) may be rewritten in the form

$$\frac{dX}{dt} = f(X, Y + h(X)), \quad X(0) = X^0, \tag{1.39}$$

$$\mu\frac{dY}{dt} = g(X, Y + h(X)) - \mu\frac{\partial h}{\partial X}f(X, Y + h(X)),\ Y(0) = Y^0, \tag{1.40}$$

where $Y(0) = Z(0) - h(X(0))$. After introducing a new time scale $t_0 = t/\mu$ into (1.39)–(1.40), we have

$$\frac{dX}{dt_0} = \mu f(X, Y + h(X)), \quad X(0) = X^0, \tag{1.41}$$

$$\frac{dY}{dt_0} = g(X, Y + h(X)) - \mu\frac{\partial h}{\partial X}f(X, Y + h(X)),\ Y(0) = Y^0. \tag{1.42}$$

From (1.41)–(1.42) it is easy to see that in the new time scale t_0 we have $dX/dt_0 \to 0$; that is, the rate of transients of $X(t)$ decreases as $\mu \to 0$. As a result, if μ tends to zero then from (1.41)–(1.42) the equation of a boundary-layer system

$$\frac{dY}{dt_0} = g(X, Y + h(X)), \quad Y(0) = Y^0 \tag{1.43}$$

follows as an asymptotic limit, where X is the frozen variable, i.e., $X \approx$ const.

1.2.4 Stability analysis

The investigation of conditions under which the trajectories of the full singularly perturbed system (1.30)–(1.31) approximate to the trajectories of the reduced model (1.37) is important both from a theoretical viewpoint and for practical applications in control system analysis and design. These conditions were considered in [Tikhonov (1948); Tikhonov (1952)] and [Vasileva (1963)] for a bounded time interval $t \in [0, t_1]$, and then in [Krasovskii (1963); Klimushchev and Krasovskii (1962)] and [Hoppensteadt (1966)] for an infinite time interval $t \in [0, \infty)$.

The simplified version of stability analysis of the singularly perturbed systems is provided below, while more detailed analysis may be found, for instance, in [Khalil (2002)].

Consider the singularly perturbed system (1.30)–(1.31)

$$\dot{X} = f(X, Z), \quad X(0) = X^0,$$
$$\mu \dot{Z} = g(X, Z), \quad Z(0) = Z^0,$$

where the following assumptions are satisfied:

- $f(0,0) = 0, \quad g(0,0) = 0.$
- The equation $g(X, Z) = 0$ has an unique isolated root $\bar{Z} = h(X)$ such that $h(0) = 0$ and

$$\|h(X)\| \leq m_1 \|X\|, \ \forall \ X \in B_{\rho_x} = \{X \in \mathbb{R}^n \mid \|X\| \leq \rho_x\}, \ m_1 > 0.$$

- The functions f, g, and h, along with their partial derivatives up to order 2, are bounded for all $Y = Z - h(X) \in B_{\rho_y}$, where

$$B_{\rho_y} = \{Y \in \mathbb{R}^m \mid \|Y\| \leq \rho_y\}.$$

In addition, we assume that a Lyapunov function $V(X)$ of the reduced system (1.37) exists such that

$$c_1 \|X\|^2 \leq V(X) \leq c_2 \|X\|^2, \tag{1.44}$$
$$\frac{d}{dt} V(X) = \frac{\partial V}{\partial X} f(X, h(X)) \leq -c_3 \|X\|^2, \tag{1.45}$$
$$\left\| \frac{\partial V}{\partial X} \right\| \leq c_4 \|X\|, \tag{1.46}$$

for all $X \in B_{\rho_x}$, where c_i are some positive constants. Therefore, in accordance with Theorem 1.1, the origin of the reduced system (1.37) is exponentially stable.

By introducing the new variable $Y = Z - h(X)$, let us rewrite equations (1.30)–(1.31) in the form (1.39)–(1.40) and consider the boundary-layer system (1.43). Assume that a Lyapunov function $W(Y)$ of (1.43) exists such that

$$b_1 \|Y\|^2 \leq W(Y) \leq b_2 \|Y\|^2, \tag{1.47}$$
$$\frac{d}{dt_0} W(Y) = \frac{\partial W}{\partial Y} g(X, Y + h(X)) \leq -b_3 \|Y\|^2, \tag{1.48}$$
$$\left\| \frac{\partial W}{\partial Y} \right\| \leq b_4 \|Y\|, \tag{1.49}$$

for all $Y \in B_{\rho_y}$, where the b_i are positive constants. Hence, by Theorem 1.1, the origin of the boundary-layer system (1.43) is exponentially stable.

Let us consider the function

$$\nu(X,Y) = (1-d)V(X) + dW(Y) \tag{1.50}$$

as a Lyapunov function candidate for the singularly perturbed system (1.39)–(1.40), where $0 < d < 1$. Then the derivative of (1.50) along the trajectories of (1.39)–(1.40) is given by

$$\frac{d\nu}{dt} = (1-d)\frac{\partial V}{\partial X} f(X, Y + h(X))$$
$$+\frac{d}{\mu}\frac{\partial W}{\partial Y} g(X, Y + h(X)) - d\frac{\partial W}{\partial Y}\frac{\partial h}{\partial X} f(X, Y + h(X)). \tag{1.51}$$

Because the function f and its partial derivatives up to order 2 are bounded for all $Y \in B_{\rho_y}$, and because $f(0,0) = 0$, the Taylor expansion of $f(X, Y + h(X))$ yields

$$f(X, Y + h(X)) = f(X, h(X)) + \frac{\partial f}{\partial Y} Y + O(\|Y\|^2),$$

where

$$\|f(X, Y + h(X))\| \leq l_0\|X\|, \quad \left\|\frac{\partial f}{\partial Y}\right\| \leq l_1, \quad \|O(\|Y\|^2)\| \leq l_2\|Y\|^2,$$

and the l_i are some positive constants.

By taking into account the above assumptions, we obtain from (1.51) the inequality

$$\frac{d\nu}{dt} \leq -(1-d)c_3\|X\|^2 + [(1-d)c_4 l_1 + d b_4 m_1 l_0]\|X\|\|Y\|$$
$$+\left[(1-d)c_4 l_2\|X\| + d b_4 m_1 l_1 + d b_4 m_1 l_2\|Y\| - \frac{d}{\mu} b_3\right]\|Y\|^2. \tag{1.52}$$

Then (1.52) can be represented as

$$\frac{d\nu}{dt} \leq -\eta^T \Gamma \eta,$$

where $\eta = \{\|X\|, \|Y\|\}^T$ and

$$\Gamma = \begin{bmatrix} \gamma_{11} & \gamma_{12} \\ \gamma_{21} & \gamma_{22} \end{bmatrix}$$

in which

$$\begin{aligned}\gamma_{11} &= (1-d)c_3,\\ \gamma_{12} &= \gamma_{21} = -0.5[(1-d)c_4 l_1 + d b_4 m_1 l_0],\\ \gamma_{22} &= \frac{d}{\mu} b_3 - (1-d)c_4 l_2\|X\| - d b_4 m_1 l_1 - d b_4 m_1 l_2 \|Y\|.\end{aligned}$$

Since $X \in B_{\rho_x}$, $Y \in B_{\rho_y}$, and $0 < d < 1$, there exists some small $\mu = \mu^\star > 0$ such that the matrix Γ is positive definite:

$$\Gamma > 0.$$

Then

$$\lambda_{\min}(\Gamma)\|\eta\|_2^2 \leq \eta^T \Gamma \eta \leq \lambda_{\max}(\Gamma)\|\eta\|_2^2.$$

From (1.44) and (1.47) it follows that some constants d_1, d_2 exist such that

$$d_1\|\eta\|^2 \leq \nu(\eta) \leq d_2\|\eta\|^2.$$

As a result we have

$$\frac{d\nu}{dt} \leq -\eta^T\Gamma\eta \leq -\lambda_{\min}(\Gamma)\|\eta\|_2^2 \leq -\frac{\lambda_{\min}(\Gamma)}{d_2}\nu.$$

Hence

$$\nu(X(t), Y(t)) \leq \nu(X(0), Y(0)) \exp\left(-\frac{\lambda_{\min}(\Gamma)}{d_2} t\right)$$

and, accordingly, we obtain

$$\|\eta(t)\| \leq \left[\frac{d_2}{d_1}\right]^{1/2} \|\eta(0)\| \exp\left(-\frac{\lambda_{\min}(\Gamma)}{2d_2} t\right).$$

Note that $\|h(X)\| \leq m_1\|X\|$ and $Y = Z - h(X)$; then there exists $m > 0$ such that

$$\|\hat{\eta}(t)\| \leq m\|\hat{\eta}(0)\| \exp\left(-\frac{\lambda_{\min}(\Gamma)}{2d_2} t\right),$$

where $\hat{\eta} = \{\|X\|, \|Z\|\}^T$ or, in other words, the origin of (1.30)–(1.31) is exponentially stable. The result may be formulated as the following theorem.

Theorem 1.3 *Consider the singularly perturbed system (1.30)–(1.31)*

$$\dot{X} = f(X, Z), \quad X(0) = X^0,$$
$$\mu\dot{Z} = g(X, Z), \quad Z(0) = Z^0,$$

under the following assumptions.

- $f(0,0) = 0, \quad g(0,0) = 0.$
- *The equation* $g(X, Z) = 0$ *has a unique isolated root* $\bar{Z} = h(X)$ *such that* $h(0) = 0$ *and* $\|h(X)\| \leq m_1\|X\|$, *where*

$$m_1 > 0, \ X \in B_{\rho_x}, \ B_{\rho_x} = \{X \in \mathbb{R}^n \mid \|X\| \leq \rho_x\}.$$

- *The functions* f, g, h *and their partial derivatives up to order 2 are bounded for all* $Y = Z - h(X) \in B_{\rho_y}$, *where*

$$Y \in B_{\rho_y}, \ B_{\rho_y} = \{Y \in \mathbb{R}^m \mid \|Y\| \leq \rho_y\}.$$

- *The Lyapunov function* $V(X)$ *of the reduced system (1.37) exists such that (1.44)–(1.46) are satisfied for all* $X \in B_{\rho_x}$.
- *The Lyapunov function* $W(Y)$ *of the boundary-layer system (1.43) exists such that (1.47)–(1.49) are satisfied for all* $Y \in B_{\rho_y}$.

Then there exists $\mu^\star > 0$ *such that for all* $\mu \in (0, \mu^\star)$, *the origin of (1.30)–(1.31) is exponentially stable.*

1.2.5 *Fast and slow-motion subsystems*

The above procedure for obtaining the boundary-layer system may be directly applied to (1.30)–(1.31) in order to obtain equations of fast-motion subsystem (FMS) and slow-motion subsystem (SMS). First, by introducing the new time scale $t_0 = t/\mu$ into (1.30)–(1.31) we have

$$\frac{dX}{dt_0} = \mu f(X, Z), \quad X(0) = X^0,$$
$$\frac{dZ}{dt_0} = g(X, Z), \quad Z(0) = Z^0,$$

where the FMS is given by

$$\frac{dZ}{dt_0} = g(X, Z), \quad Z(0) = Z^0 \tag{1.53}$$

in the new time scale t_0 and $X(t_0) \approx$ const during the transients in the subsystem (1.53).

By returning to the primary time scale t, from (1.53) the FMS equation

$$\mu \frac{dZ}{dt} = g(X, Z), \quad Z(0) = Z^0 \tag{1.54}$$

is obtained, where $X(t)$ is the frozen variable, i.e., $X(t) \approx$ const.

Second, let us assume that there is an unique equilibrium point (1.35) of (1.54) (more precisely, quasi-equilibrium point) that satisfies $g(X, Z) = 0$. Moreover, we assume that $Z = \bar{Z}$ is an exponentially stable equilibrium point of (1.54).

Finally, on the above assumption of exponential stability of the equilibrium point $Z = \bar{Z}$, we have that $Z(t) - \bar{Z} \to 0$, $\forall\ t > 0$ as $\mu \to 0$. So if the parameter μ is small enough, then after rapid decay of transients in the FMS (1.54) we find that the condition $Z = \bar{Z}$ is satisfied. Substitution of $Z = h(X)$ into (1.30) yields the SMS equation (1.37).

1.2.6 *Degree of time-scale separation*

Let us consider a linear standard singularly perturbed system

$$\dot{X} = A_{11}X + A_{12}Y, \tag{1.55}$$

$$\mu\dot{Y} = A_{21}X + A_{22}Y, \tag{1.56}$$

where μ is a small positive parameter, $X \in \mathbb{R}^n$, $Y \in \mathbb{R}^m$, and the A_{ij} are matrices with appropriate dimensions.

In accordance with the above formal algorithm of time-scale separation, we have that

$$\mu\dot{Y} = A_{21}X + A_{22}Y \tag{1.57}$$

is the FMS equation, where $X =$ const.

Assume that $\det A_{22} \neq 0$ and, moreover, A_{22} is a Hurwitz matrix. Then it is easy to find that

$$\dot{X} = A_s X \tag{1.58}$$

is the SMS equation, where

$$A_s = A_{11} - A_{12}A_{22}^{-1}A_{21}$$

and we assume that A_s is a Hurwitz matrix as well.

From a practical standpoint it is useful to have some quantitative criteria for the degree of time-scale separation between stable fast and slow motions.

The ratio

$$\eta = t_{s,SMS}/t_{s,FMS} \tag{1.59}$$

serves as a direct estimation of such a degree, where $t_{s,SMS}$ and $t_{s,FMS}$ are the settling times of the SMS and the FMS, respectively. We may also consider indirect estimates of the degree of time-scale separation between stable fast and slow motions, where the first estimate is based on solution of the Lyapunov equation and the second one is based on roots of the FMS and SMS characteristic polynomials.

Estimate based on solution of Lyapunov equation

The lower and upper bounds (1.17) and (1.19) of the linear differential equation solution may be used to introduce a quantitative criterion for degree of time-scale separation between stable fast and slow motions.

First, because the FMS (1.57) is stable, the Lyapunov function $V_F(Y) = Y^T P_F Y$ of the FMS (1.57) may be obtained by solving the Lyapunov equation

$$P_F A_{22} + A_{22}^T P_F = -Q_F, \tag{1.60}$$

where $Q_F = Q_F^T$, $Q_F > 0$, $P_F = P_F^T$, and $P_F > 0$. Then, in accordance with (1.17), the upper bound of the fast variable Y follows:

$$\|Y(t)\|_2 \leq \left[\frac{\lambda_{\max}(P_F)}{\lambda_{\min}(P_F)}\right]^{1/2} \|Y(0)\|_2 \exp\left(-\frac{\lambda_{\min}(Q_F)}{2\mu\lambda_{\max}(P_F)}t\right). \tag{1.61}$$

Next, since the SMS (1.58) is stable, the Lyapunov function $V_S(X) = X^T P_S X$ of the SMS (1.58) may be obtained by solving the Lyapunov equation

$$P_S A_S + A_S^T P_S = -Q_S,$$

where $Q_S = Q_S^T$, $Q_S > 0$, and $P_S = P_S^T, P_S > 0$. Then, in accordance with (1.19), the lower bound of the slow variable X follows:

$$\|X(t)\|_2 \geq \left[\frac{\lambda_{\min}(P_S)}{\lambda_{\max}(P_S)}\right]^{1/2} \|X(0)\|_2 \exp\left(-\frac{\lambda_{\max}(Q_S)}{2\lambda_{\min}(P_S)}t\right). \tag{1.62}$$

The ratio of the exponents in (1.61) and (1.62),

$$\eta_1 = \frac{\lambda_{\min}(P_S)\lambda_{\min}(Q_F)}{\mu\lambda_{\max}(P_F)\lambda_{\max}(Q_S)},$$

may be used as the degree of time-scale separation between fast and slow motions. In particular, by choosing $Q_F = Q_S = I$ we obtain[2]

$$\eta_1 = \frac{\lambda_{\min}(P_S)}{\mu \lambda_{\max}(P_F)}. \tag{1.63}$$

Estimate based on roots of FMS and SMS characteristic polynomials

Let us consider stable fast and slow subsystems of the linear standard singularly perturbed system (1.55)–(1.56). Denote by

$$A_{FMS}(s,\mu) = \det\left[sI - \frac{1}{\mu}A_{22}\right] = s^m + a_{m-1}^{FMS}\frac{1}{\mu}s^{m-1} + \ldots + \frac{1}{\mu^m}a_0^{FMS}$$

and

$$A_{SMS}(s) = \det[sI - A_S] = s^n + a_{n-1}^{SMS}s^{n-1} + \ldots + a_0^{SMS}$$

the FMS and SMS characteristic polynomials, respectively. Assume that $s_1^{FMS}, \ldots, s_m^{FMS}$ and $s_1^{SMS}, \ldots, s_n^{SMS}$ are the roots of the stable FMS and SMS characteristic polynomials, respectively. Denote

$$\omega_{FMS}^{\min} = \min_{i=1,\ldots,m} |\mathrm{Re}\, s_i^{FMS}|, \quad \omega_{SMS}^{\max} = \max_{i=1,\ldots,n} |\mathrm{Re}\, s_i^{SMS}|.$$

The ratio

$$\eta_2 = \frac{\omega_{FMS}^{\min}}{\omega_{SMS}^{\max}} \tag{1.64}$$

may be used as a criterion for the degree of time-scale separation between fast and slow motions.

We may also consider the ratio of FMS natural frequency to SMS natural frequency

$$\eta_3 = \frac{(a_0^{FMS})^{1/m}}{\mu (a_0^{SMS})^{1/n}} \tag{1.65}$$

as a quantitative criterion for the degree of time-scale separation between stable fast and slow motions instead of (1.63).

The estimate η_1 is more conservative than η_2 and η_3. The direct estimation (1.59) of the degree of time-scale separation between stable fast and

[2] Here I is the identity matrix. A list of notation used throughout the book appears in Appendix B starting on p. 335.

slow motions can be based on the correlations (2.4) discussed in the next chapter. From (2.4) we get

$$t_{s,FMS} \approx \frac{4}{\omega_{FMS}^{\min}}, \quad t_{s,SMS} \geq \frac{4}{\omega_{SMS}^{\max}}.$$

Therefore, by (1.59), we obtain

$$\eta \geq \frac{\omega_{FMS}^{\min}}{\omega_{SMS}^{\max}}.$$

1.3 Discrete-time singularly perturbed systems

1.3.1 *Fast and slow-motion subsystems*

In this section the discrete-time counterpart of the singularly perturbed system (1.31) is discussed. We will deal with the system of state space difference equations given by

$$X_{k+1} = \{I_n + \mu A_{11}\}X_k + \mu A_{12}Y_k, \tag{1.66}$$

$$Y_{k+1} = A_{21}X_k + A_{22}Y_k, \tag{1.67}$$

where μ is a small parameter, $X \in \mathbb{R}^n$, $Y \in \mathbb{R}^m$, and the A_{ij} are matrices with appropriate dimensions.

When $\mu = 0$, the system (1.66)–(1.67) of dimension $n + m$ degenerates into the system of dimension m given by

$$X_{k+1} = X_k,$$
$$Y_{k+1} = A_{21}X_k + A_{22}Y_k.$$

So if $\mu \to 0$, then the rate of transients of X_k decreases and, accordingly, the fast and slow modes are revealed in the system (1.66)–(1.67), where a time-scale separation between those modes is represented by the small parameter μ. If μ is sufficiently small, then from (1.66)–(1.67) the FMS equation

$$Y_{k+1} = A_{21}X_k + A_{22}Y_k \tag{1.68}$$

results, where $X_k \approx \text{const}$ during the transients in the system (1.68).

The characteristic polynomial of the FMS (1.68) is

$$A_{FMS}(z) = \det(zI_m - A_{22}).$$

Assume that all roots of $A_{FMS}(z)$ lie inside the unit circle so that the FMS (1.68) is stable. Then the steady-state of the FMS is given by

$$Y_k = \{I_m - A_{22}\}^{-1} A_{21} X_k. \tag{1.69}$$

Substitution of (1.69) into (1.66) yields the SMS

$$X_{k+1} = \{I_n + \mu[A_{11} + A_{12}(I_m - A_{22})^{-1} A_{21}]\} X_k,$$

where the characteristic polynomial of the SMS is

$$A_{SMS}(z) = \det(zI_m - A_{SMS}),$$

where

$$A_{SMS} = \{I_n + \mu[A_{11} + A_{12}(I_m - A_{22})^{-1} A_{21}]\}.$$

1.3.2 *Degree of time-scale separation*

Since the complex variables z and s are related by $z = e^{T_s s}$, the inverse mapping of the unit circle into the primary strip in the s-plane is given by

$$s = \frac{1}{T_s} \operatorname{Ln} z, \tag{1.70}$$

where T_s is the sampling period and $\operatorname{Ln} z$ is the principal value of $\ln z$. Here $z = 0$ is omitted and there is a cut along the negative real axis.

Assume that the following conditions are satisfied:

1. $z_1^{FMS}, \ldots, z_m^{FMS}$ and $z_1^{SMS}, \ldots, z_n^{SMS}$ are the roots of the FMS and SMS characteristic polynomials, respectively.
2. All roots lie inside the unit circle as shown in Fig. 1.2(a).
3. There are no roots on the cut or at the origin.

Then, by the mapping (1.70), we can obtain the sets of roots $s_1^{FMS}, \ldots, s_m^{FMS}$ and $s_1^{SMS}, \ldots, s_n^{SMS}$ as shown in Fig. 1.2(b), and construct two polynomials

$$A_{FMS}(s) = \prod_{i=1}^{m} (s - s_i^{FMS}), \qquad A_{SMS}(s) = \prod_{i=1}^{n} (s - s_i^{SMS}),$$

to which the previous criteria can be applied.

If we assume that the last mentioned condition is not satisfied, that is, there is at least one root on the cut or at the origin, then the following approach may be used.

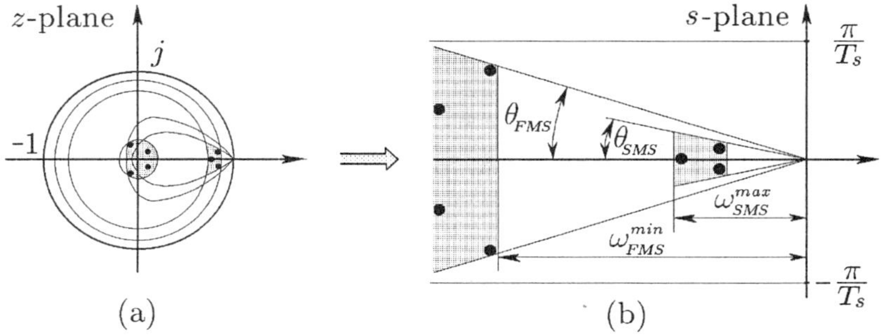

Fig. 1.2 Roots of the stable FMS and SMS characteristic polynomials in the discrete-time system (1.66)–(1.67) and their images in the primary strip on the s-plane.

Denote

$$r_{FMS} = \max_{i=1,\dots,m} |z_i^{FMS}|, \quad \text{and} \quad r_{SMS} = \min_{i=1,\dots,n} |z_i^{SMS}|,$$

where we assume that

$$0 < r_{FMS} < r_{SMS} < 1.$$

From (1.64) and (1.70) we obtain

$$\eta_2 = \frac{\ln r_{FMS}}{\ln r_{SMS}}. \tag{1.71}$$

The particular feature of the discrete-time FMS (1.68) is that a lower bound for the settling time exists, which is equal to the settling time of the deadbeat response.[3] If all roots of the FMS characteristic polynomial $A_{FMS}(z)$ are located at the origin, then the settling time of the discrete-time FMS (1.68) is equal to mT_s (the settling time of the deadbeat response for arbitrarily chosen initial conditions). However, from (2.4) and (1.70) we get

$$t_{s,FMS} \approx -\frac{4T_s}{\ln r_{FMS}}, \quad t_{s,SMS} \geq -\frac{4T_s}{\ln r_{SMS}}, \tag{1.72}$$

where $t_{s,FMS} \to 0$ as $r_{FMS} \to 0$. Therefore, from (1.72) the value $\bar{r}_{FMS}$ can be found such that the condition $t_{s,FMS}(\bar{r}_{FMS}) = mT_s$ is satisfied, where

$$\bar{r}_{FMS} = \exp(-4/m).$$

[3]The notion of the deadbeat response can be found, for instance, in [Lindorff (1965); Chen (1993); Ogata (1994)].

So the expressions (1.71) and (1.72) can be used only if the inequality $r_{FMS} > \bar{r}_{FMS}$ holds. If $r_{FMS} \leq \bar{r}_{FMS}$ then, by (1.59), we get

$$\eta \geq -\frac{4}{m \ln r_{SMS}}. \tag{1.73}$$

1.4 Notes

In this chapter we have discussed the basic principles for approximate analysis of the properties of the perturbed and singularly perturbed differential equations. The properties of the regularly and singularly perturbed differential equations that we have discussed are used throughout the book as the basis for an approximate analysis and design of nonlinear control systems.

Note that the numerical simulation of singularly perturbed differential equations has some particulars concerning the choice of step size. Usually, the higher order Runge-Kutta algorithms or Agams-Moulton methods allow us to obtain numerically stable solutions without special contrivance if the dimension of the equations is not too high.

There are many references devoted to consideration of particular details concerned with the analysis of regularly and singularly perturbed systems of differential equations. These may be found, for instance, in [Vasileva (1963); Gerashchenko (1975); Kokotović *et al.* (1986); Kokotović and Khalil (1986)] and [Sastry (1999); Khalil (2002)]. Various aspects of discrete-time singularly perturbed systems were considered in [Litkouhi and Khalil (1985); Naidu and Rao (1985); Naidu (1988)].

1.5 Exercises

1.1 The behavior of a dynamical system is described by the equation

$$x^{(2)} + 3x^{(1)} + 2x = 0, \quad x(0) = 1, \quad x^{(1)}(0) = 1.$$

Determine the lower and upper bounds for $\|X(t)\|$.

1.2 The behavior of a dynamical system is described by the equation

$$x^{(2)} + 1.5x^{(1)} + 0.5x + \mu\{2x^2 + [x^{(1)}]^2\}^{1/2} = 0.$$

Determine the region of μ such that $X = 0$ is an exponentially stable equilibrium point of the given system.

1.3 The behavior of a dynamical system is described by the equation

$$x^{(2)} + 1.5x^{(1)} + 0.5x + \mu|\sin(0.5t)| = 0.$$

Determine the parameter μ such that $\lim_{t\to\infty} \|X(t)\|_2 \leq 0.4$.

1.4 The behavior of a dynamical system is described by the equations

$$\dot{x}_1 = x_1 - x_2, \quad \mu\dot{x}_2 = 2x_1 + x_2.$$

Obtain and analyze the stability of the SMS and FMS.

1.5 The behavior of a dynamical system is described by the equations

$$\dot{x}_1 = x_1 - x_2, \quad \mu\dot{x}_2 = 2x_1 - x_2. \tag{1.74}$$

Obtain and analyze the stability of the SMS and FMS. Plot the phase portraits of the system by computer simulation for $\mu = 0.1, 0.5, 1$ and compare the results.

1.6 Consider the system (1.74). Obtain and analyze the stability of the SMS and FMS. Determine the parameter μ such that $\eta_2 = 10$.

1.7 The behavior of a dynamical system is described by the equations

$$\dot{x}_1 = x_2, \quad \dot{x}_2 = x_3 - 2x_2, \quad \mu\dot{x}_3 = x_4, \quad \mu\dot{x}_4 = -x_1 - x_3 - x_4.$$

Obtain and analyze the stability of the SMS and FMS. Determine the parameter μ, where: (a) $\eta_1 = 10$, (b) $\eta_2 = 10$, (c) $\eta_3 = 10$.

1.8 Consider the difference equations given by

$$x_1(k+1) = [1+\mu]x_1(k) + \mu x_2(k), \quad x_2(k+1) = ax_1(k) + bx_2(k).$$

Obtain and analyze the conditions for the SMS and FMS stability. Determine the parameter μ such that $\eta_2 = 10$, where $a = 0.35$, $b = 0.2$.

1.9 Consider the difference equations given by

$$\begin{aligned} x_1(k+1) &= [1-\mu]x_1(k) - \mu[x_2(k) + x_3(k)], \\ x_2(k+1) &= x_1(k) + 0.1x_2(k) + 0.2x_3(k), \\ x_2(k+1) &= 0.5x_1(k) + 0.2x_2(k) + 0.1x_3(k). \end{aligned}$$

Obtain and analyze the conditions for the SMS and FMS stability. Determine the parameter μ such that $\eta_2 = 10$.

Chapter 2

Design goal and reference model

Prior to the introduction of any specific design technique, it is appropriate to discuss performance criteria for control systems. These are usually imposed via the inclusion of some reference model in the controller, either explicitly or implicitly. We therefore use this chapter to highlight some basic correlations between the time-domain specifications of the control system output response and the pole-zero locations of the transfer function or, alternatively, the parameters of the linear differential equation.

In particular, we discuss basic step response parameters and the problem of output regulation for nonlinear time-varying control systems. We present a model of desired output behavior in the form of a differential equation, the parameters of which are based on required step response parameters (overshoot, settling time). Finally, we discuss the key role played by zeroes in the transfer function of the reference model with regard to the attainment of accuracy in the regulation problem.

2.1 Design goal

A block diagram of a general control system (GCS) appears in Fig. 2.1, where

- P is a plant,
- C is a controller,
- y is a measurable output (or controlled variable),
- u is a control (manipulated variable) of the plant,
- r is a reference input,
- w incorporates external disturbances or variable parameters unavailable for measurement.

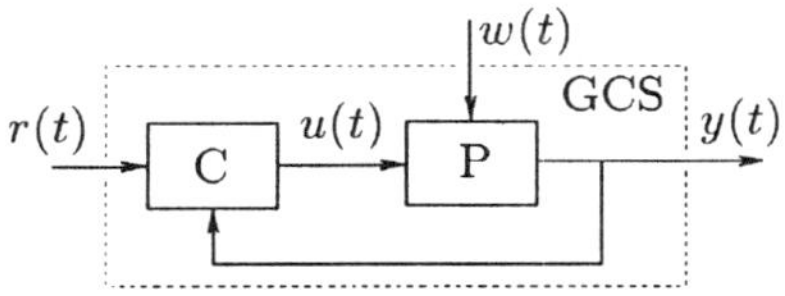

Fig. 2.1 Block diagram of the general control system.

Our goal is to design a control system subject to the condition that

$$\lim_{t\to\infty} e(t) = 0, \tag{2.1}$$

where $e(t) = r(t) - y(t)$ is the error of the reference input realization. Moreover, the controlled transients $y(t)$ should exhibit desired behaviors and should not depend on the varying parameters or external disturbances embodied in $w(t)$.

2.2 Basic step response parameters

Usually the behavior of the output variable $y(t)$ may be described by a small number of basic step response parameters. These may be directly identified by inspection of the response when a reference step input is applied to the control system (Fig. 2.2).

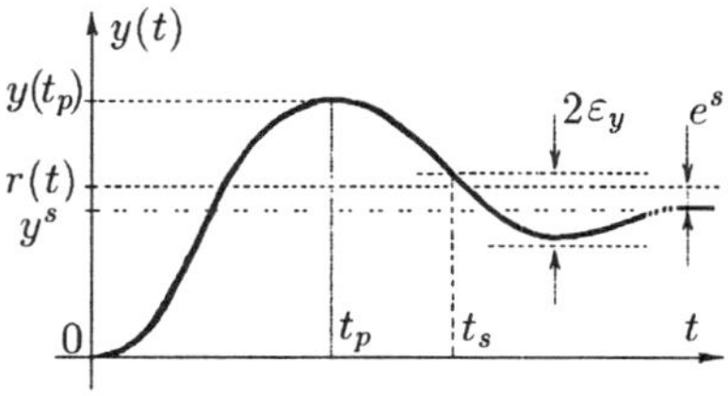

Fig. 2.2 Basic step response parameters of output variable $y(t)$.

When referring to the basic step response parameters, we have in mind the following standard concepts:

- y^s is the final value of $y(t)$, where $y^s = \lim_{t\to\infty} y(t)$.
- t_s is the settling time, which is defined as the time required for $y(t)$ to reach and remain within some neighborhood $(1 \pm \bar{\varepsilon}_y)y^s$ of its final value. We usually assume that $\bar{\varepsilon}_y = |\varepsilon_y/y^s| \in [0.01, 0.05]$.
- $y(t_p)$ is the peak value of $y(t)$, and t_p is the peak time.
- σ is the maximum percent overshoot of the output variable, where $\sigma = 100\,|(y(t_p) - y^s)/y^s|$ [%].
- e^s is the steady-state error, where $e^s = \lim_{t\to\infty} e(t)$.

We assume that the control systems discussed below will be designed to meet the following specifications:

1. The steady-state error e^s must be less than $e^s_{\max}$, i.e., $|e^s| \leq e^s_{\max}$.
2. The settling time t_s must be close to the desired value t^d_s, i.e., $t_s \approx t^d_s$.
3. The overshoot σ must be close to the desired value σ^d, i.e., $\sigma \approx \sigma^d$.

2.3 Reference model

Reference model design

Throughout the book a methodology will be discussed where the controller is designed in such a way that the closed-loop system is required to be close to some given reference model, despite the effects of varying parameters and unknown external disturbances $w(t)$ in the plant model. So, the destiny of the controller is to provide an appropriate reference input-controlled output map of the closed-loop system as shown in Fig. 2.3, where the reference model is selected based on the required output transient performance indices.

Usually, it is more convenient for practical applications to represent the reference model by a desired transfer function or appropriate desired differential equation, because the relationships between its parameters and the basic step response parameters are well known. Toward this end, the pole-zero patterns or Bode diagrams may be used to select the parameters of the reference model.

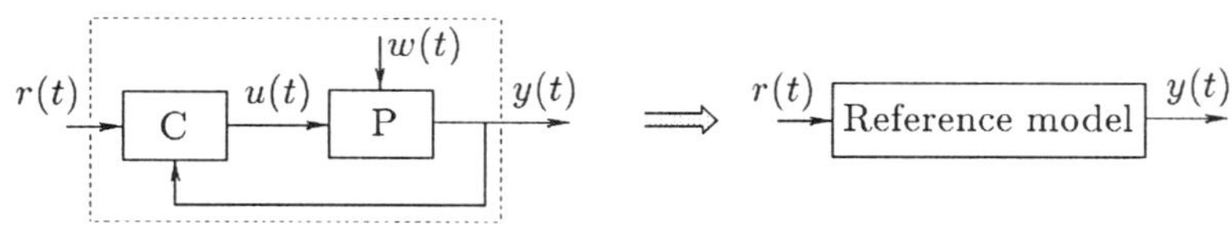

Fig. 2.3 Reference model of the closed-loop control system.

Let us now consider the reference model in the form of the rational continuous-time transfer function (desired transfer function) between the reference input r and the output y:

$$G^d_{yr}(s) = \frac{1}{A^d(s)}, \tag{2.2}$$

where

$$A^d(s) = T^n s^n + a_{n-1}^d T^{n-1} s^{n-1} + \cdots + a_1^d T s + 1. \tag{2.3}$$

A method involving a desired pole region may be used to determine the parameters $T, a_1^d, a_2^d, \ldots, a_n^d$ of the characteristic polynomial (2.3) in accordance with the requirements imposed on the step response parameters of the output transients $y(t)$. For instance, if

$$y^{(i)}(t_0) = 0 \quad \text{for all} \quad i = 1, \ldots, n-1$$

then from the second-order prototype the relationships [1]

$$t_s^d \approx \frac{4}{\omega^d}, \quad \sigma^d \approx 100 \exp\left(-\frac{\pi\zeta^d}{\sqrt{1-(\zeta^d)^2}}\right) [\%], \quad \zeta^d = \cos(\theta^d) \tag{2.4}$$

follow as a first approximation to find the desired pole region for the roots of the characteristic polynomial (2.3). This region is defined by two parameters: the angle θ^d, and the value ω^d (Fig. 2.4).

The value ζ^d is called the desired damping ratio, and ω^d is called the desired damped or actual frequency [Dorf and Bishop (2001); Kuo and Golnaraghi (2003)].

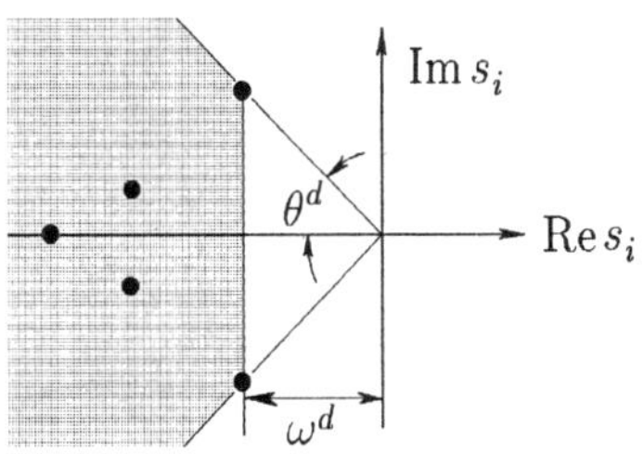

Fig. 2.4 Desired pole region of the transfer function (2.2).

So the following procedure may be applied to construct the reference model.

Step 1. From (2.4) the parameters θ^d and ω^d of the desired pole region can be found, that is

$$\theta^d \approx \tan^{-1}\left(\frac{\pi}{\ln(100/\sigma^d)}\right), \quad \omega^d \approx \frac{4}{t_s^d}. \tag{2.5}$$

[1] There are various approximate relationships in [Chen (1993); Dorf and Bishop (2001); Kuo and Golnaraghi (2003)].

Step 2. By selecting the n roots $s_1, s_2, \ldots, s_n$ inside the desired pole region, the desired characteristic polynomial

$$(s - s_1)(s - s_2) \cdots (s - s_n)$$

is obtained. Two so-called dominant poles are selected in the corners of the stated domain. This polynomial may be rewritten in the form of expression (2.3), where T is a time constant of the desired characteristic polynomial and $T = (-1)^n (s_1 s_2 \cdots s_n)^{-1/n}$.

Step 3. From the given polynomial (2.3) and the transfer function (2.2), the reference model in the form of the linear differential equation

$$T^n y^{(n)} + a_{n-1}^d T^{n-1} y^{(n-1)} + \cdots + a_1^d T y^{(1)} + y = r \tag{2.6}$$

follows. This is called the desired differential equation.

Remark 2.1 *In order to form the desired transfer function (2.2) the well known normalized step responses with non-dimensional time for the normalized transfer function*

$$G_{yr}^d(s) = \frac{1}{s^n + a_{n-1}^d s^{n-1} + \cdots + a_1^d s + 1}$$

may be used, where the denominator polynomial has its roots distributed in a Butterworth or binomial-type pattern. Such pole patterns are presented in many references [Graham and Lathrop (1953); Bosgra and Kwakernaak (2000)].

In any case, it is desirable to provide computer simulation to verify the shape of the output response.

Reference model with arbitrary system type

Let the reference input $r(t)$ be a polynomial time function of degree ρ. If $r(t) = (t^\rho/\rho!)1(t)$ (where ρ is a so-called input signal type), then the Laplace transform of the reference input is $r(s) = 1/s^{\rho+1}$.

Let us consider the desired transfer function $G_{er}^d(s)$ between reference input r and error variable e, where

$$G_{er}^d(s) = 1 - G_{yr}^d(s).$$

The Laplace transform of the error variable for the input signal of the type

ρ is given by

$$e(s) = [1 - G^d_{yr}(s)]\frac{1}{s^{\rho+1}},$$

and on the assumptions of the final value theorem (i.e., that $G^d_{yr}(s)$ is stable) the steady-state error is

$$e^s = \lim_{s \to 0} se(s).$$

Accordingly, from (2.2) and (2.3) it follows that e^s is zero if the input signal is of type 0 (unit step input).

Let $r(t)$ be the input signal of type 1 (ramp reference input, or step function of velocity):

$$r(t) = r^v t\,1(t),$$

where $r^v = \text{const}$ and $r^v \neq 0$.

The steady-state error e^s due to an input signal of type 1 is called a velocity error, i.e., $e^v_r = e^s$. Then the relative velocity error of (2.2) is given by

$$\overline{e}^v_r = \frac{e^v_r}{r^v} = \lim_{t \to \infty} \frac{r(t) - y(t)}{r^v} = a^d_1 T.$$

The system (2.2) is called a type 1 system if the velocity error e^v_r is a nonzero constant (see, for example, in [Wolovich (1994); Bosgra and Kwakernaak (2000)]).

In the next part of this section let $G^d_{yr}(s)$ be a rational strictly proper continuous-time transfer function between reference input r and output y where

$$G^d_{yr}(s) = \frac{B^d(s)}{A^d(s)}. \tag{2.7}$$

Let us assume that

$$B^d(s) = b^d_\rho \tau^\rho s^\rho + b^d_{\rho-1}\tau^{\rho-1}s^{\rho-1} + \cdots + b^d_1 \tau s + 1, \quad \rho < n.$$

Here $A^d(s)$ is the above described polynomial (2.3), whose roots lie in the stable half-plane $\text{Re}(s) < 0$, and $B^d(s), A^d(s)$ are coprime polynomials.

In accordance with the continuous-time transfer function (2.7), the reference model may be rewritten as the linear differential equation

$$\begin{aligned} T^n y^{(n)} &+ a^d_{n-1}T^{n-1}y^{(n-1)} + \cdots + a^d_1 T y^{(1)} + y \\ &= b^d_\rho \tau^\rho r^{(\rho)} + b^d_{\rho-1}\tau^{\rho-1}r^{(\rho-1)} + \cdots + b^d_1 \tau r^{(1)} + r. \end{aligned} \tag{2.8}$$

In a fashion analogous to the above, we find that the relative velocity error for the system (2.8) is given by

$$\overline{e}_r^v = \lim_{t\to\infty} \frac{r(t) - y(t)}{r^v} = a_1^d T - b_1^d \tau.$$

It is easy to see that if the condition

$$b_1^d \tau = a_1^d T \tag{2.9}$$

is satisfied, then the velocity error $\overline{e}_r^v$ equals zero for a reference model of the form (2.8).

As the next case, let us assume that (2.9) holds. Then we may consider the response of (2.8) under a parabolic reference input.

The system (2.8) is said to be of type 2 if the steady-state error is a nonzero constant due to an input signal of type 2 (parabolic reference input, step function of acceleration). In this case the steady-state error is called an acceleration error. Then the relative acceleration error is given by

$$\overline{e}_r^{acc} = \frac{e_r^{acc}}{r^{acc}} = \lim_{t\to\infty} \frac{r(t) - y(t)}{r^{acc}} = a_2^d T^2 - b_2^d \tau^2,$$

where $r(t) = r^{acc}(t^2/2!)1(t)$ and $r^{acc} = \text{const}$.

Finally, let us consider a stable linear system in the form

$$\begin{aligned} &T^n y^{(n)} + a_{n-1}^d T^{n-1} y^{(n-1)} + \cdots + a_1^d T y^{(1)} + y \\ &\quad = b_\rho^d T^\rho r^{(\rho)} + b_{\rho-1}^d T^{\rho-1} r^{(\rho-1)} + \cdots + b_1^d T r^{(1)} + r, \end{aligned} \tag{2.10}$$

where $1 < \rho < n$. Then, as opposed to (2.6), it is easy to verify that the steady-state error is zero for an input signal of type ρ if the conditions

$$b_j^d = a_j^d \quad \text{for all} \quad j = 1, \ldots, \rho$$

are satisfied.

So, it is important to note that the reference model in the form of the linear differential equation (2.10) (that corresponds to the transfer function (2.7) with zeroes) allows us to reach a higher tracking accuracy than (2.6).

For instance, the well known optimal coefficients of the transfer function

$$G_{yr}^d(s) = \frac{a_1^d s + a_0^d}{s^n + a_{n-1}^d s^{n-1} + \cdots + a_1^d s + a_0^d} \tag{2.11}$$

based on the integral of time multiplied by absolute error (ITAE) criterion

$$ITAE = \int_0^\infty t|e(t)|\, dt$$

for a ramp input may be used:

$$\begin{aligned}
&s^2 + 3.2\omega_n s + \omega_n^2,\\
&s^3 + 1.75\omega_n s^2 + 3.25\omega_n^2 s + \omega_n^3,\\
&s^4 + 2.41\omega_n s^3 + 4.93\omega_n^2 s^2 + 5.14\omega_n^3 s + \omega_n^4,\\
&s^5 + 2.19\omega_n s^4 + 6.50\omega_n^2 s^3 + 6.30\omega_n^2 s^2 + 5.24\omega_n^4 s + \omega_n^5,
\end{aligned}$$

where ω_n is the natural frequency.

Note that the approach we have discussed, of reference model design in accordance with the requirement of an appropriate system type, is widely used in various design techniques [Bosgra and Kwakernaak (2000); Dorf and Bishop (2001); Franklin *et al.* (2002); Kuo and Golnaraghi (2003)].

2.4 Notes

The main purpose of this chapter is to review some basic correlations between the time-domain specifications of the control system output response and the pole-zero locations of the transfer function or, alternatively, the parameters of the linear differential equation. The importance of such correlations to control system design technique follows from the fact that it is usually easier to formulate the requirements on control system performance in terms of time-domain specifications from an engineering point of view. But the totality of existing design procedures are based on various specifications of some reference model, usually in the form of a transfer function or differential equation. The correlations discussed above allow us to choose the reference model in accordance with the requirements placed on the desired time-domain behavior of the closed-loop system.

Note that in the problem of linear control system design, the use of a reference model in the form of desired pole and zero locations for the transfer function is usually enough. In order to solve the problem of nonlinear control system design, it is more convenient to construct the reference model in the form of a desired differential equation or desired manifold in the state space of the plant model. The correlations discussed above may be used for this purpose. This will be shown in detail in the coming chapters.

Some additional details concerned with the performance of feedback control systems may be found in such references as, for instance, [Bosgra and Kwakernaak (2000); Dorf and Bishop (2001); Franklin *et al.* (2002); Kuo and Golnaraghi (2003); Wolovich (1994)].

2.5 Exercises

2.1 Construct the reference model in the form of the 2nd order differential equation (2.6) in such a way that the step response parameters of the output meet the requirements $t_s^d \approx 6$ s, $\sigma^d \approx 0\,\%$. Plot by computer simulation the output response, and determine the steady-state error from the plot for input signals of type 0 and 1.

2.2 Construct the reference model in the form of the 3rd order differential equation (2.6) in such a way that the step response parameters of the output meet the requirements $t_s^d \approx 3$ s, $\sigma^d \approx 30\,\%$. Plot by computer simulation the output response, and determine the steady-state error from the plot for input signals of type 0 and 1.

2.3 Construct the reference model in the form of the 2nd order differential equation as the type 1 system with the following roots of the characteristic polynomial:

$$s_1 = -1 + j, \qquad s_2 = -1 - j.$$

Plot by computer simulation the output response, and determine the steady-state error from the plot for input signals of type 0, 1, and 2.

2.4 Construct the reference model in the form of the 3rd order differential equation as the type 1 system with the following roots of the characteristic polynomial:

$$s_1 = -2, \qquad s_2 = -3 + j2, \qquad s_3 = -3 - j2.$$

Plot by computer simulation the output response, and determine the steady-state error from the plot for input signals of type 0, 1, 2, and 3.

2.5 Construct the reference model in the form of the 3rd order differential equation as the type 2 system with the following roots of the characteristic polynomial:

$$s_1 = -3, \qquad s_2 = -1 + j2, \qquad s_3 = -1 - j2.$$

Plot by computer simulation the output response, and determine the steady-state error from the plot for input signals of type 0, 1, 2, and 3.

2.6 Consider the reference model in the form of the 2nd, 3rd, 4th, and 5th order transfer function (2.11) with the optimal coefficients based on the ITAE criterion. Plot by computer simulation the output response for the input signal of type 0, where: (a) $\omega_n = 1$, (b) $\omega_n = 2$.

2.7 Consider the reference model in the form of the 2nd, 3rd, 4th, and 5th order transfer function (2.11) with the optimal coefficients based on the ITAE criterion. Plot by computer simulation the output response for the input signal of type 1, where: (a) $\omega_n = 1$, (b) $\omega_n = 2$.

2.8 Consider the reference model in the form of the 2nd, 3rd, 4th, and 5th order transfer function (2.11) with the optimal coefficients based on the ITAE criterion. Plot by computer simulation the output response and determine the steady-state error from the plot for input signals of type 1 and 2, where: (a) $\omega_n = 1$, (b) $\omega_n = 2$.

Chapter 3

Methods of control system design under uncertainty

There is a broad class of methods for control system design in the presence of varying plant parameters and unknown external disturbances. Only a small subset of these is considered in this chapter. The reviewed methods are directly related to the matter of the control system design methodology in the book. In particular, a short overview of robust control synthesis techniques under uncertainty is given, where our main attention is devoted to the discussion of nonadaptive approaches such as control systems with the highest derivative of the output signal and high gain in the feedback loop, control systems with state vector and high gain in the feedback loop, and control systems with sliding motions. The interdependencies between these methods are discussed, and the main steps of the design procedures are highlighted.

3.1 Desired vector field in the state space of plant model

This section is related to the essence of the preceding chapter, with special emphasis placed on nonlinear control systems. With these systems we are naturally motivated to impose the desired performance criteria in terms of state space notions; after that, specific design techniques may be discussed. For this reason, let us select two methods in modern nonlinear control theory in order to specify the desired behavior as well as the requirement of insensitivity of the closed-loop system transients with respect to the external disturbances and varying parameters of the plant model.

The first method is based on the construction of the desired vector field in the state space of the plant model, whereas the second method is based on the construction of the desired manifold in the state space of the plant model.

Let us begin with the first method and its application to a plant model given by

$$x^{(n)} = f(X, w) + g(X, w)u, \quad y = x, \tag{3.1}$$

where $y \in \mathbb{R}^1$ is the system output, $X = \{x, x^{(1)}, \ldots, x^{(n-1)}\}^T$ is the state vector of the system (3.1), $u(t)$ is the control, $u \in \Omega_u \subset \mathbb{R}^1$ where Ω_u is a bounded set of allowable values of the control variable, and $w(t)$ is a vector of bounded external disturbances and/or varying parameters of the plant model.

Assumption 3.1 The functions $f(X, w)$, $g(X, w)$ are smooth for all $(X, w) \in \Omega_{X,w} = \Omega_X \times \Omega_w$.

Assumption 3.2 The conditions

$$|f(X, w)| \le f_{\max} < \infty, \quad 0 < g_{\min} \le |g(X, w)| \le g_{\max} < \infty \tag{3.2}$$

are fulfilled for all $(X, w) \in \Omega_{X,w}$, i.e., the functions $f(X, w)$, $g(X, w)$ are bounded for all (X, w) in the specified bounded set $\Omega_{X,w}$.

Since $g(X, w) \neq 0$, $\forall\, (X, w) \in \Omega_{X,w}$, from (3.1) it follows that the desirable value of the nth derivative $x^{(n)}(t)$, which is the highest derivative of $x(t)$ in the system (3.1), can be obtained by a proper choice of control variable $u(t)$. The set of possible values of $x^{(n)}(t)$ depends on Ω_u.

Let us assume that based on the correlations discussed in the previous chapter and in accordance with the desired time-domain specifications of the output behavior of $x(t)$, the reference model in the form of the nth order differential equation has been constructed, similar to (2.6), and is represented in the following form:

$$x^{(n)} = F(x^{(n-1)}, \ldots, x^{(1)}, x, r), \tag{3.3}$$

where $x = r$ at the equilibrium of (3.3) for $r = \text{const}$.

We call (3.3) the desired differential equation and its right member F the desired value of the highest derivative $x^{(n)}(t)$ (desired dynamics).

Let us rewrite, for short, equation (3.3) in the form

$$x^{(n)} = F(X, r). \tag{3.4}$$

Denote

$$e^F = F - x^{(n)}. \tag{3.5}$$

The value e^F is called the error of the desired dynamics realization, where the desired dynamics are assigned by (3.4).

Accordingly, if the condition

$$e^F = 0 \tag{3.6}$$

holds, then the desired behavior of $x(t)$ with the prescribed dynamics of (3.4) is fulfilled.

Expression (3.6) is the insensitivity condition of the step response parameters of output transients with respect to the external disturbances and varying parameters of the system (3.1). We can see that (3.6) allows us to simultaneously express two requirements. The first is that the output behavior is described by the desired differential equation. The second is that the output behavior exhibit insensitivity with respect to the external disturbances and varying parameters of the plant model.

Let us show that the discussed insensitivity condition corresponds to assignment of the desired vector field in the state space of the plant model.

The state space representation of the system (3.1) yields

$$\begin{aligned} \frac{d}{dt}x_i &= x_{i+1}, \qquad i = 1, \ldots, n-1, \\ \frac{d}{dt}x_n &= f(X, w) + g(X, w)u, \end{aligned} \tag{3.7}$$

where, by definition,

$$X = \{x, x^{(1)}, \ldots, x^{(n-1)}\}^T = \{x_1, x_2, \ldots, x_n\}^T.$$

For short, let us rewrite (3.7) in vector notation as

$$\frac{d}{dt}X = \hat{f}(X, w, u). \tag{3.8}$$

The vector function $\hat{f}(X, w, u)$ defines a map

$$\hat{f}(X, w, u) \colon \mathbb{R}^n \to \mathbb{R}^n$$

called the vector field. By virtue of this map, each point of $\mathbb{R}^n$ is placed into correspondence with a vector of $\mathbb{R}^n$.

Similar to the above, from the desired differential equation (3.3) we have

$$\begin{aligned} \frac{d}{dt}x_i &= x_{i+1}, \qquad i = 1, \ldots, n-1, \\ \frac{d}{dt}x_n &= F(X, r). \end{aligned} \tag{3.9}$$

Then from (3.9) we get

$$\frac{d}{dt}X = \hat{F}(X, r), \tag{3.10}$$

where the vector function $\hat{F}(X, r)$ defines a desired vector field

$$\hat{F}(X, r) \colon \mathbb{R}^n \to \mathbb{R}^n.$$

As a result, we have that the insensitivity condition (3.6) corresponds to

$$\hat{F}(X, r) = \hat{f}(X, w, u).$$

Therefore, (3.6) corresponds to assignment of the desired vector field in the state space of the plant model given by (3.1), where the desired vector field can be constructed by pattern in the form of the nth order desired differential equation in such a way as to provide for the requirement (2.1) and the desired performance indices.

3.2 Solution of nonlinear inverse dynamics

In accordance with (3.1), (3.4), and (3.5), expression (3.6) may be rewritten in the form

$$F(X, r) - f(X, w) - g(X, w)u = 0. \tag{3.11}$$

So the control problem (2.1) has been restated as the requirement to provide for the condition (3.6), or in other words, to find a solution to (3.11). Such an approach to control problem reformulation was discussed and used in [Boychuk (1966); Vostrikov (1977a)].

If the condition

$$g(X, w) \neq 0, \quad \forall\ (X, w) \in \Omega_{X,w} \tag{3.12}$$

is satisfied, then the control function $u(t) = u^{NID}(t)$ exists such that $u^{NID}(t)$ is the unique solution of (3.11):

$$u^{NID} = \{g(X, w)\}^{-1}\{F(X, r) - f(X, w)\}. \tag{3.13}$$

This is called the solution of nonlinear inverse dynamics.

Remark 3.1 *There is a broad set of publications where the function (3.13) of the nonlinear inverse dynamics solution is used as the control law in order to obtain a desired reference input-controlled output map (see, for instance, [Boychuk (1966); Popov and Krutko (1979); Petrov and Krutko*

(1980); Nijmeijer and Schaft (1990); Slotine and Li (1991); Isidori (1995); Hirschorn and Aranda-Bricaire (1998)]). Such an approach can be applied only in the presence of complete information about external disturbances and parameters of the plant model.

It is worth noting that (3.13) allows us to estimate bounds for the set

$$\Omega_u = [u_{\min}, u_{\max}],$$

such that

$$u_{\min} \leq u^{NID}(t) \leq u_{\max}, \quad \forall\ (X, w) \in \Omega_{X,w}. \tag{3.14}$$

Assume that the set Ω_u of allowable control values is assigned. Then the desired output behavior of (3.1) corresponding to (3.4) can be realized if and only if the inclusion $u^{NID}(t) \in \Omega_u$ occurs for all t.

In general, the problem of desired output transient realization depends on such properties as the invertibility and internal stability of the plant model. This problem will show up in detail below in the chapter on the realizability of a desired output behavior.

3.3 The highest derivative and high gain in feedback loop

The main subject of our consideration is the problem of control system design under the conditions when analytical expressions for the functions $f(X, w)$, $g(X, w)$ are unknown and the vector $w(t)$ of bounded external disturbances or varying parameters is unavailable for measurement as shown in Fig. 2.1.

In order to reach the discussed control goal and, as a result, to provide desired dynamical properties of $x(t)$ in the specified region of the state space of the uncertain nonlinear system (3.1), the following control law with the highest derivative of the output signal and high gain in the feedback loop

$$u = k_0\{F(X, r) - x^{(n)}\} \tag{3.15}$$

was proposed in the pioneer work [Vostrikov (1977a)], where k_0 is a high gain, $k_0 \in \mathbb{R}^1$.

Let us consider the basic correlations of the control system with the highest derivative and high gain in the feedback loop that were discussed in [Vostrikov (1977a); Utkin and Vostrikov (1978); Vostrikov (1979)]. It is easy to see that from the closed-loop system equations

$$x^{(n)} = f(X, w) + g(X, w)u, \tag{3.16}$$

$$u = k_0\{F(X, r) - x^{(n)}\}, \tag{3.17}$$

by substituting (3.17) into (3.16), we obtain

$$x^{(n)} = F(X, r) + \underbrace{\frac{1}{1 + g(X, w)k_0}}_{\to 0 \text{ as } |k_0| \to \infty} \{f(X, w) - F(X, r)\}. \tag{3.18}$$

From (3.18) it follows that the condition

$$\lim_{|k_0| \to \infty} x^{(n)} = F(X, r) \tag{3.19}$$

holds. In other words, the solution to the control problem under consideration is provided by the use of $x^{(n)}$ in the control law and the use of a high gain k_0.

Let us consider the behavior of the control variable $u(t)$ in the closed-loop system. Substituting (3.16) into (3.17) we obtain

$$u = \underbrace{\frac{k_0}{1 + g(X, w)k_0}}_{\to 1/g(X,w) \text{ as } |k_0| \to \infty} \{F(X, r) - f(X, w)\}, \tag{3.20}$$

where it was assumed that the sign of the gain k_0 is chosen in such a way that the condition

$$k_0 g(X, w) > 0, \quad \forall\ (X, w) \in \Omega_{X,w}$$

is satisfied. From (3.20) the condition

$$\lim_{|k_0| \to \infty} u = u^{NID} \tag{3.21}$$

results, where $u^{NID}(t)$ is the nonlinear inverse dynamics solution (3.13).

We can see that if $g(X, w)k_0 > 0$, $\forall (X, w) \in \Omega_{X,w}$ then conditions (3.19) and (3.21) are simultaneously satisfied despite the fact that the functions $f(X, w)$, $g(X, w)$ are unknown. So the control law (3.15) allows us to solve the control problem of interest under the condition of unknown external disturbances and varying parameters of the system (3.1).

Note that, throughout the book, the control systems discussed are as shown in Fig. 2.1 and, accordingly, the block diagram representation of the practical realization with the highest derivative and high gain in the feedback loop is as shown in Fig. 3.1. Here, in order to implement the control law, the estimations $\hat{x}^{(j)}(t)$ are used; these are received by a special dynamical system called a differentiating filter. However, the above

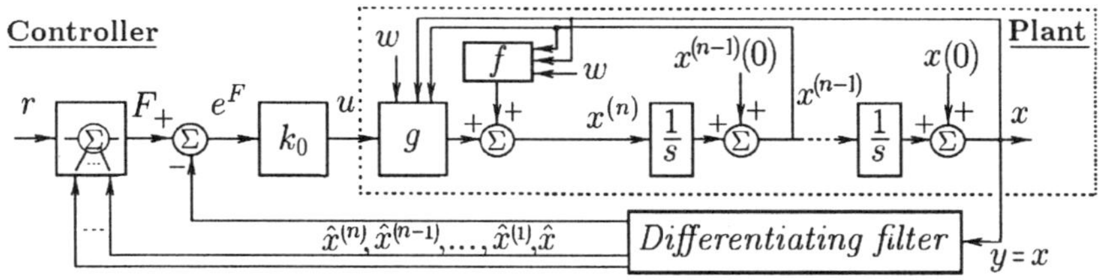

Fig. 3.1 Block diagram of the control system with the highest derivative and high gain in the feedback loop, where a real differentiating filter is used.

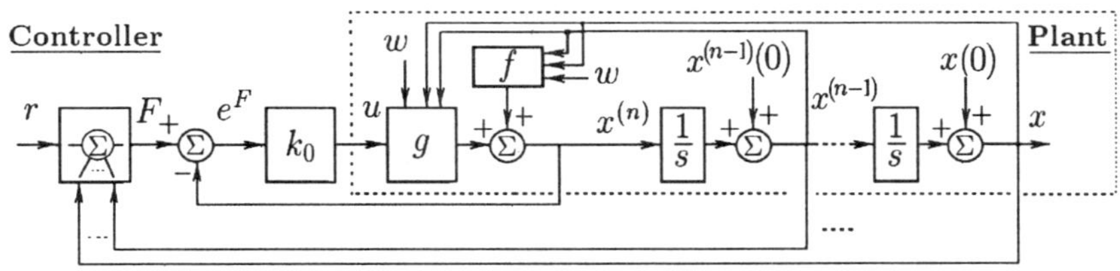

Fig. 3.2 Block diagram of the control system with the highest derivative and high gain in the feedback loop, where an ideal differentiating filter is used.

consideration was done on the assumption that the output $x(t)$ and all its derivatives, up to the highest derivative $x^{(n)}$, are measured by some ideal differentiating filter and this case corresponds to the block diagram representation of the closed-loop system equations (3.16), (3.17) as shown in Fig. 3.2. Clearly, the control law (3.17) is unrealizable in practice. Nevertheless, the above consideration is significant because the limit properties of the use of the highest derivative and high gain in the feedback loop are shown explicitly. In particular, we may represent Fig. 3.2 as shown in Fig. 3.3, where the memoryless network for the highest derivative $x^{(n)}$ is highlighted by a circuit of dots. The rejection of the influence of the unknown functions $f(X,w)$, $g(X,w)$ (and, accordingly, of the unknown bounded external disturbances and varying parameters) is provided in the memoryless network as the high gain k_0 is sufficiently large. This network is shown

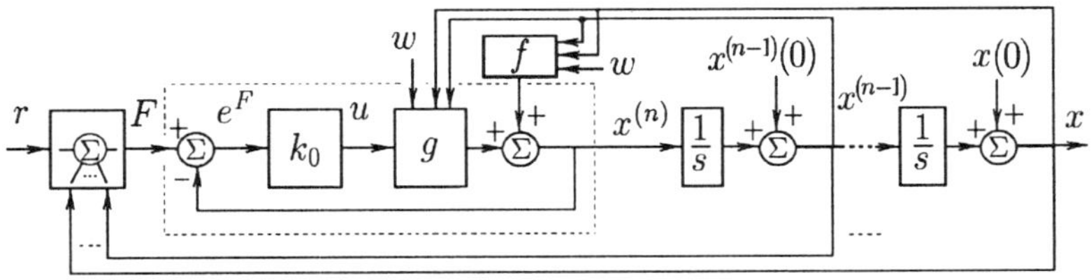

Fig. 3.3 Block diagram of the control system with the highest derivative and high gain in the feedback loop.

in Fig. 3.4 and was called a "localization circuit" in [Vostrikov (1988b); Vostrikov (1990)].

Note that the above subject matter is related to the properties of the input-output map for the memoryless system with high gain feedback as discussed, for instance, in [Bosgra and Kwakernaak (2000)].

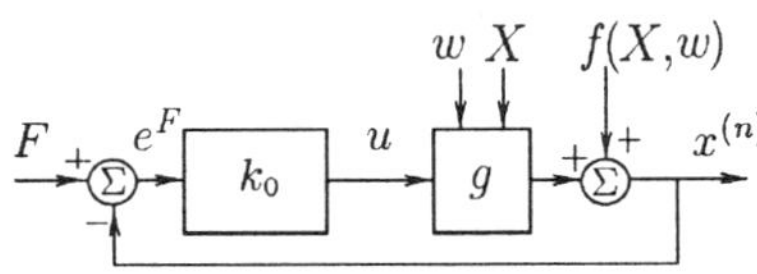

Fig. 3.4 Block diagram of the memoryless network for the highest derivative $x^{(n)}$.

3.4 Differentiating filter and high-gain observer

Now the problem of practical realization of the control law (3.15) may be discussed. In particular, in order to realize the control law in the form of (3.15) in practice, instead of the ideal derivatives $x^{(j)}(t)$ some estimations $\hat{x}^{(j)}(t)$ of these derivatives should be used as shown in Fig. 3.1.

For instance, the linear dynamical system shown in Fig. 3.5 was used in [Vostrikov (1977a)] as a real differentiating filter; its behavior is described by the equation

$$\mu^q \hat{x}^{(q)} + d_{q-1}\mu^{q-1}\hat{x}^{(q-1)} + \cdots + d_1\mu\hat{x}^{(1)} + \hat{x} = x, \quad \hat{X}(0) = \hat{X}^0. \quad (3.22)$$

Here $q \geq n$, $\hat{X} = \{\hat{x}, \hat{x}^{(1)}, \ldots, \hat{x}^{(q-1)}\}^T$ is the state vector of the system (3.22), and μ is a small positive parameter.

Note that the stability of (3.22) does not depend on μ and may be provided by selection of the parameters d_j. If μ is sufficiently small, then the variables of the state vector $\hat{X}$ may be used as the required estimations $\hat{x}^{(j)}(t)$ of the actual derivatives $x^{(j)}(t)$ since we have

$$\lim_{\mu\to 0}\{\hat{x}(t)-x(t)\}=0, \quad \forall\, t>0. \tag{3.23}$$

It is easy to see that if we denote $\mu = 1/k$, then k is a high-gain parameter where $\mu \to 0$ as $k \to \infty$. So the system (3.22) may be rewritten in the form of a system with high-gain parameters. Such dynamical systems were investigated in [Meerov (1965)].

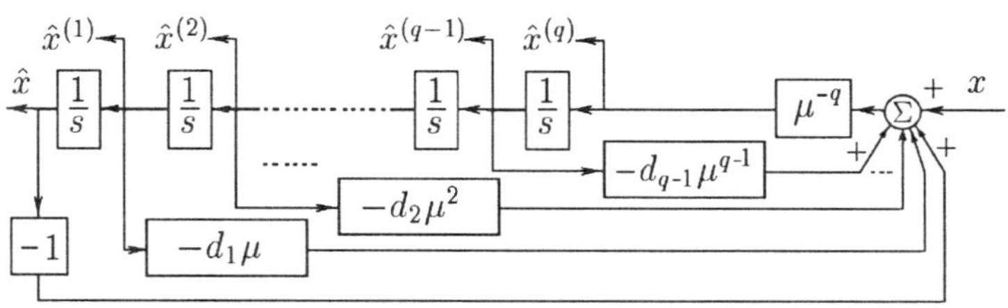

Fig. 3.5 Block diagram of the differentiating filter (3.22).

Remark 3.2 *Such dynamical systems are now usually called high-gain observers, and the various associated structures are considered in [Gauthier et al. (1991); Tornambè (1992); Bullinger and Allgöwer (1997); Levant (1998); Hammouri and Marchand (2000)]. Observer design based on sliding mode techniques is discussed, for example, in [Utkin (1977); Gauthier et al. (1991); Canudas de Wit and Slotine (1991); Hernandez and Barbot (1996); Hammouri and Marchand (2000); Perruquetti and Barbot (2002)].*

As a result of applying the real differentiating filter (3.22) in order to implement the control law (3.15) in practice, we get the singularly perturbed closed-loop system equations

$$x^{(n)} = f(X,w)+g(X,w)k_0\{F(\hat{X},r)-\hat{x}^{(n)}\},\ X(0)=X^0, \tag{3.24}$$

$$\mu^q\hat{x}^{(q)}+d_{q-1}\mu^{q-1}\hat{x}^{(q-1)}+\cdots+d_1\mu\hat{x}^{(1)}+\hat{x}=x,\ \hat{X}(0)=\hat{X}^0, \tag{3.25}$$

where slow and fast motions take place.

In particular, if $q=n=1$, then (3.24)–(3.25) assume the form

$$x^{(1)} = f(x,w)+g(x,w)k_0\{F(\hat{x},r)-\mu^{-1}(x-\hat{x})\}, x(0)=x^0, \tag{3.26}$$

$$\mu\hat{x}^{(1)}+\hat{x}=x, \quad \hat{x}(0)=\hat{x}^0. \tag{3.27}$$

A distinctive feature of (3.26)–(3.27) (as well as of (3.24)–(3.25)) is the nonstandard singular perturbation form of the closed-loop system equations. Therefore, special validation of the stability of such a system is required. For instance, it is possible to artificially transform the system (3.24)–(3.25) to the standard singular perturbation form by introducing the new variables

$$\hat{x}_1 = \hat{x}^{(1)}, \hat{x}_2 = \hat{x}^{(2)}, \ldots, \hat{x}_n = \hat{x}^{(n)};$$

by differentiation of (3.25) the system (3.24)–(3.25) is transferred to the extended form [Vostrikov (1977a); Vostrikov (1990)] given by

$$\begin{aligned}
&x^{(n)} = f(X,w) + g(X,w)k_0\{F(\hat{x}_{n-1},\ldots,\hat{x}_1,\hat{x},r) - \hat{x}_n\},\\
&\mu^q\hat{x}^{(q)} + d_{q-1}\mu^{q-1}\hat{x}^{(q-1)} + \cdots + d_1\mu\hat{x}^{(1)} + \hat{x} = x,\\
&\mu^q\hat{x}_j^{(q)} + d_{q-1}\mu^{q-1}\hat{x}_j^{(q-1)} + \cdots + d_1\mu\hat{x}_j^{(1)} + \hat{x}_j = x^{(j)}, \quad j = 1,\ldots,n-1,\\
&\mu^q\hat{x}_n^{(q)} + \cdots + d_1\mu\hat{x}_n^{(1)} + \{1 + g(X,w)k_0\}\hat{x}_n = f(X,w) + g(X,w)k_0F.
\end{aligned}$$

The behavior of $\hat{x}^{(n)}$ is described by the last differential equation of this extended system. Denote

$$\bar{D}(\mu s) = \mu^q s^q + d_{q-1}\mu^{q-1}s^{q-1} + \cdots + d_1\mu s + 1 \tag{3.28}$$

as the characteristic polynomial of system (3.22). Then the block diagram representation of the last equation may be depicted as Fig. 3.6, where the fast motions for $\hat{x}^{(n)}$ take place (fast-motion subsystem). The stability of these fast motions may be provided by choosing the controller parameters k_0, μ, d_i in accordance with the requirements placed on the admissible transient performance indices of the fast process and the degree of time-scale-separation between the fast and slow motions.

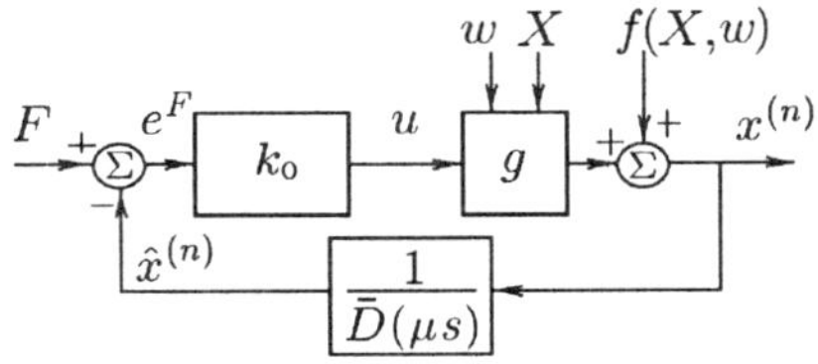

Fig. 3.6 Block diagram of the fast-motion subsystem.

If the fast motions are stable and $\mu \to 0$, then the extended system yields the SMS, which is the same as (3.18). As a result, the desired dynamical properties of $x(t)$ are provided in a specified region of the state space of the uncertain nonlinear system (3.1) in the presence of incomplete information about varying parameters of the system and external disturbances, as long as the gain k_0 is large, i.e., $|k_0| \to \infty$.

At the same time, it is necessary to say that another problem was omitted from the above consideration. Indeed, the initial conditions of the additional differential equations in the extended system depend on a small parameter μ and tend to infinity as $\mu \to 0$. So special validation of the technique for two-time-scale motion analysis is required to investigate in depth whether this is the case [Vostrikov (1979)].

Remark 3.3 *Note that this question is also related to the so-called peaking phenomenon discussed in [Sussmann and Kokotović (1991)] for systems with high-gain feedback.*

We should also note that the introduction of the differentiating filter (3.22) leads to a high pulse in the control variable $u(t)$ if a stepwise reference input $r(t)$ is applied. This results in distortion of the output from its desired behavior, and such an undesirable effect can be partly diminished by saturation in the control loop.

3.5 Influence of noise in control system with the highest derivative

The investigation of a noise influence is important in control systems with the highest derivative in feedback, since the output derivatives are used in the control law. Suppose $\bar{y}(t) = x(t) + n_s(t)$ is the sensor output corrupted by a zero-mean, high-frequency noise waveform $n_s(t)$.

First let us consider the case of the ideal differentiating filter with noisy output. Then $x(t)$ should be replaced by $\bar{y}(t)$ in (3.15) and, as a result, in the right member of (3.21) an additional term such as $g^{-1}n_s^{(n)}$ appeared. This term may be so large that saturation of the control variable occurs in real technical systems.

The block diagram representation of the practically realized control system with the highest derivative and high gain in the feedback loop is shown in Fig. 3.7, where the sensor output $y(t) = x(t)$ is corrupted by noise $n_s(t)$.

The influence of the nth derivative $n_s^{(n)}$ of the noise on the behavior of the control variable in the discussed control system with the real differen-

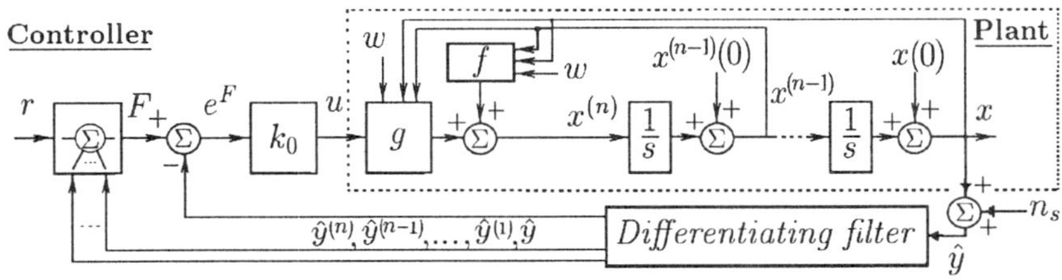

Fig. 3.7 Block diagram of the control system with the highest derivative and high gain in the feedback loop, where the real differentiating filter is used and the output $y(t) = x(t)$ is corrupted by noise $n_s(t)$.

tiating filter (3.22) may be depicted based on the block diagram shown in Fig. 3.8. It is possible to reduce the high-frequency interference by suitable choice of the degree and parameters of the differentiating filter [Vostrikov (1990)].

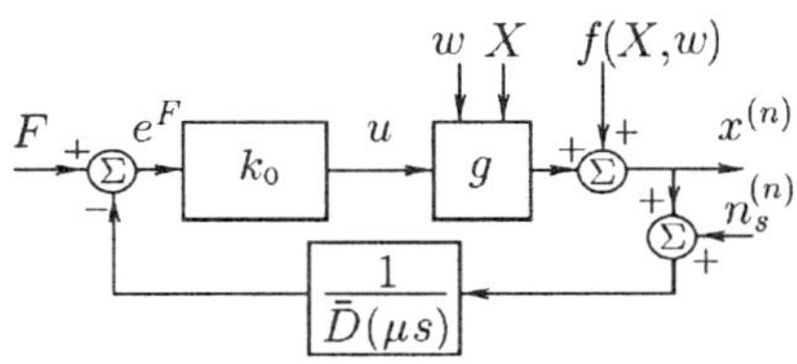

Fig. 3.8 Block diagram of the fast-motion subsystem with sensor noise.

Remark 3.4 *To conclude this section we note that there is a broad set of publications devoted to robot manipulator control by acceleration feedback, for instance, [Lun et al. (1980); Luo and Saridis (1985); Studenny and Belanger (1984); Studenny and Belanger (1986); Davis and Hirschorn (1988); Krutko (1988); Studenny et al. (1991); Krutko (1991); Krutko (1995)], where acceleration feedback control is a special case of control with the highest derivative in feedback (3.15). The problem of acceleration feedback controller design for flexible joint robots was considered in [Kotnik et al. (1988); Readman and Belanger (1991)]. An application of acceleration feedback control for flight control systems was discussed in [Batenko (1977)]. Applications of acceleration feedback control for DC drives are discussed in [Hori (1988); Schmidt and Lorenz (1990); Han et al. (2000)].*

3.6 Desired manifold in the state space of plant model

Let us consider the second method used to specify the desired behavior of the closed-loop system as well as the requirement of insensitivity of the output transients with respect to external disturbances and varying parameters of the plant model. This method is based on the construction of the desired manifold in the state space of the plant model, and is widely used in control techniques related to control systems with sliding motions [Utkin (1977); Utkin (1992)].

Let us consider, similar to the above, the plant model given by the equation (3.1)

$$x^{(n)} = f(X, w) + g(X, w)u,$$

where Assumptions 3.1, 3.2 hold.

First let us construct the desired transfer function of the $(n-1)$th order in the form

$$G^d_{xr}(s) = \frac{1}{T^{n-1}s^{n-1} + a^d_{n-2}T^{n-2}s^{n-2} + \cdots + a^d_1 Ts + 1}. \tag{3.29}$$

From (3.29) the desired differential equation

$$x^{(n-1)} = \frac{1}{T^{n-1}}\{-a^d_{n-2}T^{n-2}x^{(n-2)} - \cdots - a^d_1 Tx^{(1)} - x + r\} \tag{3.30}$$

results, where (3.30) may be rewritten concisely as

$$x^{(n-1)} = \bar{F}(x^{(n-2)}, \ldots, x^{(1)}, x, r), \tag{3.31}$$

where $r = \text{const}$. Note that $X = \{x, x^{(1)}, \ldots, x^{(n-1)}\}^T$ is the state vector of the dynamical system (3.1), and denote

$$S(X, r) = \bar{F}(x^{(n-2)}, \ldots, x^{(1)}, x, r) - x^{(n-1)}, \tag{3.32}$$

where $S(X, r)$ is the error between the actual and desired values of $x^{(n-1)}$. Accordingly, if the condition

$$S(X, r) = 0 \tag{3.33}$$

is satisfied then the desired behavior of (3.1) with prescribed dynamics of (3.31) is fulfilled in the closed-loop system.

Expression (3.33) is the insensitivity condition for the output transients with respect to the external disturbances and varying parameters of the system (3.1). This condition corresponds to assignment of the desired manifold in the state space of the plant model. So, control problem (2.1) has been reformulated as the requirement to provide the motions of the system (3.1) along the manifold (3.33).

3.7 State vector and high gain in feedback loop

Let us consider the plant model given by the equation (3.1)

$$x^{(n)} = f(X, w) + g(X, w)u,$$

where Assumptions 3.1, 3.2 hold and the control law is

$$u = k_0 S(X, r), \tag{3.34}$$

where k_0 is the high gain and $r =$ const. Denote $x_1 = x$, $x_2 = x^{(1)}$, ..., $x_n = x^{(n-1)}$. Then the closed-loop system equations (3.1) and (3.34) may be rewritten as

$$\begin{aligned} \frac{d}{dt}x_i &= x_{i+1}, \qquad i = 1, \dots, n-1, \\ \frac{d}{dt}x_n &= f(X, w) + g(X, w)u, \quad X(0) = X^0, \\ u &= k_0 S(X, r), \end{aligned} \tag{3.35}$$

where there are two-time-scale motions due to the high gain k_0. Therefore, the singular pertubation analysis can be applied [Vasileva (1963); Gerashchenko (1975); Young *et al.* (1977); Kokotović (1984); Saksena *et al.* (1984); Marino (1985)].

The equation describing the behavior of $u(t)$ in (3.35) can be obtained by differentiating (3.34) along the trajectories of (3.1) as discussed, for instance, in [Vostrikov (1988b); Vostrikov (1990)]. Then from (3.35) we obtain the extended system

$$\frac{d}{dt}x_i = x_{i+1}, \qquad i = 1, \dots, n-1, \tag{3.36}$$

$$\bar{\mu}\frac{d}{dt}x_n = \bar{\mu}f(X, w) + g(X, w)\,\mathrm{sgn}(k_0)S(X, r), \quad X(0) = X^0, \tag{3.37}$$

$$\bar{\mu}\frac{d}{dt}u = \mathrm{sgn}(k_0)\left[\sum_{i=1}^{n-1}\frac{\partial S}{\partial x_i}x_{i+1} + \frac{\partial S}{\partial x_n}\{f(X, w) + g(X, w)u\}\right], \tag{3.38}$$

$$u(0) = u^0,$$

which has the form of the standard singularly perturbed equations. Here we use the notation

$$X = \{x_1, x_2, \dots, x_n\}^T = \{x, x^{(1)}, \dots, x^{(n-1)}\}^T \tag{3.39}$$

and $\bar{\mu} = 1/|k_0|$.

If the sign of the gain k_0 is chosen in such a way that the condition

$$\mathrm{sgn}(k_0)\frac{\partial S}{\partial x_n}g(X, w) < 0, \qquad \forall\ (X, w) \in \Omega_{X,w}$$

is satisfied, then the FMS of the extended system is stable and described by

$$\bar{\mu}\frac{d}{dt}x_n = \bar{\mu}f(X, w) + g(X, w)\,\mathrm{sgn}(k_0)S(X, r), \quad X(0) = X^0,$$

$$\bar{\mu}\frac{d}{dt}u = \mathrm{sgn}(k_0)\left[\sum_{i=1}^{n-1}\frac{\partial S}{\partial x_i}x_{i+1} + \frac{\partial S}{\partial x_n}\{f(X, w) + g(X, w)u\}\right], \quad u(0) = u^0,$$

where the gain $|k_0|$ is chosen sufficiently large (i.e., $\bar{\mu} \to 0$), and X, w are the frozen variables during the transients in the FMS.

Consider the quasi-steady state of the differential equation (3.37); that is, assume $\dot{x}_n = 0$. Then, by taking into account (3.32) and (3.39), we obtain

$$\bar{\mu}f(X, w) + g(X, w)\,\mathrm{sgn}(k_0)[\bar{F}(x_{n-1}, \dots, x_1, r) - x_n] = 0. \tag{3.40}$$

Let us derive x_n from (3.40) and substitute x_n into (3.36). As a result, we obtain the SMS given by

$$\frac{d}{dt}x_i = x_{i+1}, \qquad i = 1, \dots, n-2,$$

$$\frac{d}{dt}x_{n-1} = \bar{F}(x_{n-1}, \dots, x_1, r) + [k_0 g(X, w)]^{-1}f(X, w),$$

where the choice of the high gain k_0 depends on the requirements of control accuracy and rejection of external disturbances.

Inasmuch as the control law (3.34) depends on the state vector X, equation (3.34) can be realized practically by introducing a real differentiating filter such as (3.22) or some high-gain observer. The block diagram representation of the realized control system with estimation of state vector and a high gain in the feedback loop is shown in Fig. 3.9.

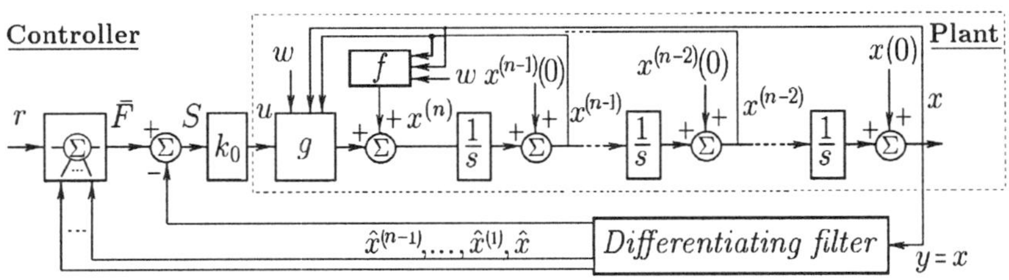

Fig. 3.9 Block diagram of the control system with state vector and a high gain in the feedback loop, where a real differentiating filter is used.

Note that the closed-loop system equations (3.35) represent the case, where the state vector $X(t)$ is measured, for instance, by some ideal differentiating filter (or ideal observer). The block diagram representation of the control system with state vector and high gain in the feedback loop (3.35) is shown in Fig. 3.10, where the part that corresponds to the plant model is encircled with dots and the remainder is the controller. Similar

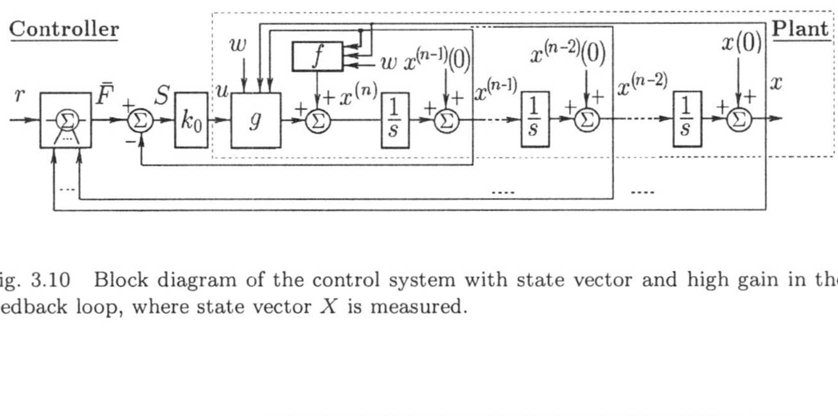

Fig. 3.10 Block diagram of the control system with state vector and high gain in the feedback loop, where state vector X is measured.

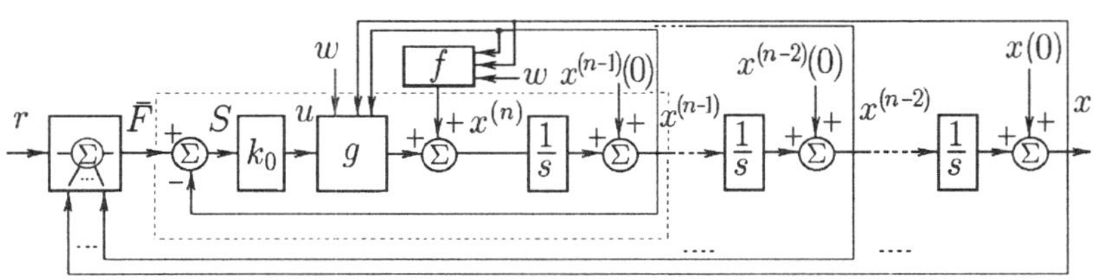

Fig. 3.11 Block diagram of the control system with state vector and high gain in the feedback loop, where the FMS is highlighted by dots.

to the above, let us represent the block diagram in Fig. 3.10 as shown in Fig. 3.11, where the portion encircled by dots corresponds to the FMS of (3.35). The inclusion of the real differentiating filter (3.22) in feedback, as shown in Fig. 3.9, yields some additional fast dynamics. Special analysis of the fast and slow motions in the closed-loop system should be performed in depth, for instance, based on the construction of the extended system. In particular, the behavior of $x^{(n-1)}$ in the FMS can be easily analysed based on the block diagram of Fig. 3.12, where X and w are the frozen variables.

It is clear to see, in contrast to the control system shown in Fig. 3.10, an upper bound for the gain k_0 appears in a control system with the real differentiating filter (3.22). This bound is determined by the requirement for FMS stability, and depends on the parameters of the differentiating filter. Hence the parameters of the differentiating filter should be chosen so that the interval $[g_{\min}, g_{\max}]$ belongs to the region of FMS stability and the FMS transients maintain the allowable performance indices.

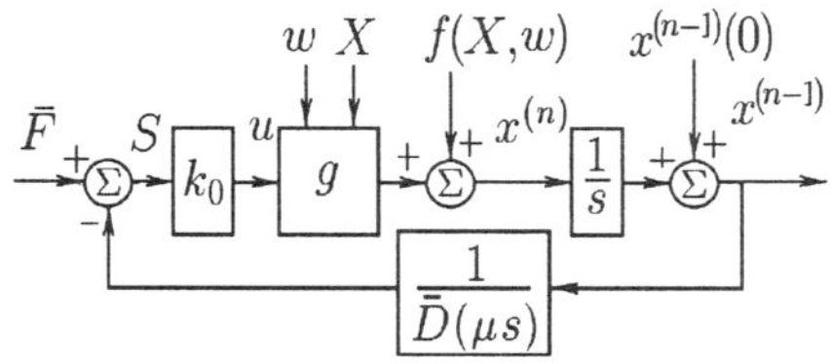

Fig. 3.12 Block diagram of the FMS in the control system with state vector and a high gain in the feedback loop, where the real differentiating filter is used.

3.8 Control systems with sliding motions

Another type of control system, which is related to the one discussed in this book and has analogous properties of high accuracy and robustness with respect to varying parameters and external disturbances, is the variable structure system and, in particular, the control system with sliding motions [Emelyanov (1963); Filippov (1964); Itkis (1976); Utkin (1977); Filippov (1988); Utkin (1992)]. The condition for disturbance rejection in variable structure systems was discussed in [Drazenovic (1969)].[1]

Let us consider the main steps of the sliding motion control system design procedure for the system given by (3.1), where Assumptions 3.1, 3.2

[1]Note that in [DeCarlo *et al.* (1988); Young *et al.* (1999)] the interested reader can find the tutorial papers devoted to sliding mode control systems.

are satisfied.

From (3.34) it follows that if $S(X,r) \neq 0$, then $|u| \to \infty$ as $|k_0| \to \infty$. Accordingly, saturation of the control variable takes place in real technical systems. It leads to consideration of the control law

$$u = u_{\max} \operatorname{sgn} S(X,r), \tag{3.41}$$

where the switching surface $S(X,r)$ is given by (3.32).

Convergence of the closed-loop system trajectories to the manifold $S(X,r)$ can be investigated through a Lyapunov function candidate, for instance, in the form $V(S) = |S(X,r)|$. We can obtain the derivative of $V(S)$ along the trajectories of (3.1) with control law given by (3.41) where $r = \text{const}$. This is

$$\frac{d}{dt}V = \operatorname{sgn}(S)\left[\sum_{i=1}^{n-1}\frac{\partial S}{\partial x_i}x_{i+1} + \frac{\partial S}{\partial x_n}f(X,w)\right] + u_{\max}\frac{\partial S}{\partial x_n}g(X,w), \tag{3.42}$$

where the sign and value of $u_{\max}$ can be chosen in such a way that the condition $dV/dt < 0$ is satisfied in the specified region of the state space of the system (3.1). Since $\partial S/\partial x_n = -1$, we have $\operatorname{sgn} u_{\max} = \operatorname{sgn} g(X,w)$.

The control law (3.41) may be realized practically by introducing a real differentiating filter such as (3.22), and the block diagram representation of such a control system is shown in Fig. 3.13. The above choice of $u_{\max}$ based on consideration of the system shown in Fig. 3.14. The part of the block diagram in Fig. 3.14 that corresponds to the plant model is encircled with dots, and the remainder is the controller.

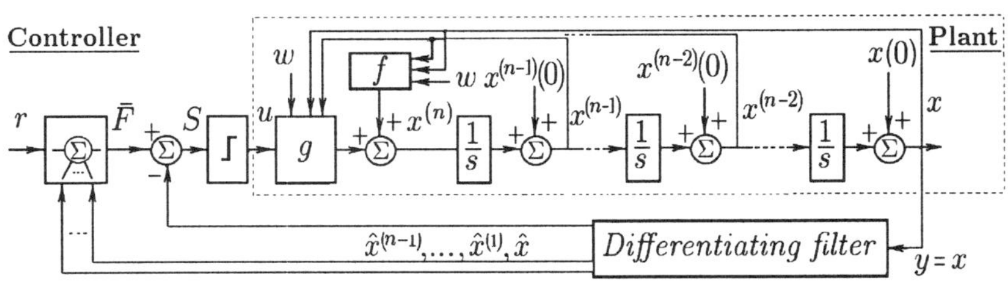

Fig. 3.13 Block diagram of the control system with sliding motions, where a differentiating filter is used.

In accordance with [Vostrikov (1977b)], the block diagram in Fig. 3.14 may be represented as in Fig. 3.15. The portion encircled by dots is where fast transients occur. The inclusion of the real differentiating filter (3.22)

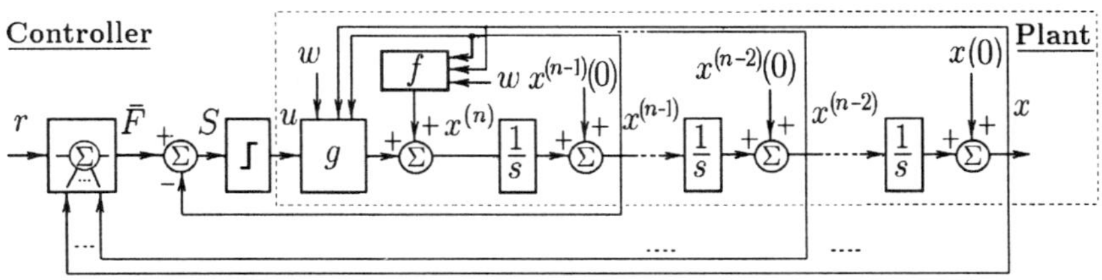

Fig. 3.14 Block diagram of the control system with sliding motions, where the state vector X is measured.

in feedback as shown in Fig. 3.13 yields some additional fast dynamics; then the two-time-scale technique should be used to determine the closed-loop properties. In particular, the behavior of $x^{(n-1)}$ in the FMS can be analysed, using the block diagram shown in Fig. 3.16, via the describing function method, where X is the frozen variable.

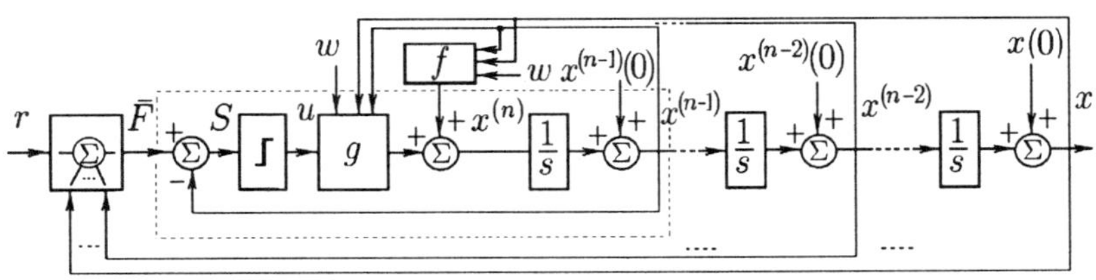

Fig. 3.15 Block diagram of the control system with ideal sliding motions in the circuit highlighted by dots.

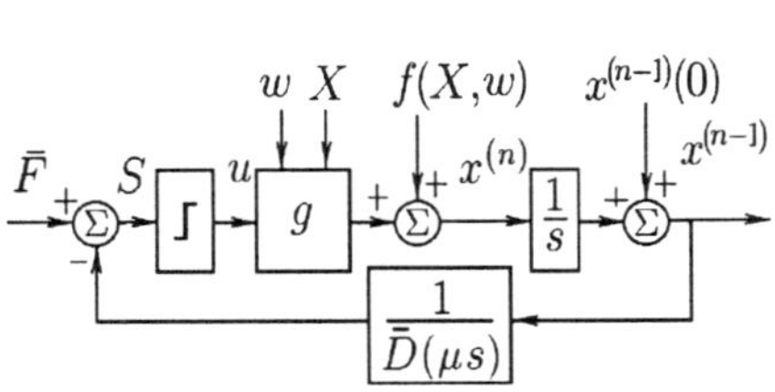

Fig. 3.16 Block diagram of the FMS in the system with sliding motions and real differentiating filter.

Remark 3.5 *Control systems with sliding motions have stimulated active research in various directions. Systems with high-order sliding modes, in* *particular, currently receive a great deal of attention; see [Emelyanov et al. (1993); Fridman and Levant (1996); Bartolini et al. (2000); Perruquetti and Barbot (2002)]. Some recent results related to sliding mode control systems are represented in [Young and Özgüner (1999)].*

3.9 Example

Let us consider the SISO nonlinear continuous-time system given by

$$x^{(2)} = -x + (1 - x^2)x^{(1)} + gu + w, \tag{3.43}$$

where $g = 1 + 0.8\sin(t)$ and we assume that the inequalities

$$|x(t)| \leq 1, \ |x^{(1)}(t)| \leq 1, \ \text{and} \ |r(t)| \leq 0.5 \tag{3.44}$$

hold for all $t \in [0, \infty)$. Consider the control law (3.41) with the real differentiating filter (see Fig. 3.13)

$$\mu^2 \hat{x}^{(2)} + d_1 \mu \hat{x}^{(1)} + \hat{x} = x. \tag{3.45}$$

Let the reference model for $x(t)$ be assigned by

$$x^{(1)} = \frac{1}{T}[r - x].$$

Then we may employ the following control law:

$$\begin{aligned} u &= u_{\max} \operatorname{sgn} S(r, x, \hat{x}^{(1)}) \\ &= u_{\max} \operatorname{sgn}(T^{-1}[r - x] - \hat{x}^{(1)}). \end{aligned} \tag{3.46}$$

From (3.42) and (3.44) we find that $dV/dt < 0$ for $u_{\max} \geq 22.25$.

Consider the FMS represented by the block diagram in Fig. 3.16. Let us assume that there is a limit cycle in the FMS. Determine the limit cycle parameters via the describing function method and, for simplicity, under the assumption that $\bar{F} = 0$, $f = 0$. In accordance with the given parameters we have

$$G_l(j\omega) = \frac{g}{j\omega(-\mu^2\omega^2 + jd_1\mu\omega + 1)}$$

as the frequency-domain transfer function (frequency response) of the linear part in Fig. 3.16. Let $G_l(j\omega)$ reveals a low-pass filtering property and

$$S_1(t) = A\sin(\omega t)$$

be the first harmonic of the Fourier series for the actual periodic signal $S(t)$ in the FMS.

The describing function of the ideal relay nonlinearity is

$$G_n(j, A) = \frac{4u_{\max}}{\pi A}. \tag{3.47}$$

Hence, the solution of the first-order harmonic balance equation

$$1 + G_n(j, A)G_l(j\omega) = 0 \tag{3.48}$$

yields the frequency ω and amplitude A of the stationary oscillations in the FMS:

$$\omega = \frac{1}{\mu} \quad \text{and} \quad A = \frac{4\mu g u_{\max}}{\pi d_1}.$$

The oscillations in the FMS induce the oscillations in the output $x(t)$ of the system (3.43). Let e_{osc} be the amplitude of the stationary oscillations of the output $x(t)$. Consider the stationary oscillation signal $x(t)$ represented by its Fourier series

$$x(t) = x_0 + \sum_{k=1}^{\infty} e_k \sin(k\omega t + \phi_k). \tag{3.49}$$

Then, the effect of chattering in the FMS on the output oscillations can be estimated by the first harmonic in (3.49). Hence, $e_{osc} \approx e_1$. If ω is sufficiently large, it is readily found that

$$\begin{aligned} e_1 &= \frac{|G_n(j, A)g|}{\omega^n} A \\ &= \frac{|\bar{D}(j\mu\omega)|}{\omega^{n-1}} A. \end{aligned} \tag{3.50}$$

From (3.45), (3.50) and taking $n = 2$, $\omega = 1/\mu$, we get

$$e_{osc} \approx e_1 = d_1 \mu A.$$

Let $\mu = 0.02$ s, $u_{\max} = 25$, and $d_1 = 2$. Since $g \in [0.2, 1.8]$, we obtain $\omega \approx 50$ rad/s, $A \in [0.06, 0.6]$, and $e_{osc} \in [0.0024, 0.024]$.

The simulation results of the system (3.43) controlled by the algorithm (3.46) for a step reference input $r(t)$ and a step disturbance $w(t)$ are displayed in Figs. 3.17–3.18, where the initial conditions are $x(0) = 0.5$, $x^{(1)}(0) = 0.5$, $\hat{x}(0) = 0$, $\hat{x}^{(1)}(0) = 0$, and $T = 1$ s. We see that the simulation results confirm the analytical calculations.

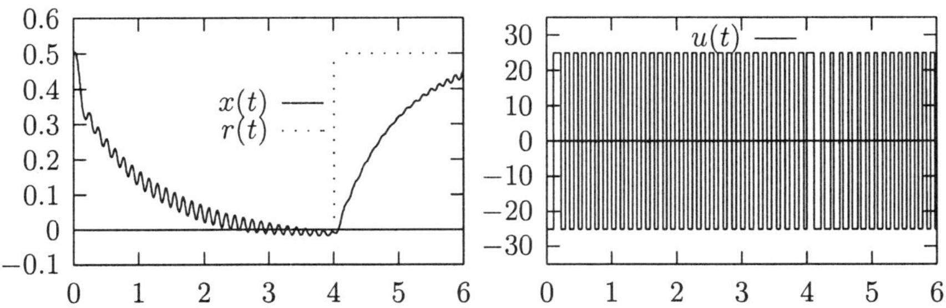

Fig. 3.17 Simulation results for $r(t), x(t)$, and $u(t)$ in the system (3.43) and (3.46).

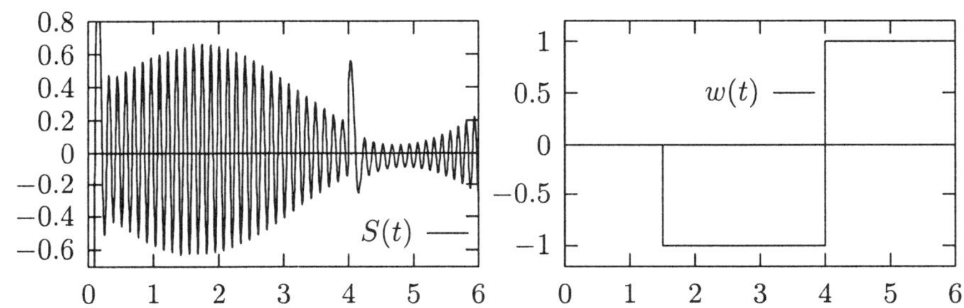

Fig. 3.18 Simulation results for $S(r(t), x(t), \hat{x}^{(1)}(t))$, and $w(t)$ in the system (3.43) and (3.46).

3.10 Notes

From the above considerations it is easy to see that the analysis of practically realizable control systems with the highest derivative of the output signal in the feedback loop, control systems with state vector and high gain in the feedback loop, and control systems with sliding motions, all lead to the singularly perturbed equations of the closed-loop system, where slow and fast motions occur.

Note that in [Vostrikov and Yurkevich (1993b)] the interested reader can find a survey of the results pertaining to control systems with the highest derivative of the output signal and high gain in the feedback loop, based on an approach known as a "localization method". Design of control systems with the highest derivative in feedback are discussed in such references as [Vostrikov and Sarycheva (1982); Vostrikov *et al.* (1982); Vostrikov and Yurkevich (1991); Vostrikov and Yurkevich (1993a); Yurkevich *et al.* (1991); Blachuta *et al.* (1992)], which are also readily available.

The main motivations of further development of control system theory with the highest derivative in the feedback loop, are the following:

- Nonstandard singular perturbation form of the closed loop system equations (3.24)–(3.25).
- Nonzero steady-state error of the reference input realization if the gain in the feedback loop has a finite value.
- If a stepwise reference input $r(t)$ is applied, then a high pulse of the control variable $u(t)$ appears; this distorts the output away from its desired behavior.
- Reference models in the form of the type 2, 3, and higher systems cannot be formulated.
- It is not easy to see how such an approach might be extended to cover digital control system design.

The above mentioned problems may be overcome through the construction of slightly different control-law structures with the highest derivative in the feedback loop. The most important purpose achieved in this way is the development of the uniform approach to continuous as well as digital control system design, and this will be demonstrated in subsequent chapters.

3.11 Exercises

3.1 The differential equation of a plant is given by

$$x^{(2)} = x^2 + |x^{(1)}| + \{1.2 - \cos(t)\}u. \tag{3.51}$$

The reference model for $x(t)$ is assigned by

$$x^{(2)} = -1.2x^{(1)} - x + r. \tag{3.52}$$

Consider the control law (3.15) with real differentiating filter (3.22) where $k_0 = 40$, $q = 2$, $\mu = 0.1$ s, $d_1 = 3$ (see Fig. 3.1). Determine the FMS and SMS equations. Perform a numerical simulation.

3.2 A system is given by (3.51). Consider the control law in the form of (3.15) with the desired dynamics given by (3.52) and the real differentiating filter (3.22) where $k_0 = 40$ and $q = 2$ (see Fig. 3.1). Determine the parameters μ and d_1 of (3.22) such that the damping ratio exceeds 0.5 in the FMS and the degree of time-scale separation between FMS and SMS exceeds 10. Compare with simulation results.

3.3 Consider the system (3.51). Design the control law (3.34) (see Fig. 3.10) such that the step response parameters of the output $x(t)$ meet the requirements $t_s^d \approx 3$ s, $\sigma^d \approx 0\,\%$. Determine the gain k_0 such that the degree of time-scale separation between FMS and SMS the closed-loop system (3.51) and (3.34) exceeds 10. Perform a simulation.

3.4 Consider the system (3.51). Design the control law (3.34) with real differentiating filter (3.22) where $q = 1$ (see Fig. 3.9). Provide that the step response parameters of the output $x(t)$ meet the requirements $t_s^d \approx 10$ s, $\sigma^d \approx 0\,\%$. Determine the gain k_0 and μ of (3.22) such that the degree of time-scale separation between FMS and SMS exceeds 10. Determine the relationship between damping ratio in the FMS and k_0. Compare with simulation results.

3.5 Consider the system (3.51). Design the control law (3.41) (see Fig. 3.14) such that the step response parameters of the output $x(t)$ meet the requirements $t_s^d \approx 6$ s, $\sigma^d \approx 0\,\%$. Determine the sign and magnitude of $u_{\max}$ such that the condition $dV/dt < 0$ is satisfied within the region specified by the inequalities $|x(t)| \leq 2$, $|x^{(1)}(t)| \leq 2$, and $|r(t)| \leq 1$.

3.6 Consider the system (3.51). Design the control law (3.41) and real differentiating filter (3.22) where $q = 2$ (see Fig. 3.13). Provide that the step response parameters of the output $x(t)$ meet the requirements $t_s^d \approx 3$ s, $\sigma^d \approx 0\,\%$. Determine the sign and magnitude of $u_{\max}$ such that the condition $dV/dt < 0$ holds within the region specified by the inequalities $|x(t)| \leq 2$, $|x^{(1)}(t)| \leq 2$, and $|r(t)| \leq 1$. Determine the relationship between frequency oscillations and parameters $u_{\max}$, μ, and d_1 based on the describing function method under the assumption that $\bar{F} = 0$, $f = 0$. Compare with simulation results.

3.7 Replace the sign function in the control law (3.41) by each of the following nonlinear functions, and determine the conditions for limit cycle existence and the limit cycle parameters in the FMS based on the input data of Exercise 3.6, under the assumption that $\bar{F} = 0$, $f = 0$: (a) saturation, (b) relay with dead zone, (c) hysteresis. Compare with simulation results.

Chapter 4

Design of SISO continuous-time control systems

The problem of output regulation of SISO nonlinear time-varying control systems is discussed in this chapter. The control system is designed to provide robust zero steady-state error of the reference input realization. Moreover, the controlled output transients should have a desired behavior. These transients should not depend on the external disturbances and varying parameters of the plant model. The model of the desired output behavior in the form of a desired differential equation is considered, with parameter selection based on the required output step response parameters (overshoot, settling time). Then an insensitivity condition of the output transient behavior with respect to the external disturbances and varying parameters of the system is introduced. The main particularity of the discussed control law lies in the use of the highest derivative in the feedback loop. The closed-loop system properties are analyzed on basis of the two-time-scale technique and, as a result, slow and fast motion subsystems are analyzed separately.

4.1 Controller design for plant model of the 1st order

4.1.1 *Control problem*

In this section we consider a nonlinear system of the form

$$\frac{dx}{dt} = f(x, w) + g(x, w)u, \quad x(0) = x_0, \tag{4.1}$$

where t denotes time, $t \in [0, \infty)$, $y = x$ is the measurable output of the system (4.1), $x \in \mathbb{R}^1$, u is the control, $u \in \Omega_u \subset \mathbb{R}^1$, w is the vector of unknown bounded external disturbances or varying parameters, $w \in \Omega_w \subset \mathbb{R}^l$, and $\|w(t)\| \leq w_{\max} < \infty$, $w_{\max} > 0$.

We assume that dw/dt is bounded for all its components,

$$\|dw/dt\| \leq \bar{w}_{\max} < \infty, \tag{4.2}$$

and that the conditions

$$0 < g_{\min} \leq |g(x, w)| \leq g_{\max} < \infty, \quad |f(x, w)| \leq f_{\max} < \infty \tag{4.3}$$

are satisfied for all $(x, w) \in \Omega_{x,w}$, where $f(x, w)$, $g(x, w)$ are unknown continuous bounded functions of $x(t)$, $w(t)$ on the bounded set $\Omega_{x,w}$ and $\bar{w}_{\max} > 0$, $g_{\min} > 0$, $g_{\max} > 0$, $f_{\max} > 0$.

A control system is being designed so that

$$\lim_{t \to \infty} e(t) = 0, \tag{4.4}$$

where $e(t)$ is an error of the reference input realization: $e(t) = r(t) - x(t)$ and $r(t)$ is the reference input.

Moreover, the output transients should have the desired performance indices. These transients should not depend on the external disturbances and varying parameters $w(t)$ of the system (4.1).

4.1.2 *Insensitivity condition*

The first point of the discussed approach is that the control problem is restated as a problem of determining the root of an equation by introducing a reference model equation whose structure is formed in accordance with the structure of the plant model equations [Boychuk (1966); Vostrikov (1977a)].

From (4.1) it follows that the first derivative of $x(t)$ depends on the control variable $u(t)$ explicitly, and so $x^{(1)}(t)$ is called the highest derivative of the system (4.1).

As any desirable value of the first derivative $x^{(1)}(t)$ may be maintained by a proper choice of the control $u(t)$, let us construct the reference model for (4.1) in the form of the 1st order desired stable differential equation

$$\frac{dx}{dt} = F(x, r). \tag{4.5}$$

For example, let us suppose that (4.5) is the linear differential equation

$$\frac{dx}{dt} = \frac{1}{T}(r - x), \tag{4.6}$$

where $x = r$ at the equilibrium for $r = \text{const}$.

We call (4.5) the desired differential equation, and its right member $F(x, r)$ a desired value of the highest derivative $x^{(1)}(t)$ (desired dynamics). Let us denote

$$e^F = F(x, r) - \frac{dx}{dt}, \tag{4.7}$$

where e^F is an error of the desired dynamics realization. Accordingly, if the condition

$$e^F = 0 \tag{4.8}$$

holds, then the behavior of $x(t)$ with prescribed dynamics of (4.5) is fulfilled.

The expression (4.8) is an insensitivity condition for the behavior of the output $x(t)$ with respect to the external disturbances and varying parameters of the system (4.1).

4.1.3 *Control law with the 1st derivative in feedback loop*

Substitution of (4.1), (4.5), and (4.7) into (4.8) yields

$$F(x, r) - f(x, w) - g(x, w)u = 0. \tag{4.9}$$

So, (4.4) has been reformulated as a problem of finding a solution of the equation $e^F(u) = 0$ when its varying parameters are unknown.

It is easy to see that the root of equation (4.9) is given by

$$u^{NID} = \{g(x, w)\}^{-1}\{F(x, r) - f(x, w)\}, \tag{4.10}$$

where $u^{NID}(t)$ is the analytical solution of (4.9). The control function $u(t) = u^{NID}(t)$ is called a solution of the nonlinear inverse dynamics (NID) [Boychuk (1966); Porter (1970); Slotine and Li (1991)].

Remark 4.1 *Obviously, the control law in the form of (4.10) may be used only if complete information is available about the disturbances, model parameters, and state of the system (4.1).*

In contrast with the approach based on the analytical solution of the nonlinear inverse dynamics problem, let us consider a radically different approach that allows us to satisfy the requirement (4.9) on the condition of incomplete information about varying parameters of the system and external disturbances.

The first way is that, as discussed above, we may consider the system (4.1) with control law (3.15) where $n = 1$. As a result, due to the use of a differentiating filter of the first order, we will have the closed-loop system (3.26)–(3.27). But let us devote our attention here to implementation of the control law with the highest derivative in the feedback loop, which allows us to incorporate integral action in the control loop without increasing the controller's order.

First, in order to obtain some justification for the control law structures introduced below, we notice that the condition (4.8) corresponds to the minimum value of the following unimodal function:

$$V(u) = 0.5\{e^F(u)\}^2. \tag{4.11}$$

Let us consider $V(u)$ as a Lyapunov function candidate. Then the requirement

$$\frac{V(u)}{dt} = \frac{\partial V(u)}{\partial u}\frac{du}{dt} < 0$$

can be satisfied for all $\partial V(u)/\partial u \neq 0$ by the control law in the form

$$\frac{du}{dt} = -k_0 \nabla_u V(u). \tag{4.12}$$

This corresponds to the gradient descent method and, by definition, we have

$$\nabla_u V(u) = \partial V(u)/\partial u = -g(x, w)e^F. \tag{4.13}$$

In accordance with (4.3), the condition

$$\operatorname{sgn}(g(x, w)) = \text{const}$$

is satisfied; then, instead of (4.12), we can use

$$\frac{du}{dt} = k_0 e^F, \tag{4.14}$$

where we assume that

$$k_0 g(x, w) > 0, \quad \forall\ (x, w) \in \Omega_{x,w}.$$

It is easy to see that an equilibrium of (4.14) is the solution of equation (4.9).

For the next step, as a generalization of (4.12), let us consider the control law given by

$$\mu\frac{du}{dt} + d_0 u = -k_0 \nabla_u V(u).$$

In a fashion analogous to the above, the control law for the system (4.1) can be introduced by the following differential equation:

$$\mu \frac{du}{dt} + d_0 u = k_0 e^F, \tag{4.15}$$

where μ is a small positive parameter, k_0 is a high gain, and $d_0 = 1$, or $d_0 = 0$.

According to (4.6) and (4.7), the control law (4.15) may be rewritten in the following form:

$$\mu \frac{du}{dt} + d_0 u = k_0 \left\{ \frac{1}{T}(r - x) - \frac{dx}{dt} \right\}. \tag{4.16}$$

This corresponds to a proper transfer function and, therefore, may be realized without an ideal differentiation of $x(t)$.

Remark 4.2 *It is easy to see that the linear control law (4.16) may be expressed in terms of transfer functions. In particular, it may be rewritten in the form of a so-called two degree-of-freedom feedback system configuration:*

$$u = \frac{k_0}{T(\mu s + d_0)} r - \frac{k_0(Ts + 1)}{T(\mu s + d_0)} x.$$

Remark 4.3 *For purposes of numerical simulation or practical realization, the control law (4.16) may be presented in a state-space form such as*

$$\begin{aligned} \frac{du_1}{dt} &= -\frac{d_0}{\mu} u_1 + k_0 \left\{ \frac{d_0}{\mu} - \frac{1}{T} \right\} x + \frac{k_0}{T} r, \\ u &= \frac{1}{\mu} u_1 - \frac{k_0}{\mu} x. \end{aligned}$$

4.1.4 *Closed-loop system properties*

In accordance with (4.1) and (4.15), the equations of the closed-loop system are given by

$$\begin{aligned} \frac{dx}{dt} &= f(x, w) + g(x, w)u, \quad x(0) = x^0, \\ \mu \frac{du}{dt} &= -d_0 u + k_0 e^F, \quad u(0) = u^0, \end{aligned}$$

or, from (4.7), its identical form

$$\frac{dx}{dt} = f(x,w) + g(x,w)u, \quad x(0) = x^0, \tag{4.17}$$

$$\mu\frac{du}{dt} = -d_0 u + k_0 \left\{ F(x,r) - \frac{dx}{dt} \right\}, \quad u(0) = u^0, \tag{4.18}$$

where the highest output derivative $x^{(1)}(t)$ of the plant model (4.1) is used in feedback.

Substitution of (4.17) into (4.18) yields the closed-loop system equations in the form

$$\frac{dx}{dt} = f(x,w) + g(x,w)u, \quad x(0) = x^0, \tag{4.19}$$

$$\mu\frac{du}{dt} = -\{d_0 + k_0 g(x,w)\}u + k_0\{F(x,r) - f(x,w)\}, u(0) = u^0, \tag{4.20}$$

where μ is a small positive parameter.

Since μ is small, the closed-loop system equations (4.19)–(4.20) are the singularly perturbed equations and, accordingly, the singular perturbation method [Tikhonov (1952); Vasileva (1963); Gerashchenko (1975); Saksena *et al.* (1984); Kokotović *et al.* (1986); Kokotović and Khalil (1986); Sastry (1999)] may be used to analyze the closed-loop system properties; here the main point is that fast and slow transients have been analyzed separately.

First, let us obtain an equation of the FMS. After introducing a new time scale $t_0 = t/\mu$ into (4.19)–(4.20) we have

$$\frac{dx}{dt_0} = \mu\{f(x,w) + g(x,w)u\}, \ x(0) = x^0, \tag{4.21}$$

$$\frac{du}{dt_0} = -\{d_0 + k_0 g(x,w)\}u + k_0\{F(x,r) - f(x,w)\}, \ u(0) = u^0. \tag{4.22}$$

By virtue of (4.21)–(4.22) we have $dx/dt_0 \to 0$ as $\mu \to 0$. Accordingly, in the new time scale t_0 from (4.21)–(4.22) we obtain the fast-motion subsystem (FMS)

$$\frac{du}{dt_0} = -\{d_0 + k_0 g(x,w)\}u + k_0\{F(x,r) - f(x,w)\}, \ u(0) = u^0, \tag{4.23}$$

where we assume that $x(t_0) \approx \text{const}$ during the transients in the subsystem (4.23).

By returning to the primary time scale t, from (4.23) the equation of the FMS

$$\mu\frac{du}{dt} + \{d_0 + k_0 g(x, w)\}u = k_0\{F(x, r) - f(x, w)\} \tag{4.24}$$

is obtained. Here we assume that the state $x(t)$ of the plant model (4.1) and, in accordance with (4.2), the vector $w(t)$ of external disturbances and varying parameters, are constants during the transients in the FMS (4.24), i.e., $w(t) = \text{const}$ and $x(t) = \text{const}$ (frozen variables).

Note that the FMS (4.24) is described by a linear differential equation. It is easy to see that the asymptotic stability[1] and desired sufficiently small settling time of the transients of $u(t)$ can be achieved by a proper choice of the parameters μ, k_0.

Second, let us obtain an equation of the slow-motion subsystem (SMS) under the condition of FMS stability. After the rapid decay of transients in (4.24), we have the steady state (more precisely, quasi-steady state) for the FMS (4.24), where

$$u(t) = u^s(t)$$

and

$$u^s = \frac{k_0}{d_0 + k_0 g(x, w)}\{F(x, r) - f(x, w)\}. \tag{4.25}$$

From (4.10) and (4.25) it follows that

$$u^s = u^{NID} + \frac{d_0}{[d_0 + k_0 g(x, w)]g(x, w)}\{f(x, w) - F(x, r)\}.$$

If the steady state of the FMS (4.24) takes place, then the closed-loop system equations imply that

$$\frac{dx}{dt} = F(x, r) + \frac{d_0}{d_0 + k_0 g(x, w)}\{f(x, w) - F(x, r)\} \tag{4.26}$$

is the equation of the SMS, where (4.26) is an equation with nonvanishing perturbation such as (1.26) and the desired equation (4.5) plays the role of the nominal system (1.2).

As a result of (4.26), if $d_0 = 1$ and $|k_0| \to \infty$, then the transients of $x(t)$ in the SMS are close to the transients of the reference model (4.5).

[1] As the FMS (4.24) is linear, the asymptotic stability of its unique equilibrium point is the same as exponential stability. In short, the FMS is stable.

If $d_0 = 0$, then the SMS (4.26) is the same as the reference model (4.5) and the desired output behavior is fulfilled. Moreover, in this case the integral action is incorporated into the control loop without increasing the controller's order and, accordingly, the robust zero steady-state error is maintained, i.e., $e^s = 0$.

Therefore, fast motions occur in the closed-loop system such that after fast ending of the fast-motion transients, the behavior of the overall singularly perturbed closed-loop system approaches that of the SMS, which is the same as the reference model.

4.2 Controller design for an nth-order plant model

4.2.1 *Control problem*

Let us consider a SISO nonlinear time-varying continuous-time system in the form

$$x^{(n)} = f(X, w) + g(X, w)u, \quad y = x, \quad X(0) = X^0, \tag{4.27}$$

where the variables are defined as follows:

y is the output of system (4.27), available for measurement, $y \in \mathbb{R}^1$;
$X = \{x, x^{(1)}, \dots, x^{(n-1)}\}^T$ is the state vector of the system (4.27);
$X(0) = X^0$ is the initial state, $X^0 \in \Omega_X$;
Ω_X is a bounded set, $\Omega_X \subset \mathbb{R}^n$;
u is the control, $u \in \Omega_u \subset \mathbb{R}^1$;
Ω_u is a bounded set of allowable values of the control variable;
w is the vector of external disturbances or varying parameters, $w \in \Omega_w$;
Ω_w is a bounded set, $\Omega_w \subset \mathbb{R}^l$.

Note that the components $x^{(1)}, \dots, x^{(n-1)}$ of the state vector $X(t)$ and the vector $w(t)$ are unavailable for measurement.

Assumption 4.1 The nonlinear functions $f(X, w)$, $g(X, w)$ are smooth for all $(X, w) \in \Omega_{X,w} = \Omega_X \times \Omega_w$, and the analytic expressions for these functions are unknown.

Assumption 4.2 The conditions

$$|f(X, w)| \le f_{\max} < \infty, \quad 0 < g_{\min} \le |g(X, w)| \le g_{\max} < \infty \tag{4.28}$$

are satisfied for all $(X, w) \in \Omega_{X,w}$, i.e., the functions $f(X, w), g(X, w)$ are bounded for all (X, w) on the specified bounded set $\Omega_{X,w}$.

Remark 4.4 *The influence of the external disturbances and varying parameters of the system (4.27) is expressed by the dependence of the functions $f(X,w)$, $g(X,w)$ on w.*

A control system is being designed to satisfy the condition

$$\lim_{t\to\infty} e(t) = 0, \tag{4.29}$$

where $e(t)$ is the error of the reference input realization, $e(t) = r(t) - y(t)$, and $r(t)$ is the reference input. Moreover, the controlled transients should have the desired behavior. These transients should not depend on the external disturbances and varying parameters of the system (4.27).

4.2.2 *Insensitivity condition*

From (4.27) it follows that any desired value of the nth derivative $x^{(n)}(t)$ may be assigned by a proper choice of the control $u(t)$. Therefore, let us construct the reference model for (4.27) in the form of the nth-order desired stable differential equation.

In the general case the reference model of the desired output transients $x(t)$ for the system (4.27) may be assigned by some stable differential equation

$$x^{(n)} = F(x^{(n-1)}, \dots, x^{(1)}, x, r^{(\rho)}, \dots, r^{(1)}, r), \tag{4.30}$$

where $\rho < n$ and $x = r$ at the equilibrium of (4.30) for $r = \text{const}$.

We call (4.30) the desired differential equation and its right member F the desired value of the highest derivative $x^{(n)}(t)$ (desired dynamics).

Let us rewrite (4.30) in the form

$$x^{(n)} = F(X, R), \tag{4.31}$$

where $R = \{r, r^{(1)}, \dots, r^{(\rho)}\}^T$ and $x = r$ at the equilibrium for $r = \text{const}$. Denote

$$e^F = F - x^{(n)}, \tag{4.32}$$

where e^F is the deviation of $x^{(n)}$ from F.

The value e^F is called the error of the desired dynamics realization which is assigned by equation (4.31).

Accordingly, if the condition

$$e^F = 0 \tag{4.33}$$

holds, then the desired behavior of $x(t)$ with prescribed dynamics of (4.31) is fulfilled.

Expression (4.33) is the insensitivity condition for the behavior of the output $x(t)$ with respect to the external disturbances and varying parameters of the system (4.27).

In accordance with (4.27), (4.31), and (4.32), expression (4.33) may be rewritten in the form

$$F(X, R) - f(X, w) - g(X, w)u = 0. \tag{4.34}$$

So, the control problem (4.29) has been reformulated as the requirement to provide the condition (4.33) or, in other words, to find a solution to (4.34) when its varying parameters are unknown.

If the condition

$$g(X, w) \neq 0, \quad \forall\ (X, w) \in \Omega_{X,w} \tag{4.35}$$

is satisfied, then the control function

$$u(t) = u^{NID}(t)$$

exists such that $u^{NID}(t)$ is the unique solution of (4.34):

$$u^{NID} = \{g(X, w)\}^{-1}\{F(X, R) - f(X, w)\}. \tag{4.36}$$

This is called the nonlinear inverse dynamics solution and, as noted before, the function $u^{NID}(t)$ may be realized in practice as the control function if and only if we have access to complete information about the disturbances, model parameters, and state of the system (4.27).

4.2.3 *Control law with the nth derivative in the feedback loop*

In order to keep hold of (4.33) under the assumption of unknown external disturbances and varying parameters of the system (4.27), let us consider the control law given by the following differential equation:

$$\begin{aligned} &\mu^q u^{(q)} + d_{q-1}\mu^{q-1}u^{(q-1)} + \cdots + d_1\mu u^{(1)} + d_0 u \\ &\quad = -k_0 \nabla_u V(u), \quad U(0) = U^0, \end{aligned} \tag{4.37}$$

where μ is a small positive parameter and

$$q \geq n,\ d_0 = 1 \text{ or } d_0 = 0,\ d_j > 0, \quad \forall\ j = 1, \ldots, q-1,$$

$$U = \{u, u^{(1)}, \ldots, u^{(q-1)}\}^T, \quad U \in \Omega_U \subset \mathbb{R}^q, \quad U(0) \in \Omega_U^0 \subset \Omega_U.$$

From (4.28) we know that the condition

$$\operatorname{sgn}(g(X, w)) = \text{const}, \quad \forall\ (X, w) \in \Omega_{X,w}$$

holds; hence, instead of (4.37) and, in accordance with (4.13), from (4.37) a control law of the form

$$\mu^q u^{(q)} + d_{q-1}\mu^{q-1}u^{(q-1)} + \cdots + d_1\mu u^{(1)} + d_0 u = k_0 e^F, \ U(0) = U^0 \quad (4.38)$$

can be constructed where it is assumed that the condition

$$k_0 g(X, w) > 0, \quad \forall\ (X, w) \in \Omega_{X,w} \quad (4.39)$$

holds.

If the desired differential equation (4.31) is assigned by (2.6), then from (4.38) the control law

$$\begin{aligned} &\mu^q u^{(q)} + d_{q-1}\mu^{q-1}u^{(q-1)} + \cdots + d_1\mu u^{(1)} + d_0 u \\ &= \frac{k_0}{T^n}\{-T^n x^{(n)} - a_{n-1}^d T^{n-1} x^{(n-1)} - \cdots - a_1^d T x^{(1)} - x + r\} \end{aligned} \quad (4.40)$$

follows, where the nth derivative of $x(t)$ in the feedback signal is used.

Accordingly, in case of (2.8) the control law (4.38) may be presented in the form

$$\begin{aligned} &\mu^q u^{(q)} + d_{q-1}\mu^{q-1}u^{(q-1)} + \cdots + d_1\mu u^{(1)} + d_0 u \\ &= \frac{k_0}{T^n}\{-T^n x^{(n)} - a_{n-1}^d T^{n-1} x^{(n-1)} - \cdots - a_1^d T x^{(1)} - x \\ &+ b_\rho^d \tau^\rho r^{(\rho)} + b_{\rho-1}^d \tau^{\rho-1} r^{(\rho-1)} + \cdots + b_1^d \tau r^{(1)} + r\}. \end{aligned} \quad (4.41)$$

It easy to see that the linear differential equation (4.41) may be expressed in terms of transfer functions, and (4.41) may be rewritten as the two degree-of-freedom feedback system configuration

$$u = \tilde{k}\frac{B^d(s)}{D(\mu s)}r - \tilde{k}\frac{A^d(s)}{D(\mu s)}x, \quad (4.42)$$

where

$$\tilde{k} = \frac{k_0}{T^n} \quad (4.43)$$

and

$$D(\mu s) = \mu^q s^q + d_{q-1}\mu^{q-1}s^{q-1} + \cdots + d_1\mu s + d_0. \quad (4.44)$$

Below we will assume that $d_0 = 1$ or $d_0 = 0$.

Remark 4.5 *Let us denote* $q = \deg D(\mu s)$, $n = \deg A^d(s)$, $\rho = \deg B^d(s)$. *If* $q \geq n$, $q \geq \rho$ *then the control law (4.42) corresponds to a proper transfer function and, therefore, may be realized without an ideal differentiation of* $x(t)$ *or* $r(t)$.

Remark 4.6 *If the cost function is assigned by (4.11), then the control law (4.37) is related to the continuous algorithm of the higher order optimization introduced in [Tsypkin (1971)]. If* $q = 1$ *and* $d_0 = 0$, *then (4.37) reduces to the differential descent equation. The problem of output regulation in the context of mechanical control systems was also discussed as the optimization problem using the criterion of minimum acceleration energy in [Krutko (1991); Krutko (1995); Krutko (1996)].*

Remark 4.7 *For numerical simulation or practical realization, the control law (4.41) can be presented in state-space form. Standard procedures for obtaining state equations from transfer functions may be found in many references, e.g., [Brogan (1991); Wolovich (1994)].*

4.2.4 *Fast-motion subsystem*

Standard singular perturbation form of closed-loop system

We now know that the behavior of the closed-loop system under consideration is described by the following differential equations:

$$x^{(n)} = f(X, w) + g(X, w)u, \quad X(0) = X^0, \tag{4.45}$$

$$\mu^q u^{(q)} + d_{q-1}\mu^{q-1}u^{(q-1)} + \cdots + d_1\mu u^{(1)} + d_0 u$$
$$= k_0\{F(X, R) - x^{(n)}\}, \; U(0) = U^0. \tag{4.46}$$

In order to analyze the closed-loop system properties, let us consider some transformations. First, substituting (4.45) into the right member of (4.46), we get

$$x^{(n)} = f(X, w) + g(X, w)u, \quad X(0) = X^0, \tag{4.47}$$

$$\mu^q u^{(q)} + d_{q-1}\mu^{q-1}u^{(q-1)} + \cdots + d_1\mu u^{(1)} + \{d_0 + k_0 g(X, w)\}u$$
$$= k_0\{F(X, R) - f(X, w)\}, \quad U(0) = U^0. \tag{4.48}$$

Next, the equations of the closed-loop system (4.47)–(4.48) may be rewritten in the following vector form:

$$\begin{aligned}
\frac{d}{dt}x_i &= x_{i+1}, \qquad i = 1, \dots, n-1, \\
\frac{d}{dt}x_n &= f + gu_1, \qquad X(0) = X^0, \qquad\qquad (4.49) \\
\mu\frac{d}{dt}u_j &= u_{j+1}, \qquad j = 1, \dots, q-1, \qquad U_1(0) = U_1^0, \\
\mu\frac{d}{dt}u_q &= -\{d_0 + k_0 g\}u_1 \cdots - d_{q-1}u_q + k_0\{F - f\}, \qquad (4.50)
\end{aligned}$$

where

$$U_1 = \{u_1, u_2, \dots, u_q\}^T$$

and

$$u_j = \mu^{j-1}u^{(j-1)}, \quad \forall\, j = 1, \dots, q. \qquad (4.51)$$

Note that f, g are functions of $X(t)$ and $w(t)$.

Since μ is a small parameter, the closed-loop system equations (4.49)–(4.50) are the singularly perturbed differential equations. If $\mu \to 0$, then fast and slow modes are forced in the closed-loop system and the time-scale separation between these modes depends on the controller parameter μ. The closed-loop system properties can be analyzed on basis of the two-time-scale technique [Tikhonov (1952); Vasileva (1963); Gerashchenko (1975); Saksena *et al.* (1984)] and, as a result, slow and fast motion subsystems are analyzed separately. This, along with Theorem 1.3 on hand, gives the justification, it will be shown in later sections, that stability conditions imposed on the fast and slow modes, and a sufficiently large mode separation rate, can ensure that the full-order closed-loop system achieves desired properties: the output transient performances are as desired, and they are insensitive to parameter variations and external disturbances.

Fast-motion subsystem equation

Let us introduce the new fast time scale $t_0 = t/\mu$ into the closed-loop system equations (4.49)–(4.50). We obtain

$$\begin{aligned} \frac{d}{dt_0}x_i &= \mu x_{i+1}, \quad i = 1, \ldots, n-1, \\ \frac{d}{dt_0}x_n &= \mu\{f + gu_1\}, \end{aligned} \tag{4.52}$$

$$\begin{aligned} \frac{d}{dt_0}u_j &= u_{j+1}, \quad j = 1, \ldots, q-1, \\ \frac{d}{dt_0}u_q &= -\{d_0 + k_0 g\}u_1 - d_1 u_2 \cdots - d_{q-1}u_q + k_0\{F - f\}, \end{aligned} \tag{4.53}$$

as the closed-loop system equations in the new time scale t_0.

It is easy to see that as $\mu \to 0$, we get the FMS equations in the new time scale t_0, that is

$$\begin{aligned} \frac{d}{dt_0}x_i &= 0, \quad i = 1, \ldots, n, \\ \frac{d}{dt_0}u_j &= u_{j+1}, \quad j = 1, \ldots, q-1, \\ \frac{d}{dt_0}u_q &= -\{d_0 + k_0 g\}u_1 - d_1 u_2 \cdots - d_{q-1}u_q + k_0\{F - f\}. \end{aligned}$$

Then, returning to the primary time scale $t = \mu t_0$, we obtain the following FMS:

$$\begin{aligned} x_i &= \text{const}, \quad i = 1, \ldots, n, \\ \mu\frac{d}{dt}u_j &= u_{j+1}, \quad j = 1, \ldots, q-1, \\ \mu\frac{d}{dt}u_q &= -\{d_0 + k_0 g\}u_1 - d_1 u_2 \cdots - d_{q-1}u_q + k_0\{F - f\}. \end{aligned} \tag{4.54}$$

These may be rewritten as

$$\begin{aligned} \mu^q u^{(q)} + d_{q-1}\mu^{q-1}u^{(q-1)} + \cdots + d_1\mu u^{(1)} + \{d_0 + k_0 g(X, w)\}u \\ = k_0\{F(X, R) - f(X, w)\}, \quad U(0) = U^0, \end{aligned} \tag{4.55}$$

where the vector $X(t)$ and, in accordance with (4.2), the vector $w(t)$ of external disturbances and varying parameters are constants during the transients in (4.55), i.e., $X(t) = \text{const}$ and $w(t) = \text{const}$.

Note that the FMS (4.55) is a linear differential equation with frozen parameters; hence, linear control system methods can be applied for stability analysis of (4.55). We summarize the main results in the following theorem.

Theorem 4.1 *If $\mu \to 0$, then in the closed-loop system (4.27) and (4.38) the two-time-scale motions appear where the FMS is described by equation (4.55).*

Remark 4.8 *If μ is sufficiently small, the condition $d_0 + k_0 g(X, w) \approx$ const holds during the transients of the FMS (4.55). Accordingly, the FMS (4.55) may be examined as the linear differential equation, where $F(X, R)$ and $f(X, w)$ play the role of disturbances.*

Let us denote

$$\gamma(X, w) = d_0 + k_0 g(X, w).$$

Remark 4.9 *As the value of $g(X, w)$ is unknown, then in accordance with (4.28) and (4.39) we know only that the inequalities*

$$0 < \gamma_{min} \leq \gamma(X, w) \leq \gamma_{max} \tag{4.56}$$

are satisfied for all $(X, w) \in \Omega_{X,w}$, i.e., $\gamma(X, w)$ is bounded and positive-valued for all (X, w) in the specified set $\Omega_{X,w}$.

Remark 4.10 *The parameters $\mu, k_0, d_0, \ldots, d_{q-1}$ can be chosen so that the asymptotic stability and sufficiently small settling time of the FMS (4.55) are achieved for all possible values γ from the given interval (4.56).*

Steady state of the fast-motion subsystem

Suppose the FMS (4.55) is stable. Taking $\mu \to 0$ in (4.55) we get

$$u(t) = u^s(t), \tag{4.57}$$

where $u^s(t)$ is a steady state (more precisely, quasi-steady state) of the FMS (4.55) and

$$u^s = \frac{k_0}{d_0 + k_0 g(X, w)} \{F(X, R) - f(X, w)\}. \tag{4.58}$$

Note that by (4.36) the expression (4.58) may be rewritten in the following form:

$$u^s = u^{NID} + \frac{d_0}{[d_0 + k_0 g(X,w)]g(X,w)}\{f(X,w) - F(X,R)\}.$$

There are two special cases here.

(a) If $d_0 \neq 0$ (in particular, if $d_0 = 1$), then from (4.58) it follows that

$$\lim_{|k_0|\to\infty} u^s(k_0) = u^{NID}.$$

(b) If $d_0 = 0$, then from the FMS (4.55) we obtain

$$\lim_{\mu\to 0} u(\mu) = u^s = u^{NID}.$$

As a result of the above, we find that if $k_0 g/d_0 \to \infty$, then the FMS steady state $u^s(t)$ tends to the nonlinear inverse dynamics solution $u^{NID}(t)$ given by (4.36).

4.2.5 *Slow-motion subsystem*

Let us assume that the FMS (4.55) is asymptotically stable. By letting $\mu \to 0$ in (4.47)–(4.48), we find that

$$x^{(n)} = F(X,R) + \frac{d_0}{d_0 + k_0 g(X,w)}\{f(X,w) - F(X,R)\}, \quad X(0) = X^0 \tag{4.59}$$

describes the SMS.

We therefore have the following theorem.

Theorem 4.2 *If the FMS (4.55) is asymptotically stable, then, after the rapid decay of fast-motion transients, the behavior of the SMS is described by (4.59).*

Note that if $d_0 \neq 0$ (e.g., $d_0 = 1$) then from (4.59) we have

$$\lim_{|k_0|\to\infty} x^{(n)}(k_0) = F(X,R),$$

i.e., $\lim_{|k_0|\to\infty} e^F(k_0) = 0$. In other words, the equation of the SMS, (4.59), tends to the desired differential equation (4.31) if the high gain k_0 is used (more precisely, if the condition

$$k_0 g(X,w) \gg d_0 \tag{4.60}$$

holds). On the other hand, if $d_0 = 0$ then from (4.47)–(4.48) the expression

$$\lim_{\mu \to 0} x^{(n)}(\mu) = F(X, R)$$

follows, i.e., $\lim_{\mu \to 0} e^F(\mu) = 0$. In contrast to the previous case, the SMS equation is the same as the desired differential equation (4.31) even if

$$k_0 g \approx 1.$$

In this case the integral action is incorporated in the control loop without increasing the controller's order and, accordingly, the robust zero steady-state error, $e^s = 0$, of the reference input realization is maintained in the closed-loop system.

Theorem 4.3 *If $k_0 g/d_0 \to \infty$, then the SMS equation (4.59) tends to the desired differential equation (4.31), i.e., $\lim_{k_0 g/d_0 \to \infty} e^F = 0$.*

Remark 4.11 *Note, that the SMS equation tends to (4.31) despite the fact that there are varying parameters of the plant model (4.27) and unknown external disturbances. Accordingly, the desired output transients are guaranteed in the closed-loop system after fast damping of FMS transients. So, if a sufficient time-scale separation between the fast and slow modes in the closed-loop system and stability of the FMS are provided, then the output transient performance indices are insensitive to parameter variations and external disturbances in (4.27).*

4.2.6 *Influence of small parameter*

Let us consider the subsystem (4.53) in the closed-loop system equations (4.52)–(4.53) represented in the new time scale t_0, i.e., the FMS given by

$$\frac{d}{dt_0} U_1 = \mathcal{A}_{FMS} U_1 + \mathcal{B}_{FMS} k_0 \{F(X, R) - f(X, w)\}, \qquad (4.61)$$

where

$$\mathcal{A}_{FMS} = \begin{bmatrix} 0 & 1 & \cdots & 0 \\ 0 & 0 & \cdots & 0 \\ \cdots & \cdots & \cdots & \cdots \\ 0 & 0 & \cdots & 1 \\ -\gamma & -d_1 & \cdots & -d_{q-1} \end{bmatrix}, \quad \mathcal{B}_{FMS} = \begin{bmatrix} 0 \\ 0 \\ \cdots \\ 0 \\ 1 \end{bmatrix},$$

$\gamma = d_0 + k_0 g(X, w)$, and U_1 is the state vector defined by (4.51).

Assume that the FMS is stable. Then, from (4.61), we find that

$$U_1^s = -\mathcal{A}_{FMS}^{-1}\mathcal{B}_{FMS}k_0\{F - f\}$$

is the unique equilibrium point; in particular, from (4.54) it is clear that $U_1^s = [u_1^s, 0, \cdots, 0]^T$, where u_1^s is defined by (4.58). For the sake of simplicity, put $d_0 = 0$ so that $u_1^s(t) = u^{NID}(t)$. By (4.36) we have $u^{NID}(t) = \varphi(X, R, w)$. Assume that both R and w are smooth and bounded; then the function $du^{NID}(t)/dt = d\varphi(X, R, w)/dt$ is bounded as well. Note that

$$\begin{aligned}\frac{d}{dt_0}U_1^{NID} &= \mu\frac{d}{dt}U_1^{NID}\\ &= \mu\frac{d}{dt}\varphi(X, R, w)\\ &= \mu\tilde{\varphi}(X, \dot{X}(X, R, w), R, \dot{R}, w, \dot{w}).\end{aligned} \tag{4.62}$$

Denote

$$\Delta U_1 = U_1 - U_1^{NID}. \tag{4.63}$$

By differentiating (4.63) with respect to t_0 and using (4.61), we get

$$\frac{d}{dt_0}\Delta U_1 = \mathcal{A}_{FMS}\Delta U_1 + \mu\tilde{\varphi}(X, R, \dot{R}, w, \dot{w}), \tag{4.64}$$

where $\mathcal{A}_{FMS}$ is the matrix with the frozen parameter $\gamma(X, w)$. The function $\tilde{\varphi}$ plays the role of perturbation in the FMS (4.64).

The stability of the system (4.64) and an error caused by finiteness of the small parameter μ may be investigated by Lyapunov's method as was done for the system (1.26). In addition, the criteria η_1, η_2, and η_3 (see p. 15) can be used to estimate the degree of time-scale separation between stable fast and slow motions in the closed-loop system.

4.2.7 *Geometric interpretation of control problem solution*

Let X_Σ be the state vector of the closed-loop system (4.49)–(4.50), where $X_\Sigma = \{X^T, U_1^T\}^T$, $X = \{x_1, x_2, \ldots, x_n\}^T$, and $U_1 = \{u_1, u_2, \ldots, u_q\}^T$. So, the state space of the closed-loop system (4.49)–(4.50) is the Cartesian product

$$\Omega_\Sigma = \Omega_X \times \Omega_{U_1}.$$

The set of trajectories of the closed-loop system, considered as a flow, is stratified into a fast flow in Ω_{U_1} and a slow flow in Ω_X as $\mu \to 0$, where we have contractile flow of the FMS into a neighborhood of the point $U_1^s(t)$ and contractile flow of the SMS into a neighborhood of the point $\{r, 0, \ldots, 0\}^T$ for $r = \text{const}$.

Let us define the manifold M_F as the set

$$M_F = \{(x^{(n)}, X, u) \mid x^{(n)} - F(X, R) = 0\},$$

which corresponds to the desired differential equation (4.30). The manifold

$$M_p = \{(x^{(n)}, X, u) \mid x^{(n)} - f(X, w) - g(X, w)u = 0\}$$

is the set of points $(x^{(n)}, X, u)$ satisfying the plant model given by (4.27). Then the control problem solution corresponds to the motion along the intersection of M_F and M_p:

$$M_\Sigma = M_F \cap M_p,$$

as shown in Fig. 4.1.

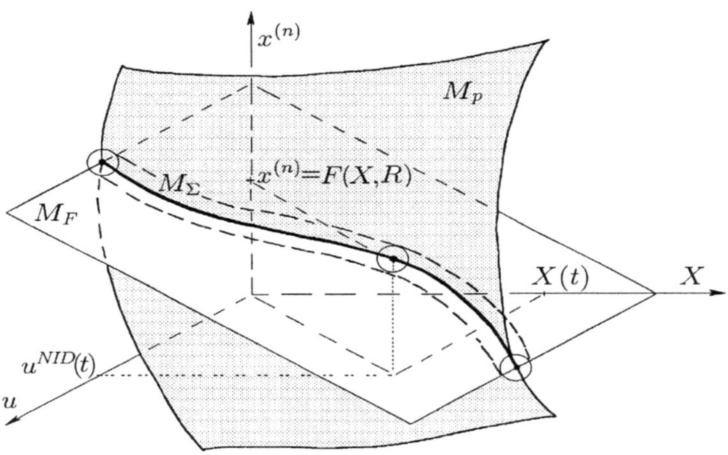

Fig. 4.1 Geometric interpretation of control problem solution in the system with the highest derivative of the output signal in the feedback loop.

4.3 Example

Consider the control law given by (4.41). By taking $q = n = 2$, we obtain

$$\mu^2 u^{(2)} + d_1 \mu u^{(1)} + d_0 u = k_0\{-x^{(2)} - \frac{a_1^d}{T} x^{(1)} + \frac{b_1^d \tau}{T^2} r^{(1)} + \frac{1}{T^2}[r - x]\}. \quad (4.65)$$

Let us rewrite the control law (4.65) in state-space form, e.g.,

$$\begin{aligned}
\dot{u}_1 &= -\frac{d_1}{\mu}u_1 + u_2 + k_0\left\{\frac{d_1}{\mu} - \frac{a_1^d}{T}\right\}x + \frac{k_0 b_1^d \tau}{T^2}r, \\
\dot{u}_2 &= -\frac{d_0}{\mu^2}u_1 + k_0\left\{\frac{d_0}{\mu^2} - \frac{1}{T^2}\right\}x + \frac{k_0}{T^2}r, \\
u &= \frac{1}{\mu^2}\{u_1 - k_0 x\}.
\end{aligned} \tag{4.66}$$

Then, from (4.66), we get the block diagram as shown in Fig. 4.2.

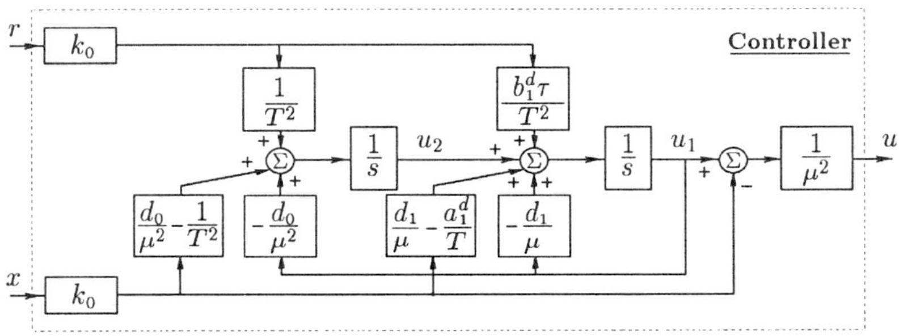

Fig. 4.2 Block diagram of the control law (4.65) represented in the form (4.66).

4.4 Notes

It has been shown that the discussed dynamical control law with the highest derivative of the output signal in the feedback loop allows us to represent the closed-loop system in the form of the standard singular perturbation system. In particular, the singularly perturbed part of this system consists of the equations of the constructed controller, where the controller corresponds to a proper transfer function and, therefore, may be realized without an ideal differentiation of the output variable or the reference variable.

As a result, two-time-scale motions are induced in the closed-loop system, where stability of the fast transients and the desired degree of time-scale separation between the fast and slow motions can be provided by choice of controller parameters. Then, after the fast motion transients have ended, the behavior of the overall closed-loop system gets close to the behavior of the slow motion subsystem. Moreover, by choice of controller parameters it is possible to ensure that the slow motion subsystem is the same as the assigned reference model.

We can now enumerate the main steps of the design procedure for nonlinear control systems with the highest derivative in the feedback loop and plant model given by (4.27).

- The model of the desired output behavior in the form of the nth-order desired differential equation (4.30) is introduced. Its parameters are selected based on the required output step response parameters (overshoot, settling time). As a result, the output regulation problem (4.29) is reformulated as the requirement to provide the insensitivity condition (4.33).
- The control law structure is chosen in the form of the differential equation (4.38), where the highest-order derivative $x^{(n)}$ is used in the feedback loop. If the desired differential equation (2.8) is used, then (4.38) has the form (4.41) and, in accordance with Remark 4.5, the control law (4.41) may be realized without an ideal differentiation of $x(t)$ or $r(t)$.
- The closed-loop system properties are analyzed on the basis of the two-time-scale technique and, as a result, the FMS equation (4.55) and the SMS equation (4.59) are obtained.
- Finally, the parameters μ, k_0, and d_j of the control law (4.38) should be selected based on the (1) required stability of the fast transients, (2) desired degree of time-scale separation between the fast and slow modes, (3) required control accuracy and rejection of external disturbances in the closed-loop system, and (4) and required high-frequency sensor noise attenuation.

Problems concerning the final step of this design procedure are discussed in detail in the next chapter. There we clarify the implementation of the control systems we have discussed, while at the same time making allowances for various practical restrictions.

4.5 Exercises

4.1 A behavior of a plant is described by the equation

$$x^{(2)} = x + x|x^{(1)}| + \{2 + \sin(t)\}u.$$

Assume that the inequalities

$$|x(t)| \leq 2, \quad |x^{(1)}(t)| \leq 10, \quad |r(t)| \leq 1, \quad |u(t)| \leq u_{\max}$$

hold for all $t \in [0, \infty)$. The reference model is chosen in the form

$$x^{(2)} = -2x^{(1)} - x + r.$$

Determine the lower bound of $u_{\max}$ for which the desired behavior of the plant can be provided in the specified region.

4.2 The differential equation of a plant is

$$x^{(2)} = x + x|x^{(1)}| + \{2 + \sin(t)\}u,$$

while that of the reference model is

$$x^{(2)} = -3.2x^{(1)} - x + 3.2r^{(1)} + r.$$

Construct the control law in the form of (4.41) where $q = 3$. Determine the FMS and SMS equations from the closed-loop system equations.

4.3 The differential equation of a plant is

$$x^{(1)} = x^2 + \{1.5 - \sin(t)\}u.$$

The step response parameters of the output $x(t)$ should meet the requirements $t_s^d \approx 3$ s, $\sigma^d \approx 0\,\%$. Construct the control law in the form of (4.40) where $q = 2$. Determine the FMS and SMS equations from the closed-loop system equations.

4.4 The differential equation of a plant is

$$x^{(2)} = xx^{(1)} + |x| + \{1.2 - \cos(t)\}u.$$

The step response parameters of the output $x(t)$ should meet the requirements $t_s^d \approx 1$ s, $\sigma^d \approx 10\,\%$. Construct the control law in the form of (4.40) where $q = 2$. Determine the FMS and SMS equations from the closed-loop system equations.

4.5 The differential equation of a plant is

$$x^{(3)} = x^{(2)} + x^{(1)} + |x|^2 + \{1.1 + \cos(x)\}u.$$

The step response parameters of the output $x(t)$ should meet the requirements $t_s^d \approx 3$ s, $\sigma^d \approx 0\,\%$. Construct the control law in the form of (4.40) where $q = 3$. Determine the FMS and SMS equations from the closed-loop system equations. Obtain a state-space representation and sketch the block diagram of the control law.

4.6 Obtain a state-space representation and sketch the block diagram of the control law (4.41) where $q = 3, n = 3, \rho = 1$.

Chapter 5

Advanced design of SISO continuous-time control systems

In Chapter 4 we presented a qualitative analysis of the properties of SISO continuous-time systems with the highest derivative in feedback control. This analysis was based on the singular perturbation method. In the present chapter we obtain relationships helpful in choosing the control law parameters of (4.38). We seek robustness of the closed-loop system properties in a specified region of the state space of the system under the assumption that we lack complete information about varying parameters of the system and external disturbances.

We begin by discussing the problem of accuracy analysis and the choice of the high gain in accordance with the requirements of control accuracy and rejection of external disturbances. As the FMS equation (4.55) is a linear differential equation, the results of linear control theory may be used to analyze the FMS properties. Therefore, in the next part of the chapter, the design of control law parameters based on desired root placement of the FMS characteristic polynomial is presented. Finally, the influence of high-frequency noisy measurements and varying parameters, as well as the application of the frequency domain approach to make a choice of the controller parameters, are considered.

5.1 Control accuracy

5.1.1 *Steady state of fast-motion subsystem*

Let us consider the closed-loop system equations (4.45)–(4.46). The relationships used to choose the high gain k_0 in the control law (4.38) can be based on the requirements placed on the error of the output behavior under either of the conditions that the steady state of the FMS or that of the SMS occurs.

Let us assume that a steady state of the FMS (4.55) occurs, i.e., that the condition (4.57) holds. Then, instead of the insensitivity condition (4.33), we may consider the requirement

$$|e^F(u^s)| \leq e^F_{\max}, \tag{5.1}$$

which should be satisfied within the bounded domains $\Omega_{X,R}$ and Ω_w. By definition,

> e^F is the error of the desired dynamics realization and the desired dynamics are assigned by (4.31),
> $e^F_{\max} = \bar{\varepsilon}_F F_{\max}$,
> $\bar{\varepsilon}_F$ is a relative error of desired dynamics realization, for example, $\bar{\varepsilon}_F \in [0.01, 0.1]$,
> $F_{\max}$ is a constant defined by $F_{\max} = \max_{\Omega_{X,R}} |F(X,R)|$,
> $\Omega_{X,R}$ is a bounded set.

In accordance with Theorem 4.3, it is easy to see that the requirement (5.1) can be provided by proper choice of the parameters k_0, d_0. In particular, from equation (4.59) for the SMS the expression

$$e^F = \left[\frac{d_0}{d_0 + k_0 g(X,w)}\right]\{F(X,R) - f(X,w)\} \tag{5.2}$$

follows, where by direct calculation we can find that (5.2) is the same as

$$e^F = \frac{d_0}{k_0} u^s.$$

Substitution of (5.2) into (5.1) gives

$$|k_0| \geq \frac{d_0}{g_{\min}} \left[\frac{\max\limits_{\Omega_{X,R,w}} |F(X,R) - f(X,w)|}{e^F_{\max}} - 1\right], \tag{5.3}$$

where k_0 corresponds to (5.1) and $g_{\min}$ is defined by (4.28).

5.1.2 *Steady state of slow-motion subsystem*

Let us consider a steady state of the SMS (4.59) by assuming that the conditions

$$X = X^s = \{x^s, 0, \ldots, 0\}^T, \; x(t) = x^s = \text{const}, \; \text{and} \; r(t) = r^s = \text{const}$$

are satisfied in (4.59). Then we have

$$e = e^s, \quad e^s = r^s - x^s.$$

Let the allowable steady-state error of the reference input realization be assigned by the requirement

$$|e^s(X^s, w)| \leq e^s_{\max}, \tag{5.4}$$

where

$e^s_{\max} = \overline{\varepsilon}_r r_{\max}$,
$r_{\max} = \max\limits_{t\in[0,\infty)} |r(t)|$,
$\overline{\varepsilon}_r$ is a relative error of the reference input realization, e.g., $\overline{\varepsilon}_r \in [0, 0.1]$.

From (4.59) and (2.8) it follows that

$$e^s = -\frac{d_0 T^n f(X^s, w)}{k_0 g(X^s, w)}. \tag{5.5}$$

Then by (5.5) and (5.4) we see that (5.4) is satisfied in a specified region assigned by the inequalities (4.28) if the condition

$$|k_0| \geq \frac{d_0 T^n f_{\max}}{e^s_{\max} g_{\min}} \tag{5.6}$$

holds.

So if $d_0 \neq 0$, then (5.3) and (5.6) may be used to choose the high gain k_0 in accordance with the requirements on admissible error of the output behavior.

If $d_0 = 0$, then the integral action is incorporated in the control loop without increasing the controller's order and, accordingly, the robust zero steady-state error of the reference input realization is provided. In this case we need not employ a large gain k_0. From a practical viewpoint the advisable selection of k_0 corresponds to the condition $k_0 g \approx 10$. Note that a decrease in $k_0 g$ leads to an increase in the dynamical error of the reference input realization under the condition of a ramp reference input or a ramp disturbance. This effect can easily be shown based on the expression for the velocity error discussed below.

5.1.3 Velocity error due to external disturbance

Let us consider a SISO linear continuous-time system in the form

$$x^{(n)} = \overline{a}X + bu + \hat{b}w, \quad X(0) = X^0, \tag{5.7}$$

where $X = \{x, x^{(1)}, \ldots, x^{(n-1)}\}^T$, $\overline{a} = \{a_0, a_1, \ldots, a_{n-1}\}$ is a row vector of real constant coefficients, $\overline{a} \in \mathbb{R}^{1\times n}$, $b = \text{const}$, $\hat{b} = \text{const}$, and $y = x$.

Substituting $d_0 = 0$ into (4.41), we obtain the control law

$$\begin{aligned} &\mu^q u^{(q)} + d_{q-1}\mu^{q-1}u^{(q-1)} + \cdots + d_1\mu u^{(1)} \\ &= \frac{k_0}{T^n}\{-T^n x^{(n)} - a_{n-1}^d T^{n-1}x^{(n-1)} - \cdots - a_1^d T x^{(1)} - x \\ &\quad + b_\rho^d \tau^\rho r^{(\rho)} + b_{\rho-1}^d \tau^{\rho-1} r^{(\rho-1)} + \cdots + b_1^d \tau r^{(1)} + r\}. \end{aligned} \tag{5.8}$$

Let us also assume that $r(t) = \text{const}$ and the external disturbance $w(t)$ is a signal of type 1 (ramp external disturbance):

$$w(t) = w^v t\, 1(t).$$

Then the steady state of the closed-loop system (5.7) and (5.8) gives

$$\lim_{t\to\infty} x(t) = x^v,$$

where $x^v = \text{const}$.

Let us introduce the following notation:

e_w^v is a velocity error due to the external disturbance $w(t)$, where $e_w^v = r - x^v$,

$\overline{e}_w^v$ is a relative velocity error due to the external disturbance $w(t)$, where $\overline{e}_w^v = e_w^v / w^v$.

It is easy to find that

$$\overline{e}_w^v = -\frac{\hat{b}T^n \mu d_1}{k_0 b}. \tag{5.9}$$

At the same time, in the closed-loop system (5.7) and (5.8) the following limit exists:

$$\lim_{t\to\infty} \{u(t) - \hat{u}(t)\} = 0,$$

where

$$\hat{u}(t) = -\frac{a_0}{b}\{r - e_w^v\} - \frac{\hat{b}}{b} w^v\, t1(t). \tag{5.10}$$

5.1.4 *Velocity error due to reference input*

Assume $w(t) \equiv 0$ and the reference input $r(t)$ is a signal of type 1 (ramp reference input):

$$r(t) = r^v t\, 1(t).$$

On one hand, we consider the reference model assigned by (2.6) and, accordingly, have the desired differential equation (4.31) in the form

$$x^{(n)} = T^{-n}\{-a_{n-1}^d T^{n-1} x^{(n-1)} - \cdots - a_1^d T x^{(1)} - x + r\}. \tag{5.11}$$

Then the steady state of the closed-loop system (5.7) and (4.40) gives

$$\lim_{t\to\infty} \{r(t) - x(t)\} = e_r^v$$

and

$$\lim_{t\to\infty} \{u(t) - \bar{u}(t)\} = 0,$$

where

$$\overline{e}_r^v = a_1^d T - \frac{a_0 T^n \mu d_1}{k_0 b}, \tag{5.12}$$

$$\bar{u}(t) = \frac{a_0 e_r^v - a_1 r^v}{b} - \frac{a_0 r^v}{b} t\, 1(t).$$

In the above expressions the following definitions are used:

e_r^v is a velocity error due to the reference input $r(t)$,
$\overline{e}_r^v$ is a relative velocity error due to the reference input $r(t)$, where $\overline{e}_r^v = e_r^v / r^v$.

On the other hand, let us consider a reference model assigned by (2.10). Then we obtain

$$\overline{e}_r^v = -\frac{a_0 T^n \mu d_1}{k_0 b}, \tag{5.13}$$

where the first term in the right member of (5.12) caused by the properties of (5.11) disappears completely. This highlights the importance of using a reference model of the form (2.10) when seeking high accuracy in output tracking control.

It is easy to see that the sufficiently small relative velocity error $\overline{e}_r^v$ due to the reference input $r(t)$ in (5.13) can be achieved by decreasing the parameter μ for the given value $k_0 b$. Note that increasing $k_0 b$ affects the FMS stability.

5.1.5 *Control law in the form of forward compensator*

A block diagram of an SISO nonlinear time-varying continuous-time system (4.27) with a control law in the form of a forward compensator is shown in Fig. 5.1. Here the control law with the nth derivative of the error $e(t)$ in the feedback signal is given by the differential equation

$$\begin{aligned} &\mu^q u^{(q)} + d_{q-1}\mu^{q-1}u^{(q-1)} + \cdots + d_1\mu u^{(1)} + d_0 u \\ &= \frac{k_0}{T^n}\{-T^n e^{(n)} - a_{n-1}^d T^{n-1} e^{(n-1)} - \cdots - a_1^d T e^{(1)} - e\}, \end{aligned} \tag{5.14}$$

where $e = r - x$ and $q \geq n$.

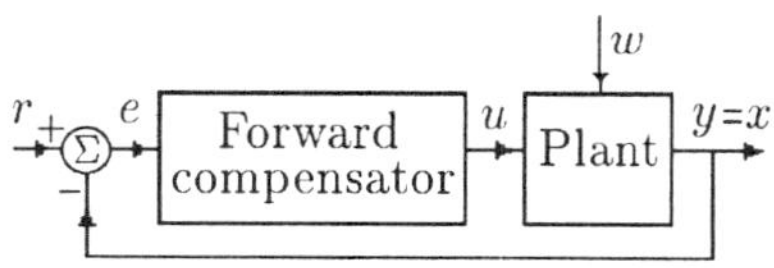

Fig. 5.1 Control law in the form of forward compensator.

We may express (5.14) in terms of the transfer function

$$u = -\tilde{k}\frac{A^d(s)}{D(\mu s)}e, \tag{5.15}$$

where $\tilde{k}$ and $D(\mu s)$ are given by (4.43) and (4.44).

In particular, let us consider the SISO linear continuous-time system (5.7). If $d_0 = 0$, $w(t) \overset{t}{\equiv} 0$, and the reference input $r(t)$ is a signal of type 1, then from (5.7) and (5.14) it follows that the relative velocity error $\bar{e}_r^v$ due to the reference input $r(t)$ is equal to (5.13).

To conclude this section, we note that the system with control law in the form of the forward compensator (5.14) is more sensitive to nonsmoothness of the reference input $r(t)$, and displays a peaking phenomenon. This will be illustrated below in an example (see Fig. 5.18).

The expressions above can be used to calculate k_0 and μ in accordance with the requirements placed on control accuracy of the desired output behavior realization for $x(t)$ if the steady-state motion of either the FMS or the SMS takes place. This does not suffice in practice, however, because some additional requirements must be maintained. These are associated with performance of the fast-motion transients. Therefore, the choice of control law parameters in accordance with requirements placed on damping and settling time of transients of the FMS (4.55) is discussed below.

5.2 Root placement of FMS characteristic polynomial

5.2.1 *Degree of time-scale separation*

From Theorems 4.2 and 4.3 (see p. 72) it follows that the desired behavior for $x(t)$ and the insensitivity of the step response parameters of the output transients with respect to the external disturbances and varying parameters of the system (4.27) occur on the condition of the steady state of the FMS (4.55). Accordingly, the stability of the FMS and a sufficient time-scale separation between the fast and slow modes in the discussed closed-loop system should be provided. Consequently, the control law parameters of (4.38) should be designed to obtain a sufficiently small settling time of FMS transients. Note that from a technical viewpoint, it is also desirable to provide an acceptable level of oscillation excited in the FMS (4.55).

Let the FMS transients of (4.55) be required to satisfy the inequality

$$t_{s,FMS} \leq t^d_{s,FMS}, \tag{5.16}$$

where $t_{s,FMS}$ is the settling time of the FMS and $t^d_{s,FMS}$ is a desired (permissible) settling time.

The desired time $t^d_{s,FMS}$ depends on the requirement placed on time-scale separation between the fast and slow modes in the closed-loop system. For example, suppose that

$$t^d_{s,FMS} \approx \frac{t_{s,SMS}}{\eta}, \tag{5.17}$$

where

$t_{s,SMS}$ is the settling time of the SMS (we usually assume that $t_{s,SMS} \approx t^d_s$),
t^d_s is a desired settling time of the output transients of (4.27),
η is a degree of time-scale separation between the fast and slow modes in the closed-loop system (we usually assume that $\eta \geq 10$).

Note that if the degree of time-scale separation is sufficiently large, then the FMS equation (4.55) may be examined as a linear differential equation and, accordingly, linear control theory may be used to analyze the FMS properties.

5.2.2 Selection of controller parameters

Let us consider the choice of control law parameters based on desired root placement of the FMS characteristic polynomial. First, assume that the degree η of time-scale separation between the fast and slow modes is sufficiently large so that the condition

$$g(X, w) \approx \text{const}$$

holds during the transients of the FMS (4.55). Then we may examine (4.55) as a linear differential equation. The resulting FMS characteristic polynomial

$$\tilde{D}_{FMS}(s) = \mu^q s^q + d_{q-1}\mu^{q-1}s^{q-1} + \cdots + d_1\mu s + \{d_0 + k_0 g\} \tag{5.18}$$

can be rewritten as

$$D_{FMS}(s) = \frac{\mu^q}{d_0 + k_0 g}s^q + \frac{d_{q-1}\mu^{q-1}}{d_0 + k_0 g}s^{q-1} + \cdots + \frac{d_1\mu}{d_0 + k_0 g}s + 1. \tag{5.19}$$

Second, we see that the assignment of a desired root placement for the FMS characteristic polynomial can provide the desired fast-motion transients. Just as the characteristic polynomial (2.3) of the reference model was formed (see p. 26), let us find the allowable root region of the FMS characteristic polynomial (5.18). This region is assigned by θ^d_{FMS} and ω^d_{FMS} (see Fig. 5.2), where

$\zeta^d_{FMS} = \cos(\theta^d_{FMS})$ is a desired damping ratio of the FMS,
ω^d_{FMS} is a desired damped or actual frequency of the FMS.

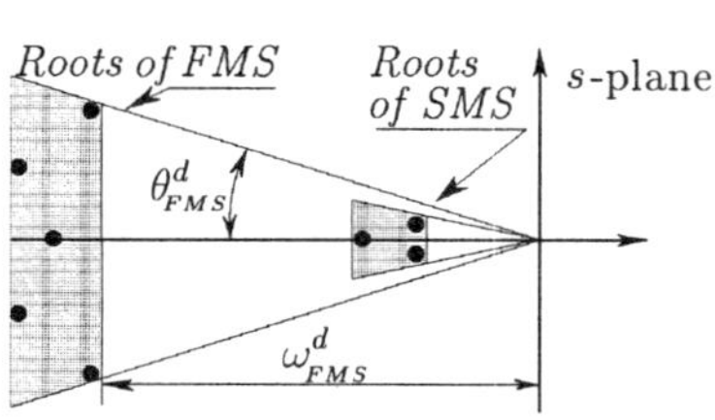

Fig. 5.2 Allowable root region of the FMS characteristic polynomial (5.19).

In order to find ω^d_{FMS} and θ^d_{FMS} for given $t^d_{s,FMS}$ and σ^d_{FMS}, the expressions (2.5) can be used. Then, by selection of the q roots s_1^{FMS}, s_2^{FMS}, $\ldots$, s_q^{FMS} in the allowable complex domain, the desired FMS characteristic polynomial

$$(s - s_1^{FMS})(s - s_2^{FMS}) \cdots (s - s_q^{FMS})$$

is assigned. Let us rewrite it in the form

$$D^d_{FMS}(s) = d^d_q s^q + d^d_{q-1}s^{q-1} + \cdots + d^d_1 s + 1. \tag{5.20}$$

Third, the requirement

$$D_{FMS}(s) = D^d_{FMS}(s) \tag{5.21}$$

is satisfied if and only if

$$\mu = \{d^d_q[d_0 + k_0 g]\}^{1/q}, \; d_j = \frac{d^d_j[d_0 + k_0 g]}{\{d^d_q[d_0 + k_0 g]\}^{j/q}}, \quad \forall\, j = 1, \ldots, q-1. \tag{5.22}$$

The relationships (5.22) allow us to choose the control law parameters in accordance with the desired root placement of the FMS characteristic polynomial.

Note that no strong restrictions exist for root placement of the FMS characteristic polynomial in the allowable region. The freedom of choice of the control law order q and the root placement of the FMS characteristic polynomial (5.20) within the allowable complex domain will be used below in order to provide the acceptable level of high-frequency sensor noise attenuation.

5.2.3 *Root placement based on normalized polynomials*

The desired FMS characteristic polynomial (5.20) can be constructed from the given normalized polynomial

$$\hat{D}^d_{FMS}(s) = s^q + \tilde{d}^d_{q-1}s^{q-1} + \cdots + \tilde{d}^d_1 s + 1, \tag{5.23}$$

where (5.23) has the desired root placement pattern.

To that end, let us rewrite (5.20) as

$$D^d_{FMS}(s) = \mu^q_d s^q + \tilde{d}^d_{q-1}\mu^{q-1}_d s^{q-1} + \cdots + \tilde{d}^d_1 \mu_d s + 1. \tag{5.24}$$

Here the time constant μ_d is defined by

$$\mu_d \le \frac{t^d_{s,FMS}}{\tilde{t}^d_s},$$

where $\tilde{t}^d_s$ is a nondimensional settling time of the system with the normalized characteristic polynomial (5.23).

It is often convenient to use the normalized polynomials for Butterworth root patterns, binomial root patterns, or ITAE standard error forms (where the integral of the time multiplied absolute error is minimal) [Graham and Lathrop (1953)]. For instance, if $\hat{D}^d_{FMS}(s)$ is the normalized polynomial for binomial root patterns, then we have

$$\tilde{d}^d_j = \frac{n!}{(n-j)!\,j!}, \quad j = 1, \ldots, q-1.$$

If the normalized coefficients $\tilde{d}^d_j$ are assigned for some root pattern, then from (5.21) and (5.24) it follows that the control law parameters are given by

$$\mu = \mu_d[d_0 + k_0 g]^{1/q}, \quad d_j = \tilde{d}^d_j[d_0 + k_0 g]^{(q-j)/q}, \tag{5.25}$$

where $j = 1, \ldots, q-1$.

5.3 Bode amplitude diagram assignment of closed-loop FMS

5.3.1 *Block diagram of closed-loop system*

The discussed closed-loop system (4.45)–(4.46)

$$\begin{aligned} x^{(n)} &= f(X,w) + g(X,w)u, \quad X(0) = X^0, \\ \mu^q u^{(q)} &+ d_{q-1}\mu^{q-1}u^{(q-1)} + \cdots + d_1\mu u^{(1)} + d_0 u \\ &= k_0\{F(X,R) - x^{(n)}\}, \; U(0) = U^0 \end{aligned}$$

corresponds to the system shown in Fig. 2.1, while the block diagram of (4.45)–(4.46) can be represented as in Fig. 5.3.

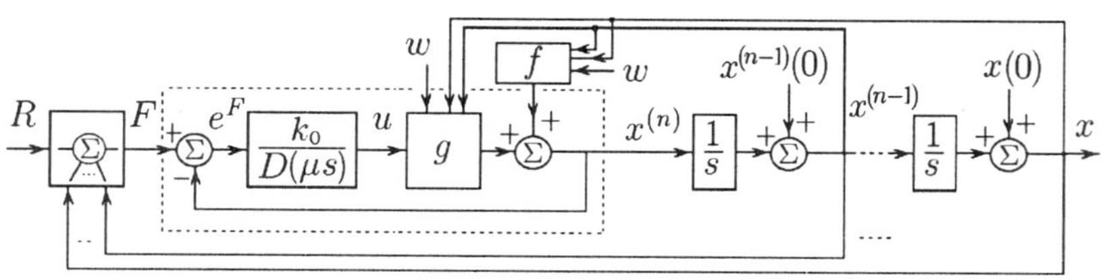

Fig. 5.3 Block diagram of the closed-loop system (4.45)–(4.46) with the highest derivative in the feedback loop.

Note that on the block diagram of Fig. 5.3 the initial conditions of the controller are omitted, and the polynomial $D(\mu s)$ has the form of (4.44).

The dotted enclosure in this block diagram, shown separately in Fig. 5.4, corresponds to the FMS (4.55). This can be verified through the use of block diagram transformations, under the assumption that w = const, X = const, F = const, f = const.

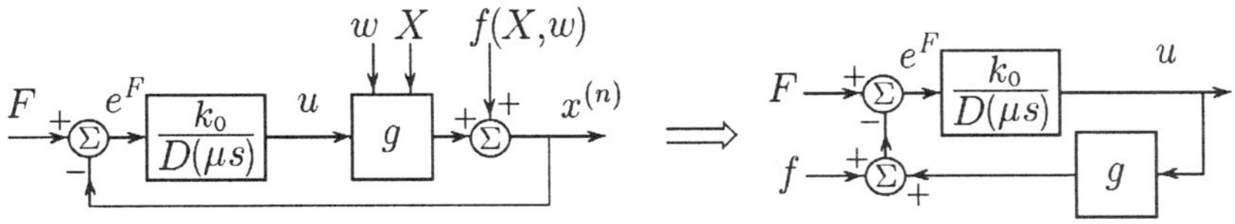

Fig. 5.4 Block diagram of the closed-loop FMS (4.55), where w = const, X = const, F = const, f = const.

5.3.2 *Bode amplitude diagram of closed-loop FMS*

Let us consider a procedure for choosing the control law parameters based on a desired Bode amplitude diagram for the closed-loop FMS of Fig. 5.4.

Assuming the degree η of time-scale separation between the fast and slow modes is sufficiently large, the FMS (4.55) will be examined as a linear differential equation in which F and f can be regarded as the external signals for the FMS (4.55).

Let us apply the Laplace transform to (4.55), given that the above assumption is satisfied and the initial conditions for (4.55) are all zero. We obtain

$$u(s) = G_{uf}(s)\{F(s) - f(s)\}, \tag{5.26}$$

where $G_{uf}(s)$ is a rational continuous-time transfer function given by

$$G_{uf}(s) = k_{uf}\,\frac{1}{D_{FMS}(s)}. \tag{5.27}$$

Here k_{uf} is the gain of the transfer function $G_{uf}(s)$ and is given by

$$k_{uf} = \frac{k_0}{d_0 + k_0 g}. \tag{5.28}$$

$D_{FMS}(s)$ is the FMS characteristic polynomial (5.19). We assume throughout the text that $d_0 + k_0 g > 0$.

Let us consider the logarithmic gain of the frequency-domain transfer function $G_{uf}(j\omega)$:

$$L_{uf}(\omega) = 20 \lg |G_{uf}(j\omega)|.$$

The plot of $L_{uf}(\omega)$ is called the Bode amplitude diagram[1] of the closed-loop FMS (4.55), and its qualitative shape is shown in Fig. 5.5.

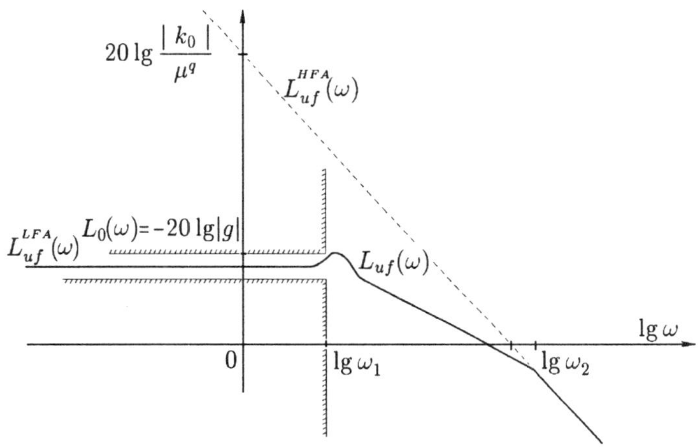

Fig. 5.5 Bode amplitude diagram $L_{uf}(\omega)$ of closed-loop FMS (4.55).

It is easy to see that

$$L_{uf}^{LFA}(\omega) = 20 \lg \frac{|k_0|}{d_0 + k_0 g} \tag{5.29}$$

is a low-frequency asymptote of $L_{uf}(\omega)$. This asymptote appears on the plot as a horizontal (zero slope) line segment.

From (4.60) and (5.28) it follows that

$$L_{uf}^{LFA}(\omega) \approx -20 \lg |g|.$$

In particular, if $d_0 = 0$ then

$$L_{uf}^{LFA}(\omega) = L_0(\omega),$$

where

$$L_0(\omega) = -20 \lg |g|.$$

[1] $\lg(x)$ denotes the logarithm of base ten, i.e., $\lg(x) = \log_{10}(x)$.

Accordingly, we can obtain

$$L_{uf}^{HFA}(\omega) = 20\lg\frac{|k_0|}{\mu^q} - 20q\lg\omega \tag{5.30}$$

as the high-frequency asymptote of $L_{uf}(\omega)$, where $L_{uf}^{HFA}(\omega)$ passes through $20\lg(|k_0|/\mu^q)$ dB at $\omega = 1$ and through 0 dB at $\omega = |k_0|^{1/q}/\mu$.

5.3.3 *Desired Bode amplitude diagram of closed-loop FMS*

From Fig. 5.5 it is clear that the condition

$$u(s) \approx k_{uf}\{F(s) - f(s)\} \tag{5.31}$$

holds for all $\omega \in [0, \omega_1]$. That is, the steady-state mode of the FMS appears if the spectra of the signals F and f lie within the frequency bandwidth $[0, \omega_1]$. Therefore, by (4.57) and (4.59), the influences of unknown disturbances and inherent properties of the plant are rejected, and the output behavior depends only on the properties of the reference model (4.31), if the low-frequency asymptote of $L_{uf}(\omega)$ belongs to some small neighborhood of $L_0(\omega)$. Note that the effective disturbance attenuation and desired output behavior are achieved only up to the frequency ω_1.

So the control law parameters may be chosen such that the Bode amplitude diagram $L_{uf}(\omega)$ of the closed-loop FMS has the desired form:

$$L_{uf}(\omega) = L_{uf}^{d}(\omega), \tag{5.32}$$

where the permissible domain (see Fig. 5.5) of $L_{uf}^{d}(\omega)$ at low-frequency bandwidth is assigned in accordance with the requirements placed on control accuracy and degree of time-scale separation.

In the frequency domain $\omega > \omega_1$ the desired Bode amplitude diagram $L_{uf}^{d}(\omega)$ should be constructed to obtain allowable FMS transients performance indices. Note that, by Remark 4.5 on p. 68, we have $q \geq n$; hence the high-frequency asymptote of $L_{uf}^{d}(\omega)$ is a straight line with a negative slope of $-20q$ dB/decade.

In accordance with the above requirements, the approximation of the actual desired Bode amplitude diagram $L_{uf}^{d}(\omega)$ may be organized as a set of straight line segments (asymptotes). These asymptotes intersect at the corner frequencies $\omega = \omega_i$.

Let us assume that $L_{uf}^{d}(\omega)$ is expressed in the corner frequency factored form

$$L^d_{uf}(\omega) = k_{uf}\frac{1}{[T_1^2 s^2 + 2\zeta_1 T_1 s + 1][T_2 s + 1]\cdots[T_j s + 1]\cdots} \tag{5.33}$$

as a product of first-order and quadratic factors. Then, from inspection of $L^d_{uf}(\omega)$ given by (5.33), the desired FMS characteristic polynomial (5.20)

$$D^d_{FMS}(s) = [T_1^2 s^2 + 2\zeta_1 T_1 s + 1][T_2 s + 1]\cdots[T_j s + 1]\cdots \tag{5.34}$$

follows.

Let us assume that $\omega_1 \leq \omega_i$, $\forall\, i$ where $\omega_i = 1/T_i$. Then the roots of quadratic factor

$$T_1^2 s^2 + 2\zeta_1 T_1 s + 1$$

are the dominant poles of $G_{uf}(s)$, where $0 < \zeta_1 < 1$. The parameter ζ_1 (damping ratio) may be calculated such that

$$\zeta_1 = \cos(\theta^d_{FMS})$$

or may be chosen through the use of the well-known Bode amplitude diagrams for a quadratic factor [Dorf and Bishop (2001); Bosgra and Kwakernaak (2000)].

The following relationship between ω_1 and $t^d_{s,FMS}$ may be used as a rule:

$$\omega_1 = \frac{1}{T_1} \approx \frac{4}{\zeta_1 t^d_{s,FMS}}. \tag{5.35}$$

5.3.4 *Selection of controller parameters*

As a result, the following procedure for choosing the control law parameters μ, d_j of (4.38) in accordance with the desired Bode amplitude diagram of the closed-loop FMS may be suggested.

Step 1. If $d_0 = 1$, then the gain k_0 is calculated based on expressions (5.3) and (5.6) in accordance with the requirement placed on control accuracy. If the robust zero steady-state error is needed, then take $d_0 = 0$.

Step 2. The low-frequency bandwidth $[0, \omega_1]$ of the Bode amplitude diagram of the closed-loop FMS is assigned in accordance with the requirement placed on the degree of time-scale separation, where ω_1 is defined, for instance, by (5.35).

Step 3. The desired Bode amplitude diagram $L^d_{uf}(\omega)$ is assigned such that its high-frequency asymptote $L^{HFA}_{uf}(\omega)$ has slope $-20q$ dB/decade where $q \geq n$.

Step 4. From inspection of $L^d_{uf}(\omega)$ the desired FMS characteristic polynomial $D^d_{FMS}(s)$ follows, which has the form (5.20).

Step 5. If $d_0 = 1$, then the parameters μ, $d_{q-1}, \ldots, d_1$ are calculated based on (5.22) as far as the gain k_0 is known. If $d_0 = 0$, then k_0 may be chosen arbitrarily — for instance, so that $1 \leq k_0 g \leq 10$. Then, by (5.21), we have

$$\mu = \{d^d_q k_0 g\}^{1/q}, \quad d_j = \frac{d^d_j [k_0 g]^{(q-j)/q}}{[d^d_q]^{j/q}}, \quad \forall\, j = 1, \ldots, q-1. \tag{5.36}$$

5.4 Influence of high-frequency sensor noise

5.4.1 *Closed-loop system in presence of sensor noise*

By choice of the degree q and root placement of the FMS characteristic polynomial (5.20) within the allowable complex domain, the solution to the problem of high-frequency measurement noise attenuation may be provided. Let us consider the solution to this problem based on the investigation of the Bode amplitude plot for the closed-loop FMS.

The SISO nonlinear time-varying continuous-time system (4.27)

$$x^{(n)} = f(X, w) + g(X, w)u, \quad \hat{y} = x + n_s, \quad X(0) = X^0 \tag{5.37}$$

is considered in this section, where the sensor output is corrupted by a zero-mean, high-frequency measurement noise $n_s(t)$ as shown in Fig. 5.6. Here $\hat{y}(t)$ is a sensor output. Then, instead of $y(t) = x(t)$, only $\hat{y}(t)$ is available for control. If we change $y(t)$ to $\hat{y}(t)$ in equation (4.38), then the control

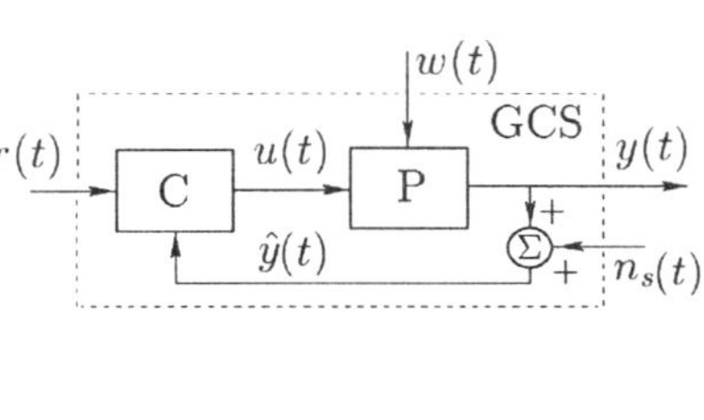

Fig. 5.6 Block diagram of the general control system with sensor noise $n_s(t)$.

algorithm becomes

$$\mu^q u^{(q)} + d_{q-1}\mu^{q-1}u^{(q-1)} + \cdots + d_1 \mu u^{(1)} + d_0 u = k_0 e^F_{\hat{y}}, \; U(0) = U^0, \tag{5.38}$$

where $e_{\hat{y}}^F = F(\hat{Y}, R) - \hat{y}^{(n)}$, $\mu > 0$, $q \geq n$, $d_0 = 1$ or $d_0 = 0$, $d_i > 0$, $\forall\, i = 1, \ldots, q-1$, and $\hat{Y} = \{\hat{y}, \hat{y}^{(1)}, \ldots, \hat{y}^{(n-1)}\}^T$.

Instead of equations (4.47) and (4.48), as a result of (5.37) and (5.38), we find that the closed-loop system equations on the condition of sensor noise are given by

$$x^{(n)} = f(X, w) + g(X, w)u, \; X(0) = X^0, \tag{5.39}$$

$$\begin{aligned}\mu^q u^{(q)} + d_{q-1}\mu^{q-1}u^{(q-1)} + \cdots + d_1\mu u^{(1)} + \{d_0 + k_0 g(X, w)\}u \\ = k_0\{F(\hat{Y}, R) - f(X, w) - n_s^{(n)}\}, \; U(0) = U^0.\end{aligned} \tag{5.40}$$

5.4.2 *Controller with infinite bandwidth*

Since μ is a small parameter, the closed-loop system equations (5.39)–(5.40) are the singular perturbed equations. Accordingly, fast and slow modes are induced in the closed-loop system as $\mu \to 0$. Then, by the time-scale separation procedure, from (5.39)–(5.40) the FMS equation

$$\begin{aligned}\mu^q u^{(q)} + d_{q-1}\mu^{q-1}u^{(q-1)} + \cdots + d_1\mu u^{(1)} + \{d_0 + k_0 g(X, w)\}u \\ = k_0\{F(\hat{Y}, R) - f(X, w) - n_s^{(n)}\}, \quad U(0) = U^0\end{aligned} \tag{5.41}$$

results. Here we assume that the state vector $X(t)$ of the subsystem (5.39) and, in accordance with (4.2), the vector $w(t)$ of external disturbances and varying parameters are constants during the transients in (5.41), i.e., that $X(t) = \text{const}$ and $w(t) = \text{const}$.

We then assume that the FMS is stable and, by finding the limit $\mu \to 0$ in (5.40), obtain $u(t) = u^s(t)$, where

$$u^s = k_0\{d_0 + k_0 g\}^{-1}\{F(X, R) - f(X, w) + F(N_s, 0) - n_s^{(n)}\}. \tag{5.42}$$

Here $N_s = \{n_s, n_s^{(1)}, \ldots, n_s^{(n-1)}\}^T$.

Substitution of (5.42) into (5.39) gives the equation of the SMS:

$$\begin{aligned}x^{(n)} = F(X, R) + F(N_s, 0) - n_s^{(n)} \\ + d_0\{d_0 + k_0 g\}^{-1}\{f(X, w) - F(X, R) - F(N_s, 0) + n_s^{(n)}\}.\end{aligned} \tag{5.43}$$

In accordance with (5.19) and (5.27), we have

$$\omega_1 \to \infty \quad \text{as} \quad \mu \to 0$$

or, in other words, the bandwidth of the FMS increases as the parameter μ decreases. Hence, for some value of μ the frequency bandwidth $[0, \omega_1]$

includes the main part of the spectral density $S_{n_s}(\omega)$ of the high-frequency noise $n_s(t)$ as shown in Fig. 5.7.

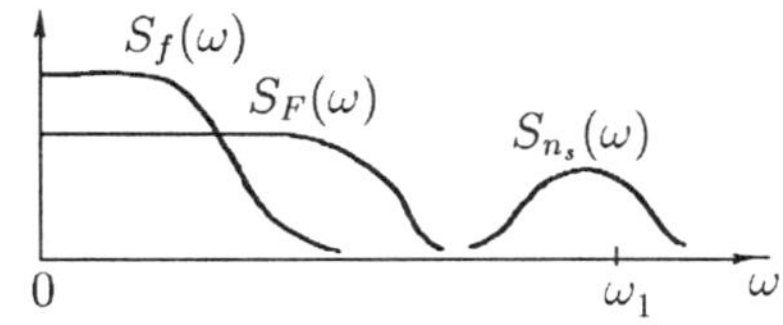

Fig. 5.7 Taking by bandwidth $[0, \omega_1]$ the main part of the noise spectral density $S_{n_s}(\omega)$.

So, the expressions (5.42) and (5.43) correspond to the controller with infinite bandwidth. We can find, by comparing (5.42) and (5.43) with (4.58) and (4.59), that there are two additional terms

$$\Delta u^s = k_0\{d_0 + k_0 g\}^{-1}\{F(N_s, 0) - n_s^{(n)}\}, \tag{5.44}$$

$$\Delta x^{(n)} = F(N_s, 0) - n_s^{(n)} + d_0\{d_0 + k_0 g\}^{-1}\{n_s^{(n)} - F(N_s, 0)\}, \tag{5.45}$$

where (5.44) and (5.45) reflect the influence of the noise $n_s(t)$.

Note that the magnitudes of the terms $\Delta u^s, \Delta x^{(n)}$ can be large and can give rise to a noise chattering effect in the control variable. In particular, if the control variable $u^s(t)$ runs to the bounds of Ω_u then saturation of $u^s(t)$ occurs. Consequently, we may have either decreasing accuracy of the desired dynamics realization or loss of closed-loop system stability.

If the condition $u^s(t) \in \Omega_u$ is satisfied, then there is small effect of the high-frequency noise $n_s(t)$ on the behavior of the output variable $y(t)$ since the system (5.37) rejects high frequencies. Therefore, the main disadvantage of the sensor noise $n_s(t)$ in the closed-loop system is that it leads to high-frequency chatter in the control variable $u^s(t)$. Accordingly, in order to attenuate sensor noise, the bandwidth $[0, \omega_1]$ of the closed-loop FMS should be bounded. Fig. 5.8 shows a case in which this requirement is met.

So the control law parameters should be chosen, first, to provide for control accuracy and, second, to attenuate the influence of high-frequency sensor noise $n_s(t)$ on the behavior of the control variable $u(t)$. In general, these two requirements are contradictory. The problem of sensor noise attenuation by assignment of the desired Bode amplitude diagram $L^d_{uf}(\omega)$ of the closed-loop FMS will be considered later.

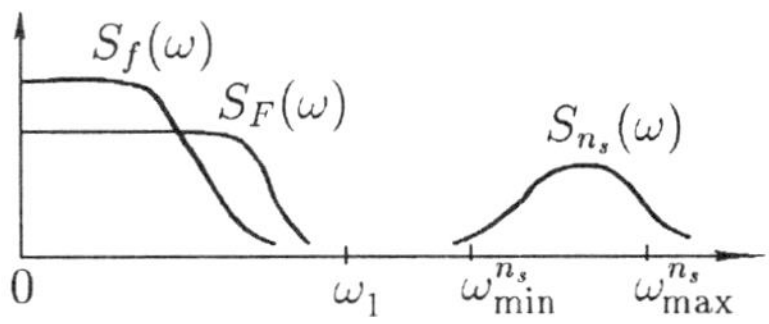

Fig. 5.8 The bandwidth $[0, \omega_1]$ is such that that the influence of the high-frequency sensor noise $n_s(t)$ is attenuated.

5.4.3 *Controller with finite bandwidth*

Selection of $G_{uf}(s)$

Let us consider a method of designing the control law (4.38) such that the influence of high-frequency sensor noise $n_s(t)$ on the behavior of the control variable $u(t)$ is attenuated. The method is based on the well-known frequency-domain approach to linear control system design under sensor noise (see, for instance, [Dorf and Bishop (2001); Bosgra and Kwakernaak (2000)]), and was used for aircraft flight controller design with the highest derivative in feedback in the presence of sensor noise in [Blachuta *et al.* (1999)].

Assume that the desired dynamics equation (4.30) has the form of the linear differential equation (2.8). Then, similar to (5.26), from (5.41) we get

$$u = G_{uf}(s)\{F - f\} - G_{un_s}(s)n_s, \tag{5.46}$$

where

$$G_{un_s}(s) = k_{un_s}\frac{A^d(s)}{D_{FMS}(s)}, \quad k_{un_s} = \frac{k_0}{[d_0 + k_0 g]T^n}, \tag{5.47}$$

and

$G_{un_s}(s)$ is a rational continuous-time transfer function between $n_s(t)$ and the control (manipulated variable) $u(t)$ of the plant,
k_{un_s} is the dc gain of $G_{un_s}(s)$, $k_{un_s} = G_{un_s}(0)$,
$A^d(s)$ is the desired characteristic polynomial (2.3) of (2.8).

The transfer function $G_{un_s}(s)$ in (5.46) determines the sensitivity of the plant input $u(t)$ to the sensor noise signal $n_s(t)$. In other words, $G_{un_s}(s)$ is an input sensitivity function with respect to noise in the closed-loop system.

Let the requirement on high-frequency sensor noise attenuation be expressed by the following inequality:

$$|G_{un_s}(j\omega)| \le \varepsilon_{un_s}(\omega), \quad \forall \quad \omega \ge \omega_{\min}^{n_s}, \tag{5.48}$$

where $\varepsilon_{un_s}(\omega)$ is an upper bound on the amplitude of the input sensitivity function with respect to noise for high frequencies.

From (5.48) it follows that an upper bound on the Bode amplitude diagram $L_{uf}(\omega)$ of the closed-loop FMS (4.55) appears for high frequencies $\omega \geq \omega_{\min}^{n_s}$. This bound is represented by a straight line $L_{\max}^{HFA}(\omega)$ in Fig. 5.9.

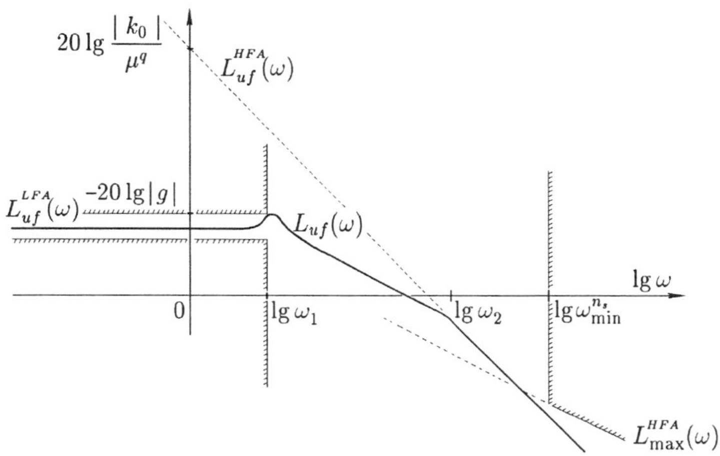

Fig. 5.9 Upper bound on the Bode amplitude diagram $L_{uf}(\omega)$ for high frequencies $\omega \geq \omega_{\min}^{n_s}$ in order to meet the requirement (5.48).

In order to find the expression for $L_{\max}^{HFA}(\omega)$, let us note that from (2.3), (5.19), and (5.47), the limit

$$\lim_{\omega\to\infty}\left[\frac{|k_0|}{\mu^q\omega^{q-n}} - |G_{un_s}(j\omega)|\right]\omega^{q-n} = 0$$

follows.

Let us assume that the condition

$$|G_{un_s}(j\omega)| \approx |G_{un_s}^{HFA}(j\omega)| = \frac{|k_0|}{\mu^q\omega^{q-n}} \tag{5.49}$$

holds for all high frequencies $\omega \geq \omega_{\min}^{n_s}$ or, in other words, that $L_{un_s}(\omega)$ is congruent with its high-frequency asymptote $L_{uf}^{HFA}(\omega)$.

From (5.49) it follows that if $q = n$, then high-frequency sensor noise attenuation is not provided as the limit

$$\lim_{\omega\to\infty}|G_{un_s}(j\omega)| = \frac{|k_0|}{\mu^q}$$

is valid.

From (5.30) and (5.49), it follows that (5.48) holds if the inequality

$$L_{uf}^{HFA}(\omega) \leq 20\lg(\varepsilon_{un_s}(\omega)) - 20n\lg\omega \tag{5.50}$$

is satisfied for all high frequencies $\omega \geq \omega_{\min}^{n_s}$. For instance, let the function $\varepsilon_{un_s}(\omega)$ be assigned by the expression

$$\varepsilon_{un_s}(\omega) = \frac{\varepsilon_{n_s}}{\omega^{\vartheta}}, \tag{5.51}$$

where $\varepsilon_{n_s} = \text{const} > 0$ and $\vartheta = \text{const}$. Then (5.50) may be rewritten in the form

$$L_{uf}^{HFA}(\omega) \leq L_{\max}^{HFA}(\omega), \quad \forall\, \omega \geq \omega_{\min}^{n_s}, \tag{5.52}$$

where

$$L_{\max}^{HFA}(\omega) = 20\lg\varepsilon_{n_s} - 20[n+\vartheta]\lg\omega.$$

The expression for $L_{\max}^{HFA}(\omega)$ is the straight line that assigns the upper bound of $L_{uf}(\omega)$ for all $\omega \geq \omega_{\min}^{n_s}$ if the assumption (5.49) is satisfied.

From inspection of Fig. 5.9 it is easy to see that upper attenuation of the high-frequency sensor noise is provided if $\omega_i = \omega_1, \ \forall\, i = 2, \ldots, q$. So, the desired FMS characteristic polynomial (5.34) should be chosen such that $T_i = T_1, \quad \forall\, i = 2, \ldots, q$.

Selection of $G_{un_s}(s)$

The high-frequency sensor noise attenuation can be investigated directly by examination of the Bode amplitude diagram for $G_{un_s}(s)$. Toward this end, let us rewrite (5.47) in the form

$$G_{un_s}(s) = \frac{k_0}{\mu^q}\frac{\hat{A}^d(s)}{\hat{D}_{FMS}(s)}, \tag{5.53}$$

where, by (2.3) and (5.18), we have

$$\hat{A}^d(s) = s^n + \frac{a_{n-1}^d}{T}s^{n-1} + \cdots + \frac{a_1^d}{T^{n-1}}s + \frac{1}{T^n},$$

$$\hat{D}_{FMS}(s) = s^q + \frac{d_{q-1}^d}{\mu}s^{q-1} + \cdots + \frac{d_1^d}{\mu^{q-1}}s + \frac{d_0 + k_0 g}{\mu^q}.$$

Denote by

$$L_{un_s}(\omega) = 20\lg|G_{un_s}(j\omega)|$$

the Bode amplitude diagram of the input sensitivity function with respect to noise in the closed-loop system. For simplicity, let us assume binomial root patterns for the polynomials $\hat{A}^d(s)$ and $\hat{D}_{FMS}(s)$ and consider the qualitative shape of $L_{un_s}(\omega)$ as shown in Fig. 5.10. Here

$$\omega_n^d = \frac{1}{T}, \quad \omega_n^{FMS} = \frac{[d_0 + k_0 g]^{1/q}}{\mu}, \quad \hat{\omega} = \frac{|k_0|^{1/(q-n)}(d_0 + k_0 g)^{1/q}}{\mu^{q/(q-n)}},$$

$$m_1 = \frac{|k_0|}{(d_0 + k_0 g)T^n}, \quad m_2 = \frac{|k_0|}{\mu^q}, \quad m_3 = \frac{|k_0|(d_0 + k_0 g)^{(q-n)/q}}{\mu^q}.$$

Recall that, by definition, $d_0 + k_0 g > 0$, and that ω_n^d and ω_n^{FMS} are called the natural frequencies of $\hat{A}^d(j\omega)$ and $\hat{D}_{FMS}(j\omega)$, respectively. Note that by Step 2 (see p. 92) ω_n^{FMS} should be chosen so that $\omega_n^{FMS} \geq \omega_1$.

From inspection of Fig. 5.10 it is easy to see that the requirement on high-frequency sensor noise attenuation given by (5.48) can be maintained by a proper choice of ω_n^{FMS} and q.

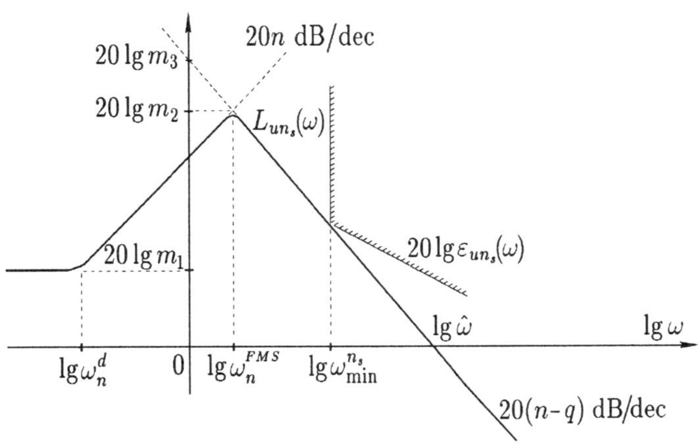

Fig. 5.10 Bode amplitude diagram $L_{un_s}(\omega)$ of the input sensitivity function with respect to noise in the closed-loop system.

5.5 Influence of varying parameters

5.5.1 *Influence of varying parameters on FMS and SMS*

Let us consider the influence of varying parameters of the plant model (4.27) on the properties of the closed-loop system (4.45)–(4.46). Then, taking into account this influence, we shall investigate some of the particularities of controller design with the highest derivative in feedback.

From (4.55) it follows that stability of the FMS depends on such plant parameters as $g(X, w)$. Regarding the SMS equation (4.59), it is easy to see that vanishing influence of varying parameters of the plant model (4.27) on the properties of (4.59) takes place if $k_0 g(X, w) \gg d_0$. So the main factor in the closed-loop system (4.45)–(4.46) is the influence of the varying parameter $g(X, w)$ on the stability of the FMS (4.55).

Note that the above consideration of the controller design in the form of (4.38) was conducted in the previous sections on the assumption that $g =$ const during the transients in the FMS (4.55). But the peculiarity of (4.27) is that the current value of $g(X, w)$ is unknown while, by Assumption 4.2 (see p. 64), it is known that

$$g(X, w) \in [g_{\min}, g_{\max}], \quad \forall\, X \in \Omega_X, \ \forall\, w \in \Omega_w. \tag{5.54}$$

So the controller parameters of (4.38) should be chosen such that stability and the requirement (5.16) are satisfied for any possible value of g from the given interval.

Taking into account the effect of the varying parameter $g(X, w)$ on the stability of the FMS (4.55), let us consider below certain particulars of controller design with the highest derivative in feedback.

5.5.2 *Michailov hodograph for FMS*

The influence of the parameter g on the stability of the FMS (4.55) may be investigated via well-known stability criteria such as the Routh–Hurwitz criterion, the Nyquist criterion, the root locus method, or the Michailov stability criterion (see, e.g., [Kolmanovskii and Nosov (1986)]). Note that Nyquist and Michailov criteria are based on the argument principle from the complex analysis.

In particular, substitution of $s = j\omega$ into (5.18) yields

$$\tilde{D}_{FMS}(j\omega) = \mu^q (j\omega)^q + d_{q-1}\mu^{q-1}(j\omega)^{q-1} + \ldots + d_1 \mu j\omega + \gamma, \tag{5.55}$$

where $\gamma = d_0 + k_0 g$. The function (5.55) is called the Michailov frequency function corresponding to the characteristic polynomial of the FMS (4.55).

Let us plot (5.55) in the complex s-plane for the frequency range $0 \leq \omega < \infty$. This plot is called the Michailov hodograph. Then, in accordance with the Michailov stability criterion, the FMS (4.55) is stable if and only if the following conditions are satisfied:

- $\tilde{D}_{FMS}(0) > 0$.
- $\tilde{D}_{FMS}(j\omega) \neq 0, \ \forall\, \omega \in [0, \infty)$.
- The Michailov hodograph encircles the origin and the traversal is in the counterclockwise direction through q quadrants in strict sequence.

In particular, the FMS (4.55) is marginally stable if some frequency $\hat{\omega} \in [0, \infty)$ satisfies $\tilde{D}_{FMS}(j\hat{\omega}) = 0$.

For example, a qualitative view of the Michailov hodograph (5.55) for a stable FMS (4.55) is shown in Fig. 5.11. Here the degree of the characteristic polynomial (5.20) is given by $q = 4$, and

$$\gamma_{\min} = d_0 + k_0 g_{\min}, \quad \gamma_{\max} = d_0 + k_0 g_{\max}.$$

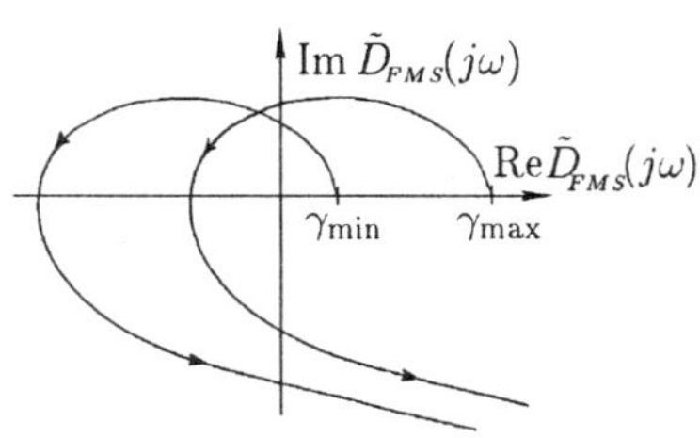

Fig. 5.11 The Michailov hodograph (5.55) of a stable FMS (4.55), where $q = 4$.

In accordance with the example shown in Fig. 5.11, we see that an increase in g leads to an oscillating margin of FMS stability, whereas a decrease in g leads to a neutral margin of FMS stability. The region of FMS stability for g can be found by inspection of the Michailov hodograph, by the Routh–Hurwitz criterion (see, for instance, [Dorf and Bishop (2001)]) or by method of D-subdivision (see, for instance, [Neimark (1947); El'sgol'ts and Norkin (1973); Kolmanovskii and Nosov (1986)]).

Hence the parameters of the control law (4.38) should be chosen so that the interval $[g_{\min}, g_{\max}]$ belongs to the region of FMS stability. Moreover, the desired relative stability of the FMS should be provided for all $g \in [g_{\min}, g_{\max}]$ such that the FMS transients maintain the allowable performance indices.

5.5.3 *Variation of FMS bandwidth*

Let us consider the influence of variations in the parameter g on the bandwidth of the FMS (4.55) (see Fig. 5.5).

Assume that the gain k_0 is chosen in accordance with the inequalities (5.3) and (5.6). Then the control accuracy requirements (5.1) and (5.4) are satisfied for all $g \in [g_{\min}, g_{\max}]$.

In addition to (5.1), (5.4), and in accordance with the 2nd step of the design procedure discussed on p. 92, the bandwidth of the FMS (4.55) should include the assigned frequency interval $[0, \omega_1]$ for all $g \in [g_{\min}, g_{\max}]$. Note that the frequency ω_1 depends on the requirement placed on the degree of time-scale separation between the fast and slow modes, and that this requirement is given by (5.16), (5.17), and (5.35).

In order to estimate the bandwidth of the FMS (4.55), let us introduce $\tilde{\omega}$ as the frequency at which the low- and high-frequency asymptotes of $L_{un_s}(\omega)$ intersect. In accordance with (5.29), (5.30), and from the condition

$$L_{uf}^{LFA}(\omega) = L_{uf}^{HFA}(\omega),$$

it follows that

$$\tilde{\omega} = \frac{[d_0 + k_0 g]^{1/q}}{\mu}. \tag{5.56}$$

Then from (5.54) we find that the bounds for $\tilde{\omega}$ are given by

$$\tilde{\omega} \in [\tilde{\omega}_{\min}, \tilde{\omega}_{\max}], \quad \forall\, X \in \Omega_X,\ \forall\, w \in \Omega_w, \tag{5.57}$$

where

$$\tilde{\omega}_{\min} = \frac{[d_0 + k_0 g_{\min}]^{1/q}}{\mu} \quad \text{and} \quad \tilde{\omega}_{\max} = \frac{[d_0 + k_0 g_{\max}]^{1/q}}{\mu}. \tag{5.58}$$

So $\tilde{\omega}$ can vary over the interval

$$[\tilde{\omega}_{\min}, \tilde{\omega}_{\max}]. \tag{5.59}$$

Hence the controller parameters of (4.38) should be chosen so that the requirement

$$\tilde{\omega}_{\min} \geq \omega_1 \tag{5.60}$$

holds, where ω_1 is assigned by (5.35).

The influence of variations in g on the bandwidth of the FMS (4.55) is illustrated by Fig. 5.12. Here, as an example, we assume that the parameters of the control law (4.38) were chosen so that if $g = g_{\min}$, then the condition (5.32) (see p. 91) holds and the desired FMS characteristic polynomial (5.34) is given by

$$D^d_{FMS}(s) = [T_1 s + 1]^q. \tag{5.61}$$

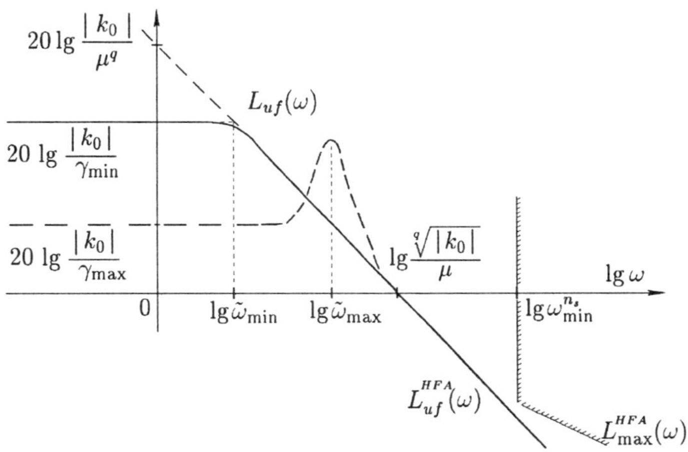

Fig. 5.12 The influence of variations in the parameter g on the bandwidth of the FMS (4.55).

5.5.4 *Degree of control law differential equation*

First, by inspection of Fig. 5.12 it is easy to see that the requirement (5.60) may be provided by choice of location of the high-frequency asymptote $L^{HFA}_{uf}(\omega)$, given by (5.30). At the same time, the parameter k_0 was assigned beforehand such that the control accuracy requirements (5.1) and (5.4) are satisfied for all $g \in [g_{\min}, g_{\max}]$. Then there is only one possible way to assign the location of $L^{HFA}_{uf}(\omega)$: by choice of the parameter μ and the order q of the control law differential equation (4.38).

From (5.58) and (5.60) it follows that an upper bound on μ is given by

$$\mu \leq \frac{[d_0 + k_0 g_{\min}]^{1/q}}{\omega_1}. \tag{5.62}$$

Second, from the requirement placed on high-frequency sensor noise attenuation (5.48) and, in accordance with (5.49), (5.51), and (5.52), it follows that a lower bound on μ is given by

$$\mu \geq \left[\frac{|k_0|}{\varepsilon_{n_s} [\omega_{\min}^{n_s}]^{q-n-\vartheta}} \right]^{1/q}, \tag{5.63}$$

where $q \geq n + \vartheta$.

A simultaneous solution exists for both inequalities (5.62) and (5.63) if the degree of the control law differential equation (4.38) is chosen such that the condition

$$q \geq n + \vartheta + \left[\ln \frac{|k_0|\omega_1}{\varepsilon_{n_s}[d_0 + k_0 g_{\min}]} \right] [\ln \omega_{\min}^{n_s}]^{-1} \tag{5.64}$$

holds.

5.5.5 *Root placement of FMS characteristic polynomial*

If $q \geq 3$, then from (5.18) we see that an increase in g leads first of all to underdamping and a decrease in the relative stability of the FMS. Further increase leads to marginal stability, and finally to instability, of the FMS (4.55). The qualitative change of the Bode amplitude diagram caused by increasing g from $g_{\min}$ to $g_{\max}$ is shown in Fig. 5.12 as a broken curve.

Suppose root placement of the FMS characteristic polynomial is used to obtain the control law parameters of (4.38) with a desired root pattern. Then, in order to avoid underdamping of the FMS caused by variations of g, we can use the following procedure instead of (5.25).

Step 1. The parameter μ is determined on the condition when $g = g_{\min}$ by the relationship

$$\mu = \mu_d[d_0 + k_0 g_{\min}]^{1/q}. \tag{5.65}$$

This choice allows us to provide the desired lower bound for relative stability or, in other words, the damped frequency of the FMS for all $g \in [g_{\min}, g_{\max}]$.

Step 2. The parameters d_j are determined on the condition $g = g_{\max}$ by the relationship

$$d_j = \tilde{d}_j^d [d_0 + k_0 g_{\max}] \left[\frac{\mu_d}{\mu} \right]^j, \quad \forall\, j = 1, \ldots, q-1. \tag{5.66}$$

This choice allows us to provide the desired lower bound for the FMS damping ratio (see Fig. 5.2, p. 86) or, in other words, the requirement

$\zeta_{FMS} \geq \zeta^d_{FMS}$ for all $g \in [g_{\min}, g_{\max}]$.

5.6 Bode amplitude diagram assignment of open-loop FMS

Let us consider a procedure for choosing the control law parameters of (4.38), based on the formation of a desired Bode amplitude diagram for the open-loop FMS. Therefore, in contrast to Fig. 5.4 (see p. 89), let us consider a block diagram representation of the FMS in which the loop is broken as in Fig. 5.13. It is easy to see that the Bode diagram of the open-loop FMS

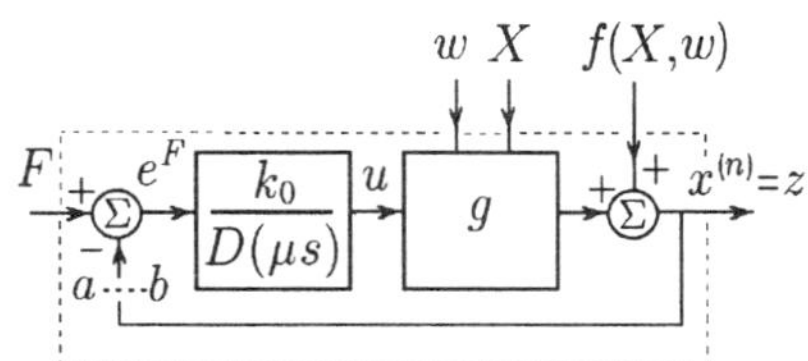

Fig. 5.13 Block diagram of the open-loop FMS (4.55), where w = const, X = const, F = const, f = const.

is given by

$$\overset{o}{L}_{zF}(\omega) = 20 \lg |\overset{o}{G}_{zF}(j\omega)|, \tag{5.67}$$

where $z = x^{(n)}$ and

$$\overset{o}{G}_{zF}(s) = \frac{k_0 g}{D(\mu s)}. \tag{5.68}$$

Then, in order to obtain allowable FMS transient performance indices, the desired Bode diagram of the open-loop FMS can be designed by choosing the control law parameters k_0, μ, and d_j in (4.38). So the standard design procedure using Bode diagrams [Chen (1993); Kuo and Golnaraghi (2003)] can be used. The main steps of the design procedure using the Bode amplitude diagram of the open-loop FMS are highlighted below.

Step 1. Determine the frequency interval $[0, \hat{\omega}_1]$ and the prohibited area of the low-frequency part of $\overset{o}{L}_{zF}(\omega)$ (crosshatched region at lower left in Fig. 5.14) by taking into account the requirements (5.1) and (5.4). Take $d_0 = 1$ or, if robust zero steady-state error is required, $d_0 = 0$.

Step 2. Determine the prohibited area of the high-frequency part of $\overset{o}{L}_{zF}(\omega)$ (crosshatched region at upper right in Fig. 5.14) by taking into account the requirement (5.48). To find this area, note first that from

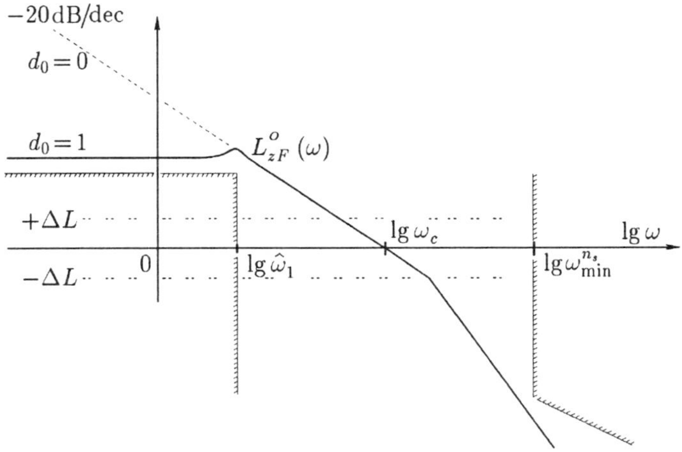

Fig. 5.14 Bode amplitude diagram $\overset{o}{L}_{zF}(\omega)$ of the open-loop FMS.

(5.68) we have

$$\lim_{\omega \to \infty} \left[\frac{k_0 g}{\mu^q \omega^q} - |\overset{o}{G}_{zF}(j\omega)| \right] \omega^q = 0.$$

Second, assume that

$$|\overset{o}{G}_{zF}(j\omega)| \approx \frac{k_0 g}{\mu^q \omega^q} \tag{5.69}$$

for all $\omega \geq \omega_{\min}^{n_s}$. Then, from (5.67) and (5.69), it follows that the requirement (5.48) holds if

$$\overset{o}{L}_{zF}(\omega) \leq 20 \lg |g \varepsilon_{un_s}(\omega)| - 20 n \lg \omega, \quad \forall \quad \omega \geq \omega_{\min}^{n_s}. \tag{5.70}$$

In particular, if $\varepsilon_{un_s}(\omega)$ is given by (5.51), then by (5.70) the prohibited area of the high-frequency part of $\overset{o}{L}_{zF}(\omega)$ is defined by the inequality

$$\overset{o}{L}_{zF}(\omega) \leq 20 \lg |g \varepsilon_{n_s}| - 20(n + \vartheta) \lg \omega, \quad \forall \, \omega \geq \omega_{\min}^{n_s}. \tag{5.71}$$

Step 3. The desired shape of the Bode amplitude diagram $\overset{o}{L}_{zF}(\omega)$ of the open-loop FMS is constructed in the allowable domain in such a way that it meets the desired gain (ΔL) and phase margins. The crossover frequency ω_c should be chosen to satisfy the requirement (5.16).

Step 4. From direct inspection of the desired shape of $\overset{o}{L}_{zF}(\omega)$, the gain k_0 and parameters of the polynomial $D(\mu s)$ result, where $D(\mu s)$ is given by (4.44) (see p. 67).

So, if the equation of the desired dynamics is assigned by (4.30), then calculation of the remaining parameters of the control law (4.38) may be examined as the design of a linear compensator of the form

$$C_c(s) = \frac{k_0}{D(\mu s)}$$

in the unity-feedback system as shown in Fig. 5.13. Note that, in general, the discussed procedure allows us to design the linear compensator given by

$$C_c(s) = k_0 \frac{K(\mu s)}{D(\mu s)}, \tag{5.72}$$

where

$$K(\mu s) = k_l \mu^l s^l + k_{l-1} \mu^{l-1} s^{l-1} + \cdots + k_1 \mu s + 1.$$

Then, instead of (5.38), we have that the resulting general control law is given by

$$\begin{aligned} &\mu^q u^{(q)} + \mu^{q-1} d_{q-1} u^{(q-1)} + \cdots + \mu d_1 u^{(1)} + d_0 u \\ &= k_0 \{ k_l \mu^l [e_{\hat{y}}^F]^{(l)} + k_{l-1} \mu^{l-1} [e_{\hat{y}}^F]^{(l-1)} + \cdots + k_1 \mu [e_{\hat{y}}^F]^{(1)} + e_{\hat{y}}^F \}. \end{aligned} \tag{5.73}$$

We can realize (5.73) without an ideal differentiation of $\hat{y}(t)$ or $r(t)$ if $q \geq l + n$. Taking into account the additional requirements of the high-frequency sensor noise attenuation given by (5.48) and (5.51), we see that the degree of the control law differential equation (5.73) should be chosen so that the condition $q \geq l + n + \vartheta$ is satisfied.

5.7 Relation with PD, PI, and PID controllers

Since the proportional-derivative (PD), proportional-integral (PI), and proportional-integral-derivative (PID) controllers are widely used (see, for instance, [Chen (1993); Kuo and Golnaraghi (2003)]), we now discuss the relationship between these and the above controller with the highest derivative in the feedback loop. It will be shown that the latter leads to the former types under certain conditions.

Let us consider the control law with the nth derivative of the error $e(t)$ in the feedback signal, given by (5.14). At noted above, (5.14) can be represented in terms of the transfer function (5.15).

From (5.15) the controller

$$u = -\tilde{k}\{T^n s^n + a_{n-1}^d T^{n-1} s^{n-1} + \cdots + a_1^d Ts + 1\}e \tag{5.74}$$

follows as $\mu = 0$ and $d_0 = 1$. In particular, if $n = 1$ then (5.74) reduces to the structure of the PD controller

$$u = -\tilde{k}\{Ts + 1\}e. \tag{5.75}$$

Note that (5.74) is described by an improper transfer function. Hence the control law (5.74) corresponds to the use of an ideal differentiating filter in feedback (or, from a technical point of view, to the use of special sensors capable of measuring all derivatives up to the nth order).

From (5.15) the controller with the proper transfer function of the form

$$u = -\tilde{k}\frac{Ts + 1}{\mu s + 1}e \tag{5.76}$$

results as $q = 1$, $n = 1$, and $d_1 = d_0 = 1$. It is easy to see that (5.76) reduces to the PD controller (5.75) as $\mu \to 0$.

On the other hand, assume that $q = 1$, $n = 1$, $\mu = 1$, $d_1 = 1$, and $d_0 = 0$. Then from (5.15) the equation of the PI controller

$$u = -\tilde{k}\left[T + \frac{1}{s}\right]e \tag{5.77}$$

follows.

Let us consider the next case. By substitution of $q = 1$, $n = 2$, $\mu = 1$, $d_1 = 1$, and $d_0 = 0$ into (5.15), the PID controller

$$u = -\tilde{k}\left[T^2 s + a_1^d T + \frac{1}{s}\right]e \tag{5.78}$$

results. Since (5.78) is described by an improper transfer function, to realize the PID controller in practice (i.e., without a sensor for the derivative of $e(t)$) let us take $q = 2$, $n = 2$, $d_2 = d_1 = 1$, and $d_0 = 0$ in (5.15). As a result, the proper PID controller

$$u = -\frac{\tilde{k}}{\mu}\frac{(T^2 s^2 + a_1^d Ts + 1)}{(\mu s + 1)s}e \tag{5.79}$$

follows.

So, the above design methodology gives a clear procedure for calculation of PD or PI controller parameters for a plant model of the first degree, and the parameters of a PID controller for a plant model of the second degree.

The advantage of this methodology is that the desired performance criteria are guaranteed under conditions of incomplete information regarding varying parameters of the plant model and unknown external disturbances.

5.8 Example

Let us consider a SISO nonlinear continuous-time system in the form

$$x^{(2)} = [1 + \sin(x^{(1)})]x + [1 - 0.5\sin(x)]u + w(t), \tag{5.80}$$

where $y(t) = x(t)$ and the desired dynamics of $y(t)$ are assigned by

$$y^{(2)} = -a_1^d \frac{1}{T} y^{(1)} - \frac{1}{T^2} y + \frac{1}{T^2} b_1^d \tau r^{(1)} + \frac{1}{T^2} r. \tag{5.81}$$

Expression (5.81) corresponds to (4.31). We assume that $T = 1$ s. Let $q = 2$; then, from (4.41), we obtain the control law structure given by

$$\mu^2 u^{(2)} + d_1 \mu u^{(1)} + d_0 u = k_0\{-y^{(2)} - \frac{a_1^d}{T} y^{(1)} + \frac{b_1^d \tau}{T^2} r^{(1)} + \frac{1}{T^2}[r - y]\}. \tag{5.82}$$

Hence, the FMS characteristic polynomial is

$$\tilde{D}_{FMS}(s) = \mu^2 s^2 + d_1 \mu s + d_0 + k_0 g,$$

where $g \in [0.5, 1.5]$.

By taking $\mu = 0.1$ s, $k_0 = 10$, $d_0 = 0$, and $d_1 = 2$, we find that as the parameter g is varied from 0.5 to 1.5, the degree of the time-scale separation $\eta_3 = [d_0 + k_0 g]^{0.5} T/\mu$ varies from 22 to 38 and the damping ratio ζ_{FMS} varies from 0.44 to 0.25.

Usually, in order to perform a computer simulation, we must represent the control law (5.82) in state-space form, e.g.,

$$\begin{aligned}
\dot{u}_1 &= -\frac{d_1}{\mu} u_1 + u_2 + k_0 \left\{ \frac{d_1}{\mu} - \frac{a_1^d}{T} \right\} y + \frac{k_0 b_1^d \tau}{T^2} r, \\
\dot{u}_2 &= -\frac{d_0}{\mu^2} u_1 + k_0 \left\{ \frac{d_0}{\mu^2} - \frac{1}{T^2} \right\} y + \frac{k_0}{T^2} r, \\
u &= \frac{1}{\mu^2} \{u_1 - k_0 y\}.
\end{aligned} \tag{5.83}$$

The simulation results for the output of the system (5.80) controlled by the algorithm (5.83) in response to a step reference input $r(t)$ and a step disturbance $w(t)$ are displayed in Fig. 5.15, where the initial conditions are zero and $T = \tau = 1$ s, $a_1^d = 1.4$, $b_1^d = 0$, and $t \in [0, 8]$ s.

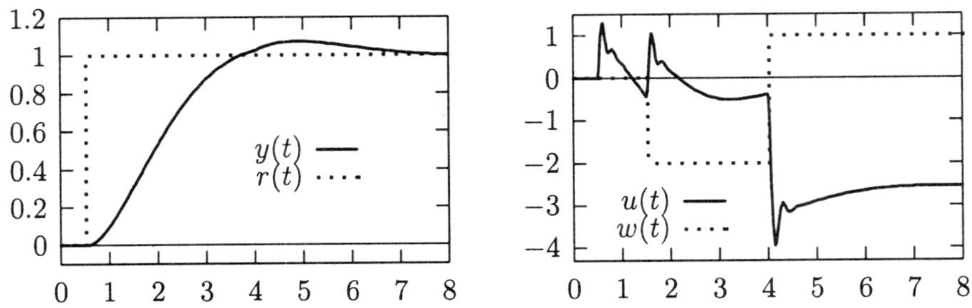

Fig. 5.15 Output response in the system (5.80) and (5.83) for a step reference input $r(t)$ and a step disturbance $w(t)$, where $T = \tau = 1$ s, $a_1^d = 1.4$, $b_1^d = 0$.

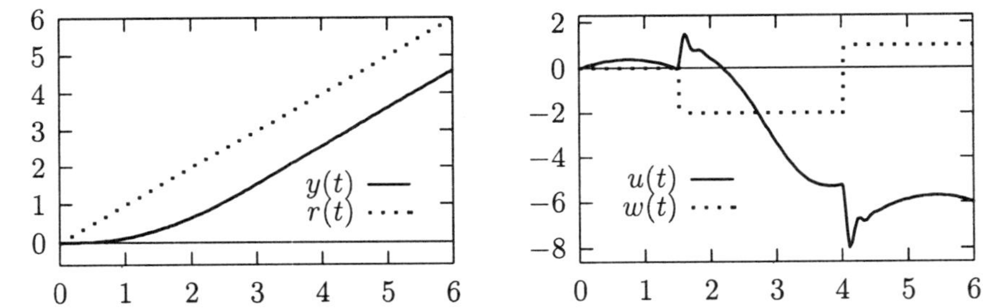

Fig. 5.16 Output response in the system (5.80) and (5.83) for a ramp reference input $r(t)$ and a step disturbance $w(t)$, where $T = \tau = 1$ s, $a_1^d = 1.4$, and $b_1^d = 0$.

The output response in the system (5.80) and (5.83) for a ramp reference input $r(t)$ and a step disturbance $w(t)$ is shown in Fig. 5.16, where $r(t) = r^v t1(t)$, $r^v = 1$, $T = \tau = 1$ s, $a_1^d = 1.4$, $b_1^d = 0$, and $t \in [0, 6]$ s.

If, in accordance with (2.10) (see p. 29), we assume that $a_1^d = b_1^d = 1.4$, then the reference model is a type 2 system. The simulation results for the output response in the system (5.80) controlled by the algorithm (5.83) to a ramp reference input $r(t)$ and a step disturbance $w(t)$ are shown in Fig. 5.17, where $t \in [0, 6]$ s.

The simulation results for the output response in the system (5.80) controlled by the algorithm in the form of the forward compensator (5.14) to a ramp reference input $r(t)$ and a disturbance $w(t) \equiv 0$ are shown in Fig. 5.18, where $e(t) = r(t) - y(t)$, $r(t) = r^v t1(t)$, $r^v = 1$, $T = 1$ s, $a_1^d = 1.4$, $\mu = 0.1$ s, $k_0 = -10$, $d_0 = 0$, $d_1 = 2$, and $t \in [0, 6]$ s.

The sign of k_0 undergoes a change here in accordance with the requirement of FMS stability if the control law (5.14) is used. In comparison with

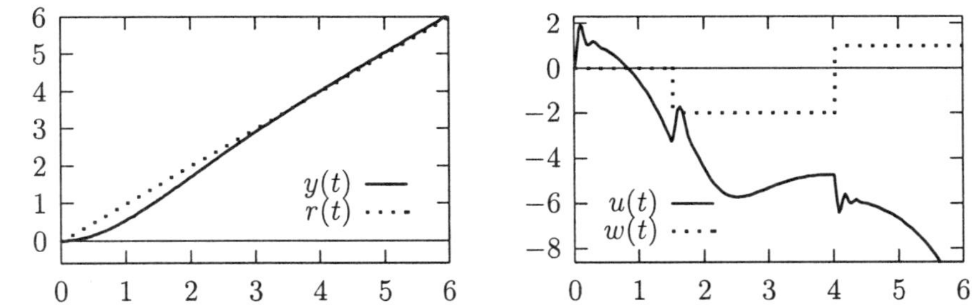

Fig. 5.17 Output response in the system (5.80) and (5.83) for a ramp reference input $r(t)$ and a step disturbance $w(t)$, where $T = \tau = 1$ s and $a_1^d = b_1^d = 1.4$ (reference model is a system of type 2).

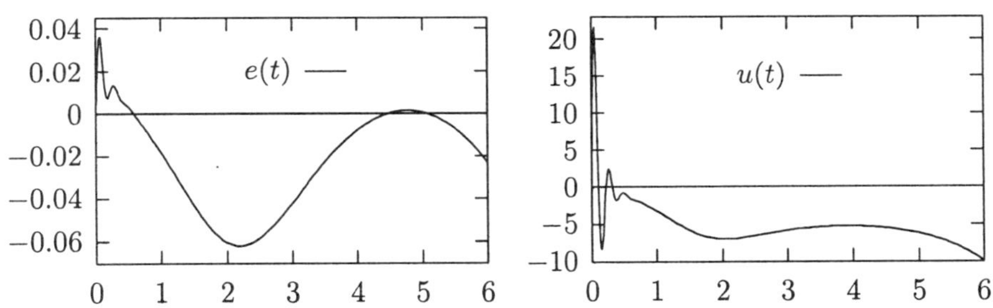

Fig. 5.18 Output response in the system (5.80) and (5.14) for a ramp reference input $r(t)$ and disturbance $w(t) = 0$.

Fig. 5.17, we can see that the closed-loop system with control law (5.14) is more sensitive with respect to nonsmoothness of the reference input $r(t)$ and reveals a peaking phenomenon. So the reference signal $r(t)$ should be a continuously differentiable function in order to avoid a high pulse in the control variable $u(t)$ in Fig. 5.18.

5.9 Notes

Our main purpose has been to explain the various design procedures for choosing the parameters of the control law with the highest derivative in the feedback loop for SISO systems, such that the robustness of the closed-loop system properties is provided in a specified region of the state space of the system — despite incomplete information regarding varying parameters of the system and external disturbances.

We have shown that for the given structure of the control law with the highest derivative in the feedback loop, the design procedure for controller parameters may lead to linear compensator design by linear control system methods, given that the degree of time-scale separation between induced fast and slow motions is sufficiently large in the closed-loop system. We have also shown that the desired degree of time-scale separation can be provided by proper choice of the controller parameters.

The relationships used to choose the control law parameters of (4.38) in accordance with the requirements on control accuracy and disturbance rejection, as well as high-frequency sensor noise attenuation, have been obtained in this chapter. These make it possible to employ control systems with highest derivative in feedback while making allowance for various practical restrictions.

Finally, the relationship between the controller with the highest derivative in the feedback loop and the PD, PI, and PID controller types has been established. The resulting design methodology can be used in order to obtain PD or PI controller parameters for a first-degree plant model, and PID controller parameters for a second-degree plant model. The advantage of this approach to PID controller design is the guaranteed performance of the output response for nonlinear systems in the presence of unknown external disturbances and varying parameters.

Note that there is a broad set of publications devoted to problem of integral controller design for linear as well as nonlinear systems, where various types of output feedback controllers with observers are discussed, e.g., [Davison (1976); Francis (1977); Isidori and Byrnes (1990); Huang and Rugh (1990); Mahmoud and Khalil (1996); Khalil (2000)].

5.10 Exercises

5.1 The differential equation of a plant model is given by

$$x^{(2)} = x + x|x^{(1)}| + \{1.5 + \sin(t)\}u. \tag{5.84}$$

Assume that the specified region is given by the inequalities $|x(t)| \leq 2$, $|x^{(1)}(t)| \leq 10$, and $|r(t)| \leq 1$, where $t \in [0, \infty)$. The reference model for $x(t)$ is chosen as $x^{(2)} = -2x^{(1)} - x + r$. Determine the parameters of the control law (4.40) to meet the requirements: $\overline{\varepsilon}_F = 0.05$, $\overline{\varepsilon}_r = 0.02$, and $\zeta_{FMS} \geq 0.5$; $\eta_3 \geq 20$. Compare simulation results with the assignment. Note that η_3 is the degree of time-scale separation between stable fast and slow motions defined by (1.65).

5.2 The behavior of a plant model is described by the equation

$$x^{(2)} = 2x^{(1)} + x + 2u. \tag{5.85}$$

Assume that the inequalities $|x(t)| \leq 4$, $|x^{(1)}(t)| \leq 20$, and $|r(t)| \leq 2$ hold for all $t \in [0, \infty)$. The reference model for $x(t)$ is assigned by

$$x^{(2)} = -3x^{(1)} - x + r. \tag{5.86}$$

Determine the parameters of the control law (4.40), where $d_0 = 1$, to meet the following requirements: $\overline{\varepsilon}_F = 0.1$, $\overline{\varepsilon}_r = 0.05$, $\zeta_{FMS} \geq 0.2$; $\eta_3 \geq 20$, $q = 2$. Determine the equation for the steady-state error due to a ramp reference input $r(t)$ if $d_0 = 0$. Modify (5.86) in order to obtain the reference model as the system of type 2. Compare simulation results with the assignment.

5.3 The differential equation of a plant model is given by

$$x^{(2)} = x^{(1)} + x + 5u. \tag{5.87}$$

Assume that the inequalities $|x(t)| \leq 3$, $|x^{(1)}(t)| \leq 20$, $|r(t)| \leq 1$ hold for all $t \in [0, \infty)$. The reference model for $x(t)$ is assigned by

$$x^{(2)} = -1.4x^{(1)} - x + r. \tag{5.88}$$

Determine the parameters of the control law (4.40), where $d_0 = 1$, to meet the following requirements: $\overline{\varepsilon}_F = 0.1$, $\overline{\varepsilon}_r = 0.05$, $\zeta_{FMS} \geq 0.3$; $\eta_3 \geq 20$, $q = 2$. Determine the equation for the steady-state error due to a ramp disturbance $w(t)$ if $d_0 = 0$. Compare simulation results with the assignment.

5.4 The plant model and reference models are given by (5.87)–(5.88), and the control law has the form (4.40) where $k_0 = 8$, $q = 3$, $\mu = 0.1$ s, $d_2 = 5$, $d_1 = 9$, and $d_0 = 0$. Determine the steady-state error due to a ramp disturbance $w(t)$. Compare simulation results with the assignment.

5.5 Determine the region of stability for g of the FMS (4.55), where $k_0 = 8$, $q = 3$, $\mu = 0.1$ s, $d_2 = 5$, $d_1 = 9$, and $d_0 = 0$.

5.6 The plant model and reference models are given by (5.87)–(5.88), and the control law has the form (4.40) where $k_0 = 10$, $q = 2$, $\mu = 0.1$ s, $d_1 = 2$, and $d_0 = 0$. Determine the equation of the steady-state error due to a ramp reference input $r(t)$. Compare simulation results with the assignment.

5.7 Determine the region of variations (5.59) of the frequency $\tilde{\omega}$ of intersection between the low- and high-frequency asymptotes of $L_{un_s}(\omega)$. Also determine the region of variations of the damping ratio of the FMS, where $g \in [0.5, 10], q = 2$, $k_0 = 10$, $\mu = 0.1$ s, $d_1 = 2$, and $d_0 = 1$. Find the root loci of the FMS for $g \geq 0$.

5.8 Find the root loci of the FMS (4.55) for $g \geq 0$, where $q = 2$, $k_0 = 10$, $\mu = 0.1$ s, $d_1 = 3$, and $d_0 = 0$.

5.9 The differential equation of a plant model is given by (5.84). Determine the parameters of the control law (4.40) such that $\bar{\varepsilon}_r = 0$, $t_s^d \approx 2$ s, $\sigma^d \approx 5\,\%$, $\zeta_{FMS} \geq 0.2$, $\eta_3 \geq 20$, and $q = 2$. Compare simulation results of the step response in the closed-loop control system with the assignment.

5.10 The differential equation of a plant model is given by (5.85). Determine the parameters of the control law (4.40) such that $\bar{\varepsilon}_r = 0$, $t_s^d \approx 1$ s, $\sigma^d \approx 10\,\%$, $\zeta_{FMS} \geq 0.3$, and $\eta_3 \geq 10$. The additional requirement (5.48) should be provided such that $\varepsilon_{un_s}(\omega) = 10^3$ and $\omega_{\min}^{n_s} = 10^3$ rad/s. Compare simulation results with the assignment.

5.11 The plant model is given by (5.85). Determine the parameters of the control law (4.40) to meet the following requirements: $\bar{\varepsilon}_r = 0$, $t_s^d \approx 3$ s, $\sigma^d \approx 5\,\%$, $\zeta_{FMS} \geq 0.5$, $\eta_3 \geq 8$. The requirement (5.48) should be provided such that $\varepsilon_{un_s}(\omega) = 360$ and $\omega_{\min}^{n_s} = 10^2$ rad/s. Compare simulation results with the assignment.

5.12 The plant model is given by (5.84). Determine the parameters of the control law (4.40) to meet the requirements $\bar{\varepsilon}_r = 0$, $t_s^d \approx 2$ s, $\sigma^d \approx 5\%$, $\zeta_{FMS} \geq 0.5$, $\eta_3 \geq 8$. The requirement (5.48) should be provided such that $\varepsilon_{un_s}(\omega) = 10^3/\omega$ and $\omega_{\min}^{n_s} = 10^2$ rad/s. Compare simulation results with the assignment.

Chapter 6

Influence of unmodeled dynamics

We have considered problems of SISO continuous-time control system design with the highest derivative in feedback, in which uncertainties were caused by variations in the parameters of the model or unknown external disturbances and the degree of the plant model remained the same as in the system without uncertainties. Such uncertainties are usually called structured uncertainties [Slotine and Li (1991)]. The robustness of the output performance of the control systems with the highest derivative in feedback was discussed in the presence of bounded structured uncertainties. In particular, the design procedures for a controller of the form (4.38) for system (4.27) were given, where the bounded structured uncertainties were assigned by the conditions (4.28).

Another type of uncertainty in the model description is the so-called unstructured uncertainty. These reflect errors in the system degree (or relative degree). Taking them into account usually leads to examination of the plant model with an additional small pure time delay, as well as models in the form of regularly and/or singularly perturbed systems of differential equations.

From a practical viewpoint, it is more convenient to use some simplified (nominal) model of the system reflecting only the main qualitative and quantitative features of the system. A more detailed description leads to a change in the system order. For instance, accounting for actuator and/or sensor dynamics can lead to an increase in the system order (so-called fast unmodeled dynamics). So, the nominal model results from the premeditated neglect of small parameters in order to get a simplified model and hence a simplified controller. However, the neglected parameters affect the performance of the physical closed-loop system.

In this chapter we first examine the robustness of performance of control systems with the highest derivative in feedback in the presence of the unstructured uncertainties. Second, we examine restrictions on controller parameters caused by various types of unstructured uncertainties. Finally, we discuss the influence of a nonsmooth nonlinearity in the control loop.

6.1 Pure time delay

6.1.1 *Plant model with pure time delay in control*

After a closer examination of the real system properties, it may happen that some small pure time delay in the control loop exists and affects the stability of the closed-loop system (Fig. 6.1). Such delays usually occur in models of technical systems that involve the movement of some substance: examples include hydraulic systems, heat exchangers, chemical reactors, long current lines, and pipelines [Schneider (1988)]. Many works have dealt with systems having pure time delays; for example, problems of sliding mode control for systems with time delay are investigated in [Fridman *et al.* (1996); Gouaisbaut *et al.* (1999)].

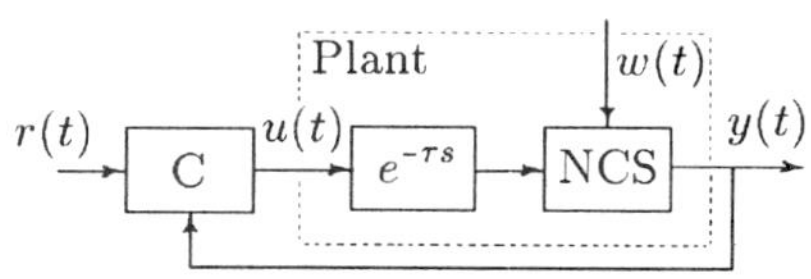

Fig. 6.1 Block diagram of the nonlinear continuous system (NCS) with a pure time delay in the control variable.

In this section we examine the peculiarities caused by a small pure time delay in control for systems with the highest derivative in feedback.

Let us consider a nonlinear time-varying system described by an nth-order differential equation with a pure time delay in the control variable:

$$x^{(n)}(t) = f(X(t), w(t)) + g(X(t), w(t))u(t-\tau), \quad X(0) = X^0. \tag{6.1}$$

Here

x is the output variable, available for measurement;
t is the time variable, $t > 0$;
τ is the pure time delay, $\tau > 0$;
$X = \{x, x^{(1)}, \ldots, x^{(n-1)}\}^T$ is the state vector;
$X(0) = X^0$ is the initial state;
$u(t-\tau)$ is the control variable with pure time delay;
$f(X, w), g(X, w)$ are functions satisfying (4.28).

Let us assume that τ is constant and belongs to a known interval:

$$\tau \in [0, \tau_{\max}]. \tag{6.2}$$

Note that the pure time delay model may be approximated by a finite dimensional transfer function of arbitrary order. Since it alters the order of the plant model, the delay block is discussed here as a particular case of unstructured uncertainty.

If $\tau = 0$, then from (6.1) the simplified model of the form (4.27) follows. Assume that the control law structure of the form (4.38) was constructed for the simplified model (4.27) based on the above design procedure. The main purpose of this section is to choose the parameters of the control law (4.38) to reduce the effect of the pure time delay on performance of the closed-loop system when the control (4.38) is applied to the true system (6.1).

6.1.2 *Closed-loop system with delay in feedback loop*

The closed-loop system equations of the plant model (6.1) and controller (4.38) are given by

$$x^{(n)}(t) = f(X(t), w(t)) + g(X(t), w(t))u(t-\tau), \ X(0) = X^0, \tag{6.3}$$

$$\mu^q u^{(q)}(t) + d_{q-1}\mu^{q-1}u^{(q-1)}(t) \cdots + d_0 u(t) = k_0 e^F, \ U(0) = U^0, \tag{6.4}$$

where μ is a small positive parameter and $q \geq n$.

In accordance with (4.32) and (6.1), equations (6.3)–(6.4) of the closed-loop system may be rewritten in the form

$$x^{(n)}(t) = f(X(t), w(t)) + g(X(t), w(t))u(t-\tau), \ X(0) = X^0, \tag{6.5}$$

$$\begin{aligned} &\mu^q u^{(q)}(t) + \cdots + d_0 u(t) + k_0 g(X(t), w(t))u(t-\tau) \\ &= k_0\{F(X(t), R(t)) - f(X(t), w(t))\}, \ U(0) = U^0, \end{aligned} \tag{6.6}$$

and then, by (4.51), in the form of the singular perturbed model

$$\begin{aligned} \frac{dx_i}{dt} &= x_{i+1}, \quad i = 1, \ldots, n-1, \\ \frac{dx_n}{dt} &= f(\cdot) + b(\cdot)u_1(t-\tau), \\ \mu\frac{du_j}{dt} &= u_{j+1}, \quad j = 1, \ldots, q-1, \\ \mu\frac{du_q}{dt} &= -d_0 u_1 + k_0 g(\cdot)u_1(t-\tau) - d_1 u_2 - \cdots - d_{q-1}u_q + k_0\{F - f\}. \end{aligned} \tag{6.7}$$

6.1.3 *Fast motions in presence of delay*

First, in order to enable usage of the above standard technique for two-time-scale motions analysis, we must represent the time delay τ in the normalized form

$$\tau = \tau_0 \mu \tag{6.8}$$

where τ_0 is the normalized time delay.

Second, let us introduce the differentiation operator

$$p = \frac{d}{dt}$$

and rewrite (6.7) in the following operator form:

$$\begin{aligned}
px_i &= x_{i+1}, \quad i = 1, \dots, n-1, \\
px_n &= f(\cdot) + g(\cdot)e^{-\tau_0 \mu p} u_1, \\
\mu\, pu_j &= u_{j+1}, \quad j = 1, \dots, q-1, \\
\mu\, pu_q &= -\{d_0 + k_0 g(\cdot) e^{-\tau_0 \mu p}\} u_1 - d_1 u_2 - \cdots - d_{q-1} u_q + k_0\{F - f\}.
\end{aligned}$$

By introducing the new time scale $t_0 = t/\mu$ and, accordingly, the new differentiation operator $p_0 = d/dt_0$ where $p = \mu^{-1} p_0$, we obtain

$$\begin{aligned}
p_0 x_i &= \mu x_{i+1}, \quad i = 1, \dots, n-1, \\
p_0 x_n &= \mu\{f(\cdot) + g(\cdot)e^{-\tau_0 p_0} u_1\}, \\
p_0 u_j &= u_{j+1}, \quad j = 1, \dots, q-1, \\
p_0 u_q &= -\{d_0 + k_0 g(\cdot) e^{-\tau_0 p_0}\} u_1 - d_1 u_2 - \cdots - d_{q-1} u_q + k_0\{F - f\}.
\end{aligned}$$

By setting $\mu = 0$, we find that the FMS equations in the time scale t_0 are described by

$$\begin{aligned}
p_0 x_i &= 0, \quad i = 1, \dots, n, \\
p_0 u_j &= u_{j+1}, \quad j = 1, \dots, q-1, \\
p_0 u_q &= -\{d_0 + k_0 g e^{-\tau_0 p_0}\} u_1 - d_1 u_2 - \cdots - d_{q-1} u_q + k_0\{F - f\}.
\end{aligned}$$

From the above, we get the equations of the FMS

$$\begin{aligned}
\frac{du_j}{dt_0} &= u_{j+1}, \quad j = 1, \dots, q-1, \\
\frac{du_q}{dt_0} &= -d_0 u_1 + k_0 g u_1(t_0 - \tau_0) - d_1 u_2 - \cdots - d_{q-1} u_q \\
&\quad + k_0\{F - f\},
\end{aligned} \tag{6.9}$$

where $X(t) = \text{const}$. By (4.2), $w(t) = \text{const}$ in the system (6.9). Note that the stability of the FMS (6.9) is invariant with respect to μ by the condition (6.8).

Returning to the primary time scale $t = \mu t_0$, from (6.9) we obtain the FMS equations

$$\begin{aligned}
\mu \frac{du_j}{dt} &= u_{j+1}, \quad j = 1, \ldots, q-1, \\
\mu \frac{du_q}{dt} &= -d_0 u_1 + k_0 g u_1(t-\tau) - d_1 u_2 - \cdots - d_{q-1} u_q \\
&\quad + k_0\{F - f\},
\end{aligned} \tag{6.10}$$

where $X(t)$ and $w(t)$ are the frozen variables.

It is easy to see that (6.10) can be rewritten in the form

$$\begin{aligned}
&\mu^q u^{(q)}(t) + \cdots + d_1 \mu u^{(1)}(t) + d_0 u(t) + k_0 g(X(t), w(t)) u(t-\tau) \\
&\quad = k_0\{F(X(t), R(t)) - f(X(t), w(t))\}, \quad U(0) = U^0,
\end{aligned} \tag{6.11}$$

where we assume that $X(t) = \text{const}$ and $g(X, w) = \text{const}$ during the transients in (6.11).

Remark 6.1 *Inasmuch as the FMS (6.11) may be examined as a linear system with frozen parameter $g(X, w)$, the use of the Nyquist stability criterion [Nyquist (1932)] is a more natural and simple form of stability analysis from a practical point of view. The known method of D-subdivision [Neimark (1947); El'sgol'ts and Norkin (1973); Kolmanovskii and Nosov (1986)] may be used as well. In general, the stability analysis of FMS (6.11) based on Lyapunov functions [Krasovskii (1963); Kolmanovskii and Nosov (1986)] — in particular, the Lyapunov-Krasovskii functionals and Lyapunov-Razumikhin functions — may also be used.*

6.1.4 *Stability of FMS with delay*

If we have the linear differential equations of the FMS (6.11) with frozen parameter $g(X, w)$, then the methods of linear control theory may be used in order to analyze the FMS properties. In particular, the characteristic equation of (6.11) has the following form:

$$D(\mu s) + k_0 g e^{-\tau s} = 0.$$

The closed-loop system (6.3)–(6.4) corresponds to the system shown in Fig. 6.1, while the system (6.3)–(6.4) can be represented in block diagram form as in Fig. 6.2. Here the time delay τ is given by a block having transfer function $e^{-\tau s}$. The highlighted portion corresponds to the FMS (6.11).

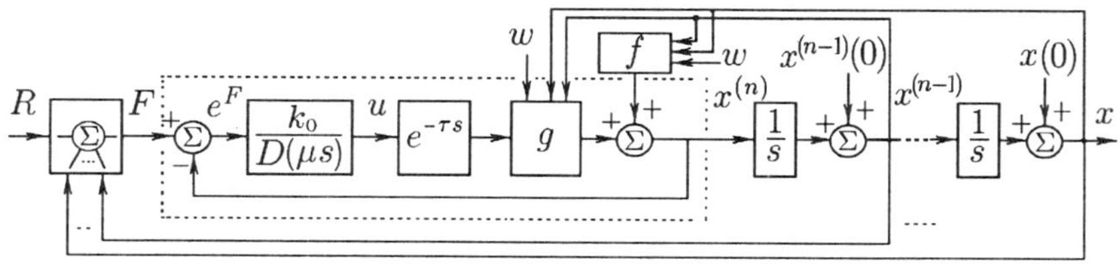

Fig. 6.2 Block diagram of the closed-loop system (6.3), (6.4).

The Nyquist criterion [Nyquist (1932)] is more convenient to use in order to analyze the stability of the FMS (6.11) with a pure time delay. From a practical viewpoint this allows us to obtain simple relationships for choice of the controller parameters. Therefore, let us select from Fig. 6.2 the part corresponding to the FMS (6.11) and consider the sinusoidal transfer function

$$G^{O}_{FMS}(j\omega, \mu) = \frac{k_0 g e^{-j\tau\omega}}{D(j\mu\omega)}, \tag{6.12}$$

where $G^{O}_{FMS}(s, \mu)$ is the transfer function of the open-loop FMS with time delay as shown in Fig. 6.3.

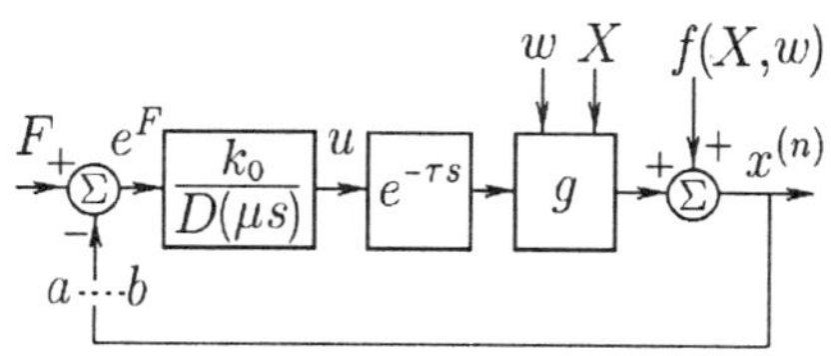

Fig. 6.3 Block diagram of the open-loop FMS with delay (6.11), where $w = \text{const}$, $X = \text{const}$, $F = \text{const}$, $f = \text{const}$.

From the expression (4.44) for the polynomial $D(\mu s)$ it follows that the Nyquist plot of (6.12) approaches the Nyquist plot of

$$G^{O}_{FMS}(j\omega, \mu = 0) = \frac{k_0 g}{d_0} e^{-j\tau\omega}$$

as $\mu \to 0$, where the Nyquist plot of $G^{O}_{FMS}(j\omega, \mu = 0)$ is a circle of radius

$$r_0 = k_0 g / d_0.$$

Let us assume that the characteristic polynomial $D(\mu s)$ is stable and $k_0 g > d_0$ as shown in Fig. 6.4. As μ is decreased it will eventually reach some value such that the Nyquist plot of (6.12) encircles the point $(-1, j0)$ in the clockwise direction. Therefore, loss of stability of the FMS (6.11) occurs when the parameter μ is decreased.

So, in contrast to the control system (4.45)–(4.46), a lower bound for the parameter μ appears in a control system with pure time delay. This bound is determined by the requirement for FMS stability, and depends on the value of the pure time delay τ.

Fig. 6.4 Nyquist plot of the FMS (6.11) with time delay.

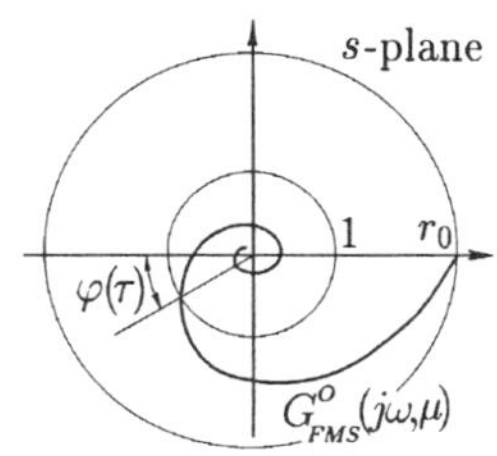

In accordance with the Nyquist stability criterion, the FMS (6.11) is marginally stable if the conditions

$$|D(j\mu_m\omega_m)| = k_0 g, \tag{6.13}$$

$$\text{Arg}D(j\,\mu_m\,\omega_m) + \tau_m\omega_m = \pi \tag{6.14}$$

hold[1]. From equations (6.13) and (6.14), the lower bound on μ is given by

$$\mu_m = \tau a_m\{\pi - \text{Arg}D(ja_m)\}^{-1}, \tag{6.15}$$

where a_m satisfies

$$|D(ja_m)| = k_0 g.$$

In accordance with (6.2), the FMS (6.11) is asymptotically stable if

$$\mu > \tau_{\max} a_m\{\pi - \text{Arg}D(ja_m)\}^{-1}.$$

6.1.5 *Phase margin of FMS with delay*

The main question regards conditions under which the influence of the pure time delay is negligible.

It is obvious that a closed-loop system having a marginally stable FMS will burn out, disintegrate, or saturate in practice. In practical control

[1] $\text{Arg}D(j\,\mu_m\,\omega_m)$ denotes the principal value of $\arg D(j\,\mu_m\,\omega_m)$.

systems it is required that for all $\tau \in [0, \tau_{\max}]$ the FMS (6.11) be asymptotically stable with the required settling time (5.16), that is,

$$t_{s,FMS} \leq t_{s,FMS}^{d}$$

where $t_{s,FMS}^{d}$ is defined by the degree of time-scale separation between the fast and slow modes in the closed-loop system. Usually, it is also desirable to provide an acceptable level of oscillation excited in the FMS (6.11).

In accordance with the above procedure for designing the controller (4.38), we assume that the polynomial $D(\mu s) + k_0 g$ is stable and that the condition $k_0 g > d_0$ holds. Therefore, we must verify (5.16) only in the case when $\tau = \tau_{\max}$. It is more convenient to evaluate the performance of the FMS transients in the presence of a pure time delay based on the calculation of phase margin and gain margin. The settling time can be estimated by means of the crossover frequency ω_c on the Nyquist plot of the FMS (6.11), where from (6.12) we find that ω_c is defined by the equation

$$|G_{FMS}^{o}(j\omega_c, \mu)| = 1.$$

See Fig. 6.4. In particular, the phase margin $\varphi(\tau)$ of the FMS (6.11) is given by

$$\varphi(\tau) = \pi - \mathrm{Arg}D(j\mu\omega_c) - \tau\omega_c. \tag{6.16}$$

Finally, the gain margin should also be taken into account in order to obtain the allowable performance specifications of the FMS transients.

6.1.6 *Control with compensation of delay*

Let us consider the special case when τ is a known constant. Then, in order to reduce the influence of τ on the stability of the FMS, let us modify (4.38) and consider the control law given by

$$\begin{aligned} &\mu^q u^{(q)}(t) + d_{q-1}\mu^{q-1}u^{(q-1)}(t) + \cdots + d_1\mu u^{(1)}(t) + d_0 u(t) \\ &+\gamma[u(t) - u(t-\tilde{\tau})] = k_0\{F(X(t), R(t)) - x^{(n)}(t)\},\ U(0) = U^0. \end{aligned} \tag{6.17}$$

The plant model (6.1) and the new control law (6.17) can be rewritten in the operator form

$$p^n x(t) = f(X(t), w(t)) + g(X(t), w(t))e^{-\tau p}u(t), \tag{6.18}$$

$$\{D(\mu p) + \gamma(1 - e^{-\tilde{\tau}p})\}u(t) = k_0\{F(X(t), R(t)) - p^n x(t)\}. \tag{6.19}$$

Let us assume that the conditions

$$\gamma = k_0 g, \quad \tilde{\tau} = \tau$$

hold. Next, we can use the above operator procedure to obtain the FMS equation; substitution of (6.18) into the right member of (6.19) yields

$$p^n x(t) = f(X(t), w(t)) + g(X(t), w(t))e^{-\tau p} u(t), \tag{6.20}$$

$$\{D(\mu p) + k_0 g\} u(t) = k_0 \{F(X(t), R(t)) - f(X(t), w(t))\}. \tag{6.21}$$

In accordance with the above procedure for two-time-scale motion analysis with the operator form of the differential equations, and on the assumption (6.8), it is easy to see that in the closed-loop system equations (6.20)–(6.21) the characteristic polynomial of the FMS is given by

$$D(\mu s) + k_0 g = 0. \tag{6.22}$$

From (4.44) and (6.22) it follows that the stability of the FMS in this case does not depend on μ. This is similar to the case of a system with no time delay and, accordingly, the lower bound on μ has disappeared completely.

It may be shown that as $\mu \to 0$, equations (6.21)–(6.20) yield

$$\lim_{\mu \to 0} e^F(\mu) = \left\{1 - \frac{k_0 g}{d_0 + k_0 g} e^{-\tau p}\right\} \{F(X, R) - f(X, w)\}. \tag{6.23}$$

With $d_0 = 0$, the time domain description of (6.23) yields

$$\lim_{\mu \to 0} e^F(\mu) = \sum_{i=1}^{\infty} \frac{(-\tau)^i}{i!} \frac{d^i}{dt^i} [f(X, w) - F(X, R)]. \tag{6.24}$$

So, on one hand, application of the control law (6.17) allows us to provide compensation for the delay in the FMS; on the other hand, such a control law structure leads to the additional error (6.24) of the desired dynamics realization.

Remark 6.2 *It is clear that the modification of the controller (4.38) to a control law of the form (6.17) is related to the main idea of the time delay compensation scheme now known as the Smith predictor [Smith (1957)]. This idea is widely used in controller design for processes with time delays [Palmor (1996)].*

6.1.7 *Velocity error with respect to external disturbance*

Consider the LTI system (5.7), where the pure time delay τ is introduced into the control variable:

$$x^{(n)} = \bar{a}X(t) + bu(t-\tau) + \hat{b}w(t), \quad X(0) = X^0. \tag{6.25}$$

Let us assume that the control law (6.17) with compensation of delay is applied, where $\gamma = k_0 b$ and $\tilde{\tau} = \tau$.

If $d_0 = 0$, $r(t) = \text{const}$, and $w(t) = w^v t 1(t)$, then from (6.25) and (6.17) it follows that the relative velocity error due to a ramp external disturbance $w(t)$ is given by

$$\bar{e}_w^v = \lim_{t\to\infty} \frac{r(t) - x(t)}{w^v} = -\hat{b}T^n \frac{\mu d_1 + \tilde{\tau}\gamma}{k_0 b} \tag{6.26}$$

and

$$\hat{u}(t) = -\frac{\hat{b}}{b}w^v\tau - \frac{a_0}{b}\{r - \bar{e}_w^v w^v\} - \frac{\hat{b}}{b}w^v\, t. \tag{6.27}$$

By comparing expressions (5.9) and (5.10) with (6.26) and (6.27), respectively, we can see that the additional terms caused by the pure time delay τ exist, and this fact corresponds to (6.24).

6.1.8 *Example*

Consider the second-order system

$$x^{(2)}(t) = [1 + \sin(x^{(1)}(t))]x(t) + [1 - 0.5\sin(x(t))]u(t-\tau) + w(t). \tag{6.28}$$

The desired dynamics of $y(t) = x(t)$ is assigned by the equation

$$T^2 y^{(2)} + a_1^d T y^{(1)} + y = r \tag{6.29}$$

and the control law is given by

$$\mu^2 u^{(2)} + d_1\mu u^{(1)} + d_0 u = k_0\left\{-y^{(2)} - \frac{a_1^d}{T}y^{(1)} + \frac{1}{T^2}[r - y]\right\}, \tag{6.30}$$

where $\mu = 0.1$ s, $k_0 = 10$, $d_0 = 0$, and $d_1 = 2$.

The simulation results for the closed-loop system equations (6.28) and (6.30) are displayed in Fig. 6.5, where $\tau = 0.022$ s, $a_1^d = 1.4$, $T = 1$ s, and $t \in [0, 12]$ s. Note that the parameter g depends on $x(t)$ where $g \in [0.5, 1.5]$. Then, by (6.13)–(6.14), we can find that $\tau_m(g = 0.5) \approx 0.045$

s and $\tau_m(g = 1.5) \approx 0.014$ s. We can see the loss of FMS stability for $t \in [10, 12]$ s, caused by increasing the parameter g.

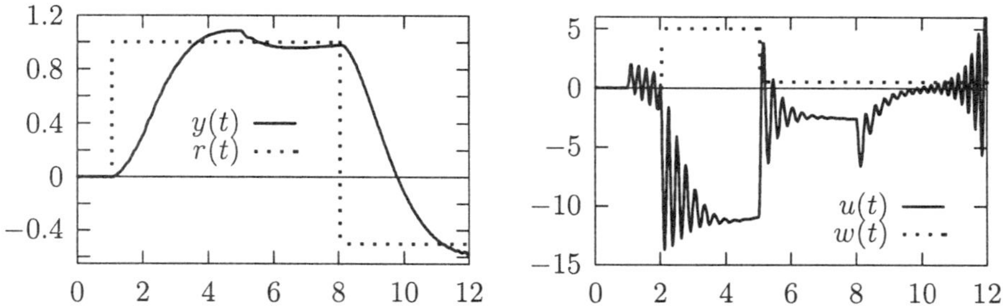

Fig. 6.5 Output response of the system (6.28), (6.30) for a step reference input $r(t)$ and a step disturbance $w(t)$ without compensation of delay, where $\tau = 0.022$ s.

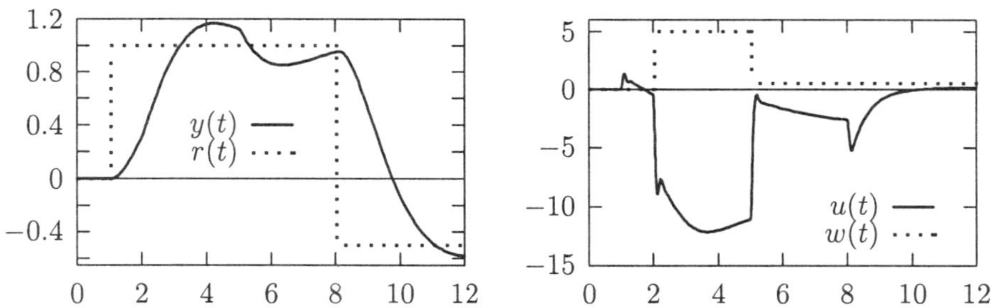

Fig. 6.6 Output response of the system (6.28), (6.31) with compensation of delay, where $\tau = 0.022$ s.

In order to provide compensation for the time delay τ, let us consider the control law in the form (6.17). If $q = n = 2$, then from (6.17) and (6.29) the control law

$$\mu^2 u^{(2)}(t) + d_1 \mu u^{(1)}(t) + d_0 u(t) + \gamma[u(t) - u(t - \tilde{\tau})]$$
$$= k_0 \left\{ -y^{(2)}(t) - \frac{a_1^d}{T} y^{(1)}(t) + \frac{1}{T^2}[r(t) - y(t)] \right\} \tag{6.31}$$

results, where $\mu = 0.1$ s, $k_0 = 10$, $d_0 = 0$, $d_1 = 2$, $\tau = 0.022$ s, $T = 1$ s, $a_1^d = 1.4$, $\gamma = k_0$, and $\tilde{\tau} = \tau$.

The simulation results for the control system (6.28), (6.31) with compensation of delay are shown in Fig. 6.6. Stability of the FMS is maintained under variations of the parameter g.

Note that the control law (6.31) can be rewritten in the form

$$\begin{aligned}\dot{u}_1(t) &= -d_1\mu u(t) + u_2(t) - \frac{k_0 a_1^d}{T} y(t),\\ \dot{u}_2(t) &= -d_0 u(t) - \gamma[u(t) - u(t-\tau)] + \frac{k_0}{T^2}[r(t) - y(t)],\\ u(t) &= \frac{1}{\mu^2}[u_1(t) - k_0 y(t)].\end{aligned} \tag{6.32}$$

Then, from (6.32), we get the block diagram as shown in Fig. 6.7.

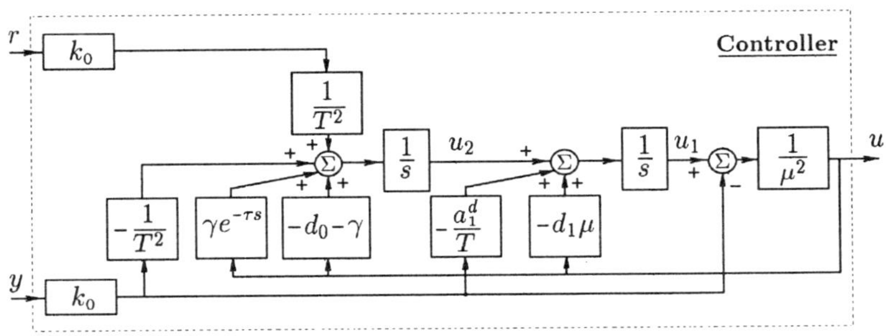

Fig. 6.7 Block diagram of the control law (6.31) represented in the form (6.32).

6.2 Regular perturbances

6.2.1 *Regularly perturbed plant model*

This section is devoted to the case when the unstructured uncertainty leads to examination of the plant model in the form of regularly perturbed differential equations and the relative degree of the system is modified. Such a mathematical model occurs, for instance, in applications to the planar vertical takeoff and landing (PVTOL) aircraft [Hauser *et al.* (1992)]. The properties of regularly perturbed systems have been discussed in such references as [Sastry *et al.* (1989); Isidori *et al.* (1992); Barbot *et al.* (1994); Sastry (1999)] in the context of approximate linearization and the relationship between regularly perturbed nonlinear systems and singularly perturbed zero dynamics. Here we concern ourselves with the peculiarities caused by regular perturbances in control systems with the highest derivative in feedback.

Consider the plant model given by

$$x^{(n)} = f(X, w) + \epsilon_m u^{(m)} + \cdots + \epsilon_1 u^{(1)} + g(X, w)u. \tag{6.33}$$

We assume that the following conditions are satisfied.

(i) The inequalities

$$|\epsilon_j| \le \epsilon \; g_{\min} \quad \forall\, j = 1, \ldots, m \tag{6.34}$$

hold, where ϵ is a small positive parameter.

(ii) $f(X, w), g(X, w)$ are continuous bounded functions where the inequalities (4.28) are satisfield for all $(X, w) \in \Omega_{X,w}$.

The system given by (6.33) is called a regular perturbance of the nominal plant model (4.27). It is easy to see that the relative degree of the system (6.33) is $n - m$ if $\epsilon_m \neq 0$.

From the plant model (6.33), the simplified model of the form (4.27) follows as $\epsilon \to 0$. We assume that the control law (4.38) was constructed based on the simplified model (4.27). The main purpose of this section is to choose the parameters of the control law (4.38) to reduce the effect of perturbances ϵ_j on the performance of the closed-loop system when the control (4.38) applied to the true system (6.33).

6.2.2 *Fast motions in presence of regular perturbances*

Note that in the general case, a decrease in μ in the closed-loop system (6.33) and (4.38) may result in loss of FMS stability due to finiteness of the parameters ϵ_j. Therefore, as with control systems having a pure time delay and, to enable us to use the standard technique of two-time-scale motion analysis [Tikhonov (1952)], we must normalize the ϵ_j to μ:

$$\epsilon_j = \epsilon_j^0 \mu^j, \quad \forall\, j = 1, \ldots, m. \tag{6.35}$$

Then (6.33) can be rewritten in the form

$$x^{(n)} = f(X, w) + \epsilon_m^0 \mu^m u^{(m)} + \cdots + \epsilon_1^0 \mu u^{(1)} + g(X, w)u. \tag{6.36}$$

If $\mu = 0$, then from (6.36) the simplified model (4.27) follows. Assume that the control law in the form (4.38) is constructed, and let us consider the effect of regular perturbances on the performance of the closed-loop system

$$x^{(n)} = f(X, w) + \epsilon_m^0 \mu^m u^{(m)} + \cdots + \epsilon_1^0 \mu u^{(1)} + g(X, w)u, \tag{6.37}$$

$$\mu^q u^{(q)} + d_{q-1}\mu^{q-1} u^{(q-1)} + \cdots + d_1 \mu u^{(1)} + d_0 u = k_0 e^F, \tag{6.38}$$

where $X(0) = X^0$ and $U(0) = U^0$.

From (4.32) and by substituting (6.37) in (6.38), we obtain

$$x^{(n)} = f + \epsilon_m^0 \mu^m u^{(m)} + \cdots + \epsilon_1^0 \mu u^{(1)} + gu, \quad X(0) = X^0, \tag{6.39}$$

$$\mu^q u^{(q)} + \cdots + d_{m+1}\mu^{m+1}u^{(m+1)} + \{d_m + k_0\epsilon_m^0\}\mu^m u^{(m)} + \cdots$$
$$+\{d_1 + k_0\epsilon_1^0\}\mu u^{(1)} + \{d_0 + k_0 g\}u = k_0\{F - f\}, \ U(0) = U^0. \tag{6.40}$$

From (6.39)–(6.40) we get the FMS equations

$$\mu^q u^{(q)} + \ldots + d_{m+1}\mu^{m+1}u^{(m+1)} + \{d_m + k_0\epsilon_m^0\}\mu^m u^{(m)} + \cdots$$
$$+\{d_1 + k_0\epsilon_1^0\}\mu u^{(1)} + \{d_0 + k_0 g\}u = k_0\{F - f\}, \ U(0) = U^0, \tag{6.41}$$

where $g(X, w)$ is the frozen parameter during the transients in (6.41).

Assume that the FMS (6.41) is stable. Then the SMS of the form (4.59) results from (6.39)–(6.40) as $\mu \to 0$. Note that this occurs because of (6.35). So the desired output transients are guaranteed fully in the closed-loop system after damping of the FMS transients.

6.2.3 *Selection of controller parameters*

Let us assume that the parameters ϵ_j are unknown and belong to the known intervals

$$\epsilon_j \in [\underline{\epsilon}_j, \overline{\epsilon}_j], \quad \forall\, j = 1, \ldots, m, \tag{6.42}$$

where $\underline{\epsilon}_j \leq \overline{\epsilon}_j$. Then, from (6.35) and (6.42) we can find intervals normalized by μ:

$$\epsilon_j^0 \in [\underline{\epsilon}_j^0, \overline{\epsilon}_j^0], \quad \forall\, j = 1, \ldots, m. \tag{6.43}$$

From (6.41) the FMS characteristic polynomial

$$\mu^q s^q + d_{q-1}\mu^{q-1}s^{q-1} + \cdots + d_{m+1}\mu^{m+1}s^{m+1}$$
$$+\{d_m + k_0\epsilon_m^0\}\mu^m s^m + \cdots + \{d_1 + k_0\epsilon_1^0\}\mu s + \{d_0 + k_0 g\} \tag{6.44}$$

follows, where the polynomial coefficients belong to the known intervals depending on the bounds of the intervals (6.43).

Note that the stability of the polynomial (6.44) does not depend on the value of the parameter μ, and corresponds to the stability of the normalized polynomial

$$s^q + \cdots + d_{m+1}s^{m+1} + \{d_m + k_0\epsilon_m^0\}s^m + \cdots + \{d_1 + k_0\epsilon_1^0\}s + \{d_0 + k_0 g\}. \tag{6.45}$$

So we have the following result.

Suppose:

(i) The plant model with regular perturbances has the form (6.33).
(ii) The conditions (6.35) and (6.43) hold.
(iii) The equation of the desired output dynamics is assigned by (4.30).
(iv) The parameters μ, d_0, k_0 of the control law (4.38) are assigned such that the required control accuracy and time scale-separation are satisfied.

Then the calculation of the remaining parameters of the control law (4.38) is reduced to a choice of $d_1, \ldots, d_q$ in such a way that the interval polynomial (6.45) must be stable[2] for all possible values of ϵ_j^0. Moreover, some nonzero value of relative stability must be provided.

Remark 6.3 *The next step of the design procedure deals with robust stability analysis of (6.45) via interval analysis tools [Piazzi and Marro (1996); Kharitonov and Torres Munoz (2002)]. In particular, frequency-domain criteria for robust stability may be used [Tsypkin and Polyak (1991)].*

6.2.4 *Control with compensation of regular perturbances*

Consider the plant model (6.33), where we assume that the ϵ_j are known and $\epsilon_j = \text{const}$, $\forall j = 1, \ldots, m$. Then let us modify (4.38) and consider the control law given by

$$\begin{aligned}\mu^q u^{(q)} + \ldots + d_{m+1}\mu^{m+1}u^{(m+1)} + \{d_m\mu^m - \gamma_0\epsilon_m^0\}u^{(m)} + \cdots \\ + \{d_1\mu - \gamma_0\epsilon_1^0\}u^{(1)} + d_0 u = k_0\{F - x^{(n)}\},\end{aligned} \tag{6.46}$$

where (6.46) may be rewritten in the operator form

$$\{D(\mu p) - \gamma_0[\epsilon_m p^m + \epsilon_{m-1}p^{m-1} + \cdots + \epsilon_1 p]\}u = k_0 e^F. \tag{6.47}$$

Assume that $\gamma_0 = k_0$. Then from the closed-loop system equations (6.33) and (6.46), it follows that the corresponding FMS equation has characteristic polynomial of the form (6.22). As a result, compensation for the regular perturbances in the FMS occurs. At the same time, as $\mu \to 0$ in (6.33) and (6.46) we obtain the operator equation of the desired dynamics realization

[2] The polynomial is said to be stable if and only if the real part is strictly negative for all roots of the polynomial.

given by

$$\lim_{\mu\to 0} e^F(\mu) = \frac{d_0}{d_0 + k_0 g}\{F(X,R) - f(X,w)\}$$
$$+\frac{k_0}{d_0 + k_0 g}[\epsilon_m p^m + \cdots + \epsilon_1 p]\{f(X,w) - F(X,R)\}. \tag{6.48}$$

With $d_0 = 0$, the time domain description of (6.48) yields

$$\lim_{\mu\to 0} e^F(\mu) = \frac{1}{g}\sum_{i=1}^{m} \epsilon_i \frac{d^i}{dt^i}[f(X,w) - F(X,R)]. \tag{6.49}$$

So the control law (6.46) with compensation of regular perturbances gives the additional error of the desired dynamics realization. Note that (6.48) is the full counterpart of (6.23) obtained above for control systems with compensation of the pure time delay.

6.2.5 *Example*

Let us consider an SISO nonlinear continuous-time system in the form

$$\begin{aligned}\dot{x}_1 &= x_2 + \epsilon_1 u,\\ \dot{x}_2 &= [1+\sin(x_1)]x_1 + [1 - 0.5\sin(x_1)]u + w(t),\\ y &= x_1,\end{aligned}$$

where $\epsilon_1 = -0.015$. By differentiating $y(t)$ we obtain

$$y^{(2)} = [1+\sin(y)]y + \epsilon_1 u^{(1)} + [1 - 0.5\sin(y)]u + w(t), \tag{6.50}$$

where $\epsilon_1 = -0.015$.

Equation (6.50) corresponds to equation (6.33), and the approximate system (with $\epsilon_1 = 0$) is given by (5.80) (see p. 109). Assume that the desired behavior of $y(t)$ is assigned by

$$y^{(2)} = -\frac{a_1^d}{T}y^{(1)} - \frac{1}{T^2}y + \frac{b_1^d\tau}{T^2}r^{(1)} + \frac{1}{T^2}r. \tag{6.51}$$

Let $q = 2$; then for the approximate system (5.80) we may consider the control law given by (5.82):

$$\mu^2 u^{(2)} + d_1\mu u^{(1)} + d_0 u = k_0\left\{-y^{(2)} - \frac{a_1^d}{T}y^{(1)} + \frac{b_1^d\tau}{T^2}r^{(1)} + \frac{1}{T^2}[r-y]\right\},$$

where we assume that $\mu = 0.1$ s, $k_0 = 10$, $d_0 = 0$, and $d_1 = 2$.

The control law (5.82) applied to the true system (6.50) gives the FMS characteristic polynomial of the form

$$\mu^2 s^2 + \{d_1\mu + \epsilon_1 k_0\}s + \{d_0 + k_0 g\},$$

where for the given system parameters a decrease in μ leads to loss of FMS stability (since $\epsilon_1 < 0$). An increase in μ leads to loss of the degree of time-scale separation between the FMS and SMS, and so to a loss in accuracy of the desired dynamics realization. Note that the root-locus method may be useful to find an appropriate value of the parameter μ.

Simulation results for the output response in the system (6.50) controlled by the algorithm (5.82) to a step reference input $r(t)$ and a step disturbance $w(t)$ are displayed in Fig. 6.8. Here the initial conditions are zero and $T = \tau = 1$ s, $a_1^d = 1.4$, $b_1^d = 0$, and $t \in [0, 8]$ s. Results for the same system where $\mu = 0.08$ s and $\mu = 0.3$ s are shown in Figs. 6.9 and 6.10, respectively.

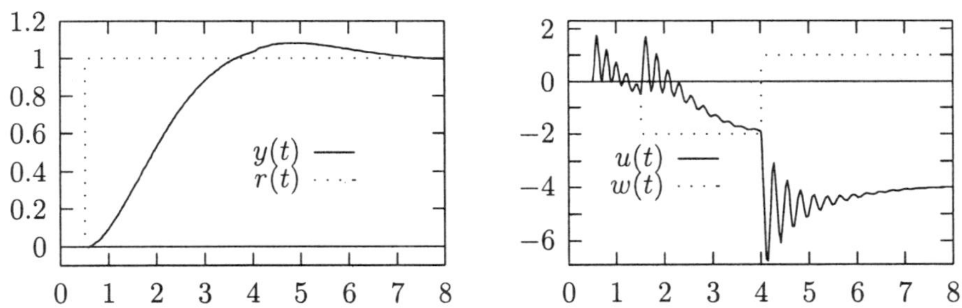

Fig. 6.8 Output response of the system (6.50) and (5.82) for a step reference input $r(t)$ and a step disturbance $w(t)$, where $\mu = 0.1$ s.

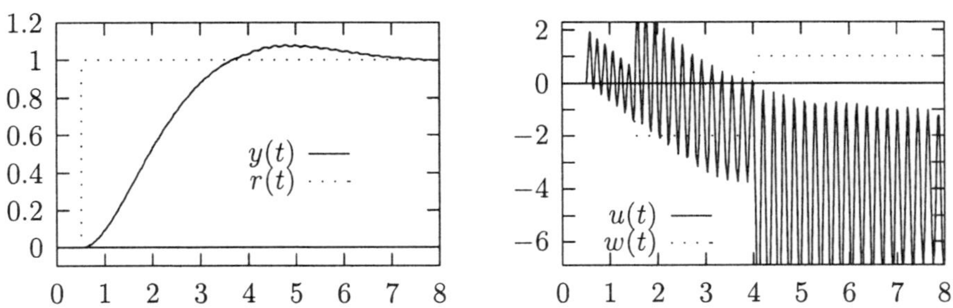

Fig. 6.9 Output response of the system (6.50) and (5.82) for a step reference input $r(t)$ and a step disturbance $w(t)$, where $\mu = 0.08$ s.

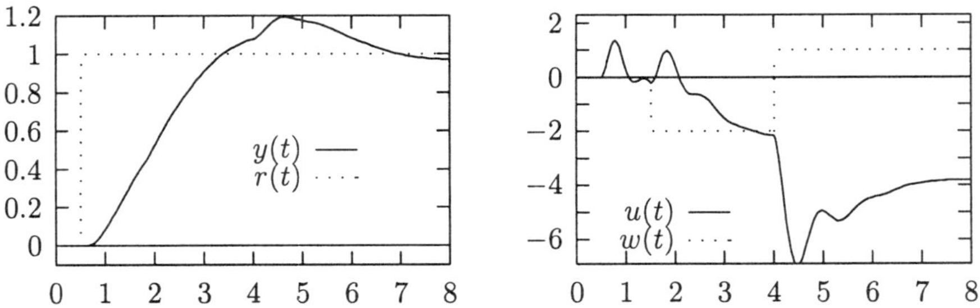

Fig. 6.10 Output response for the system (6.50), (5.82) for a step reference input $r(t)$ and a step disturbance $w(t)$, where $\mu = 0.3$ s.

6.3 Singular perturbances

6.3.1 *Singularly perturbed plant model*

Let us consider the other possible type of unstructured uncertainty that leads to a plant model in the form of singularly perturbed differential equations. In particular, taking into account the actuator and/or sensor dynamics leads to an increase in the system order and the appearance of additional fast dynamics. As a result, the plant model has the form of singularly perturbed differential equations (singular perturbances). The various classes of control systems with singularly perturbed models have been widely investigated (see, for instance, [Kokotović *et al.* (1976); Ioannou and Kokotović (1983); Riedle and Kokotović (1985); Marino (1985); Khalil (1987)]).

The main subject of this section is the performance and robustness of a control system with the highest derivative in feedback in the presence of singular perturbances. In particular, the nonlinear control system (NCS) preceded by the fast actuator (A) will be discussed as shown in Fig. 6.11.

Assume that the nominal model of the plant is the NCS governed by (4.27) (see p. 64). At the same time, taking into consideration the fast dynamics of the actuator (A), we obtain the following singularly perturbed

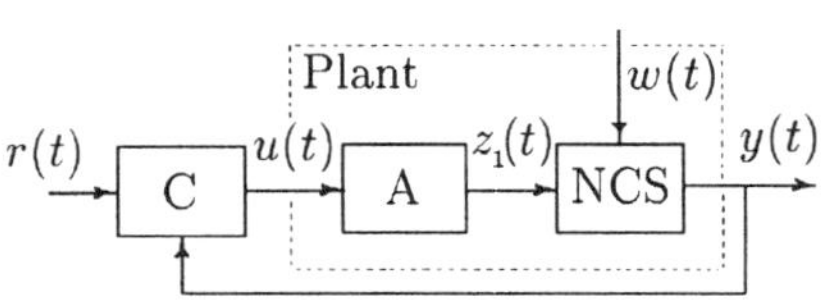

Fig. 6.11 Block diagram of the nonlinear control system (NCS) preceded by fast actuator (A).

system:

$$x^{(n)} = f(X,w) + g(X,w)z_1 \ , y = x, \ X(0) = X^0, \tag{6.52}$$
$$\nu z_i^{(1)} = z_{i+1} \ , \quad i = 1,\dots,l-1, \tag{6.53}$$
$$\nu z_l^{(1)} = \varphi(t, z_1, \dots, z_l, u) \ , \quad Z(0) = Z^0, \tag{6.54}$$

where $Z = \{z_1, z_2, \dots, z_l\}^T$ is the state vector of the additional subsystem (6.53)–(6.54) caused by the fast dynamics of the actuator, $Z(0) = Z^0$ is the initial state of subsystem (6.53)–(6.54), and ν is the small positive parameter.

If $\nu = 0$, then from (6.53)–(6.54) the equality $z_1 = u$ follows and from (6.52)–(6.54) the simplified model (4.27) results, where the degree of (4.27) is equal to n. From a practical viewpoint, we also assume that the additional fast subsystem described by (6.53)–(6.54) is stable, as only on such an assumption can the simplified model (4.27) reflect the main qualitative and quantitative performance of the whole system (6.52)–(6.54) when ν is small enough.

The additional fast subsystem described by (6.53)–(6.54) is examined here as a particular case of the unstructured uncertainty, where the degree of the system (6.52)–(6.54) is equal to $n + l$. The system of equations (6.52)–(6.54) is called the singular perturbance of the nominal plant model (4.27).

Let the control law structure of the form (4.38) be constructed based on the simplified model (4.27). The main purpose of this section is to choose the parameters of the control law (4.38) to reduce the effect of singular perturbance on performance of the closed-loop system when the control (4.38) applied to the true system (6.52)–(6.54).

6.3.2 *Fast motions in presence of singular perturbances*

The closed-loop system equations of the plant model (6.52)–(6.54) and controller (4.38) are given by

$$\begin{aligned}
x^{(n)} &= f(X,w) + g(X,w)z_1, \quad X(0) = X^0, \\
\nu z_i^{(1)} &= z_{i+1}, \quad i = 1,\dots,l-1, \\
\nu z_l^{(1)} &= \varphi(t, z_1, \dots, z_l, u), \ Z(0) = Z^0, \\
\mu^q u^{(q)} &+ d_{q-1}\mu^{q-1}u^{(q-1)} + \cdots \\
&+ d_1\mu u^{(1)} + d_0 u = k_0\{F(X,R) - x^{(n)}\}, \ U(0) = U^0.
\end{aligned}$$

Substituting the expression of $x^{(n)}$ into the right member of the second equation, we get

$$\begin{aligned} x^{(n)} &= f(X,w) + g(X,w)z_1, \quad X(0) = X^0, \\ \nu z_i^{(1)} &= z_{i+1}, \quad i = 1,\dots,l-1, \\ \nu z_l^{(1)} &= \varphi(t, z_1,\dots,z_l,u), \quad Z(0) = Z^0, \\ \mu^q u^{(q)} &+ d_{q-1}\mu^{q-1}u^{(q-1)} + \dots + d_1\mu u^{(1)} + d_0 u \\ &= k_0\{F(X,R) - f(X,w) - g(X,w)z_1\}, \ U(0) = U^0. \end{aligned} \tag{6.55}$$

First, in order to enable use of the above standard technique for two-time-scale motion analysis of (6.55), we must assume that ν and μ are interdependent. In particular, let $\nu = \nu_0\mu$ where $\nu_0 = \text{const}$.

Second, let us rewrite (6.55) in the new fast time scale $t_0 = t/\mu$ and, by setting $\mu = 0$, find the FMS equations in the t_0 scale. Then, by returning to the primary time scale $t = \mu t_0$, we obtain the FMS equations

$$\begin{aligned} \nu z_i^{(1)} &= z_{i+1}, \quad i = 1,\dots,l-1, \\ \nu z_l^{(1)} &= \varphi(t, z_1,\dots,z_l,u), \quad Z(0) = Z^0, \\ \mu^q u^{(q)} &+ d_{q-1}\mu^{q-1}u^{(q-1)} + \dots + d_1\mu u^{(1)} + d_0 u \\ &= k_0\{F(X,R) - f(X,w) - g(X,w)z_1\}, \ U(0) = U^0, \end{aligned} \tag{6.56}$$

where $X(t)$ and $w(t)$ are the frozen variables during the transients in (6.56).

The controller parameters should be selected such that to maintain allowable performance of the fast-motion transients described by (6.56).

6.3.3 *Selection of controller parameters*

The main subject matter is an additional restriction on the controller parameters of (4.38) caused by singular perturbances in the form of the subsystem (6.53)–(6.54).

Let us consider a particular case of the subsystem (6.53)–(6.54) described by the transfer function

$$G_s(s) = \frac{1}{S(\nu s)}, \tag{6.57}$$

where

$$S(\nu s) = \nu^l s^l + \tilde{s}_{l-1}\nu^{l-1}s^{l-1} + \dots + \tilde{s}_1 \nu s + 1. \tag{6.58}$$

The block diagram representation of the closed-loop system equations (6.55) is shown in Fig. 6.12. The additional fast dynamical system is represented by the block with transfer function (6.57), and some portion of this block diagram is highlighted by a circuit of dots. This part corresponds to the FMS (6.56).

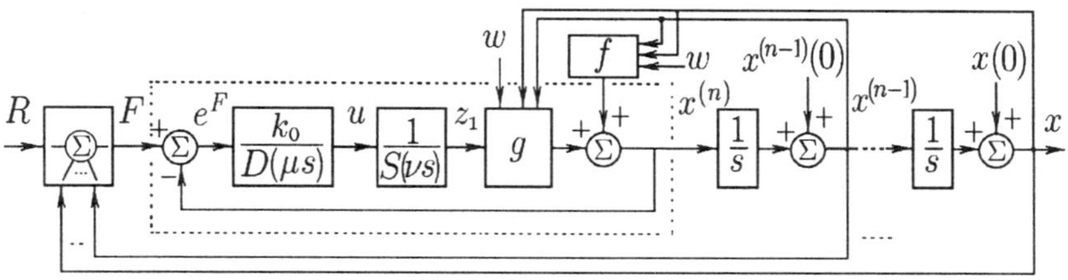

Fig. 6.12 Block diagram of the closed-loop system (6.55)-(6.57).

The FMS (6.56) is a linear system where g is the frozen parameter. Let the equation of the desired dynamics (4.30) be constructed based on the differential equation given by (2.8). Similar to (5.26) and (5.46), let us apply the Laplace transform to (6.56), given that the initial conditions of (6.56) are all zero. We get

$$\begin{aligned} u(s) = &\frac{k_0 S(\nu s)}{D(\mu s)S(\nu s) + k_0 g}\{F(s) - f(s)\} \\ &- \frac{k_0 S(\nu s) A^d(s)}{T^n[D(\mu s)S(\nu s) + k_0 g]} n_s(s), \end{aligned} \tag{6.59}$$

where

$$A^d(s) = T^n s^n + a^d_{n-1} T^{n-1} s^{n-1} + \cdots + a^d_1 T s + 1.$$

Let $\nu = \text{const}$. Then from (6.59) it follows that

$$\begin{aligned} \lim_{\mu \to 0} u(s, \mu) = &\frac{k_0 S(\nu s)}{d_0 S(\nu s) + k_0 g}\{F(s) - f(s)\} \\ &- \frac{k_0 S(\nu s) A^d(s)}{T^n[d_0 S(\nu s) + k_0 g]} n_s(s) \end{aligned} \tag{6.60}$$

as $\mu \to 0$.

In particular, if $d_0 = 0$ then

$$\lim_{\mu \to 0} u(s, \mu) = g^{-1} S(\nu s)[F(s) - f(s) - T^{-n} A^d(s) n_s(s)]. \tag{6.61}$$

By comparing (6.59), (6.60), and (6.61) with (4.36), (5.42), and (5.46), respectively, we can see that the influence of the high-frequency sensor noise $n_s(t)$ on the control $u(t)$ (manipulated variable) is increased as $\mu \ll \nu$ because of the factor $S(\nu s)$. As a result, the requirements on the allowable range of control variations are increased.

We can see that if $\mu \gg \nu$, then from (6.58) and (6.59) the expression (5.46) follows. This is the case when the influence of singular perturbances in the form of stable subsystem (6.53)–(6.54) can be neglected.

On the other hand, if $\mu \approx \nu$ then from (6.59) we can see that the parameters of the polynomial $D(\mu s)$ should be chosen in such a way that the characteristic polynomial of the FMS

$$D(\mu s)S(\nu s) + k_0 g$$

is stable.

So, in the presence of singular perturbances of the form (6.53)–(6.54), it is advisable to choose the small parameter μ of the control law (4.38) such that the additional restriction $\mu \geq \nu$ is satisfied. Note that if an increase in μ conflicts with the requirement on time-scale separation degree (5.16)–(5.17), then the control law structure should be chosen based on the nonsimplified model (6.52)–(6.54).

6.4 Nonsmooth nonlinearity in control loop

6.4.1 *System preceded by nonsmooth nonlinearity*

Nonsmooth nonlinearities are inherent in a wide set of mechanical actuators, electrical and electro-mechanical systems. This section deals with nonlinear continuous systems preceded by a nonsmooth nonlinearity, and the peculiarities caused by a nonsmooth nonlinearity in a control system with the highest derivative in feedback.

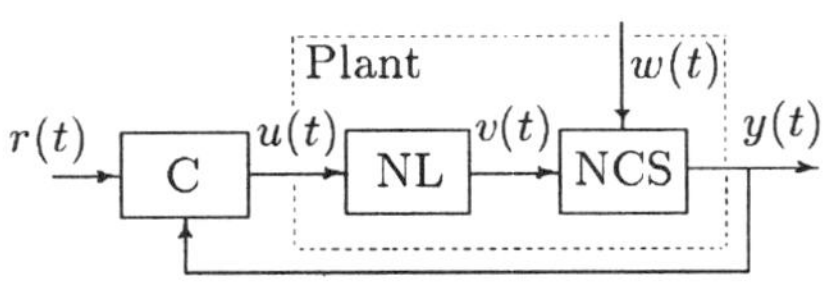

Fig. 6.13 Block diagram of the nonlinear continuous system (NCS) preceded by nonsmooth nonlinearity (NL).

A block diagram of the system under consideration, a nonlinear continuous system (NCS) preceded by a nonsmooth nonlinearity (NL), is shown in Fig. 6.13. The control system is being designed to provide the condition

(2.1) with prescribed output response specifications. Moreover, the controlled transients of $y(t)$ should have the desired behavior in the presence of varying parameters and external disturbances $w(t)$ in the plant model.

In particular, we discuss the problem of controller design for the nonlinear time-varying continuous systems governed by differential equation (4.27), assuming that parameters of the system are unknown and may vary in some bounded set, and that the conditions (4.28) are satisfied. Assume that the system (4.27) is preceded by a nonsmooth nonlinearity. This may be associated with saturation, relay with dead zone, hysteresis, or backlash hysteresis as shown in Fig. 6.14.

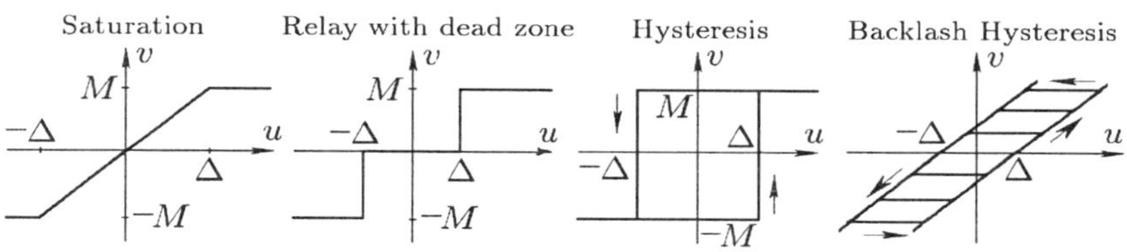

Fig. 6.14 Nonsmooth nonlinearities.

Let us consider the following system:

$$x^{(n)} = f(X, w) + g(X, w)v, \quad X(0) = X^0, \tag{6.62}$$

$$y = x, \quad v = \varphi(u, u^{(1)}), \tag{6.63}$$

where the nonlinearity is represented by the function $v = \varphi(u, u^{(1)})$; X is the state vector which is unavailable for measurement, $X = \{x, x^{(1)}, \dots, x^{(n-1)}\}^T$; w is the vector of (unavailable for measurement) external disturbances or varying parameters, $w \in \mathbb{R}^\rho$; y is the measurable output (controlled variable) of the system, $y \in \mathbb{R}^1$; u is the control variable, $u \in \mathbb{R}^1$.

The purpose of this section is to discuss the peculiarities caused by the nonsmooth nonlinearity in the system (6.62)–(6.63) with a controller of the form (4.38):

$$\mu^q u^{(q)} + d_{q-1}\mu^{q-1}u^{(q-1)} + \cdots + d_1\mu u^{(1)} + d_0 u = k_0 e^F.$$

In extended form this control law can be represented as (4.41):

$$\begin{aligned}&\mu^q u^{(q)} + d_{q-1}\mu^{q-1}u^{(q-1)} + \cdots + d_1\mu u^{(1)} + d_0 u\\ &= \frac{k_0}{T^n}\{-T^n x^{(n)} - a^d_{n-1}T^{n-1}x^{(n-1)} - \cdots - a^d_1 T x^{(1)} - x\\ &\quad + b^d_\rho \tau^\rho r^{(\rho)} + b^d_{\rho-1}\tau^{\rho-1}r^{(\rho-1)} + \cdots + b^d_1 \tau r^{(1)} + r\}.\end{aligned}$$

As a result, we have the closed-loop system given by

$$\begin{aligned}x^{(n)} = f(X,w) + g(X,w)v,\ v = \varphi(u,u^{(1)}),\ X(0) = X^0,\\ \mu^q u^{(q)} + d_{q-1}\mu^{q-1}u^{(q-1)} + \cdots + d_1\mu u^{(1)} + d_0 u = k_0 e^F,\ U(0) = U^0.\end{aligned}$$

In accordance with (4.32), the above system can be rewritten as

$$x^{(n)} = f(X,w) + g(X,w)\varphi(u,u^{(1)}), \tag{6.64}$$

$$\begin{aligned}&\mu^q u^{(q)} + d_{q-1}\mu^{q-1}u^{(q-1)} + \cdots + d_1\mu u^{(1)} + d_0 u\\ &\qquad = k_0\{F(X,R) - x^{(n)}\}.\end{aligned} \tag{6.65}$$

Substituting (6.64) into (6.65), we obtain the closed-loop system equations in the form

$$x^{(n)} = f(X,w) + g(X,w)\varphi(u,u^{(1)}), \tag{6.66}$$

$$\begin{aligned}&\mu^q u^{(q)} + d_{q-1}\mu^{q-1}u^{(q-1)} + \cdots + d_1\mu u^{(1)} + d_0 u\\ &+k_0 g(X,w)\varphi(u,u^{(1)}) = k_0\{F(X,R) - f(X,w)\}.\end{aligned} \tag{6.67}$$

Since μ is the small positive parameter, (6.66)–(6.67) are the singularly perturbed equations. A formal application of the above considered standard procedure [Tikhonov (1952)] for the time-scale separation gives the FMS equation of the following form:

$$\begin{aligned}&\mu^q u^{(q)} + d_{q-1}\mu^{q-1}u^{(q-1)} + \cdots + d_1\mu u^{(1)} + d_0 u\\ &+k_0 g(X,w)\varphi(u,u^{(1)}) = k_0\{F(X,R) - f(X,w)\},\end{aligned} \tag{6.68}$$

where X and w are the frozen variables.

6.4.2 *Describing function analysis of limit cycle in FMS*

Let us consider the block diagram representation of the closed-loop system equations (6.66)–(6.67) as shown in Fig. 6.15, where the initial conditions of the controller are omitted and the polynomial $D(\mu s)$ has the form of (4.44).

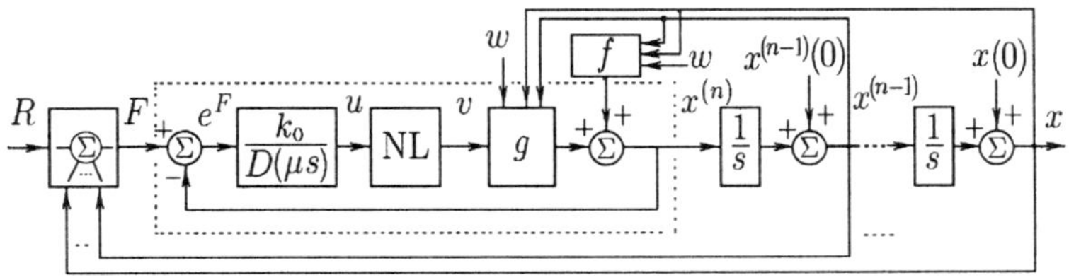

Fig. 6.15 Block diagram of the closed-loop system (6.66)–(6.67).

The highlighted portion of the diagram, shown separately in Fig. 6.16, corresponds to the FMS (6.68). From the block diagram representation it is clear that the FMS (6.68) can be examined as a sequence of a linear system and nonsmooth nonlinearity. It is well known that the describing function method is a powerful tool for the analysis of periodic solutions in such systems.

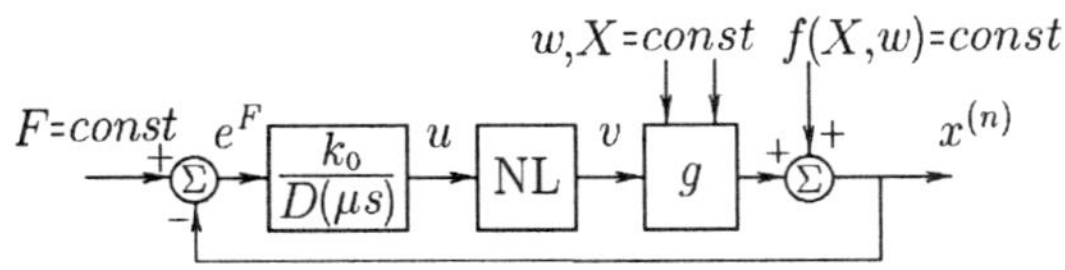

Fig. 6.16 Block diagram of the FMS (6.68) where w = const, X = const, F = const, f = const.

Remark 6.4 *Some particulars related to the describing function method can be found in a broad set of references, for instance, [Mees and Bergen (1975); Atherton (1981); Mickens (1981); Slotine and Li (1991); Taylor (2000); Khalil (2002); Vukić et al. (2003)].*

Remark 6.5 *Note that the application of the describing function method to analyze closed-loop systems with nonsmooth nonlinearity and the highest derivative in feedback of the form (3.15) was discussed in [Suvorov (1991)]. In contrast to (3.15), the discussed control law structure (4.41) allows us to include the integral action in the control loop without increasing the controller's order.*

Denote

$$G_l(j\mu\omega) = \frac{k_0 g}{D(j\mu\omega)} \tag{6.69}$$

and assume that the sinusoidal transfer function $G_l(j\mu\omega)$ reveals a low-pass filtering property. Assume that a limit cycle in the FMS (6.68) exists, and the stationary oscillation signal $u(t)$ is

$$u(t) = u_0 + A\sin(\omega t). \tag{6.70}$$

Then the output of the nonlinearity $v = \varphi(u, u^{(1)})$ can be represented by its Fourier series

$$v(t) = v_0 + b_1 \sin(\omega t) + c_1 \cos(\omega t) + \sum_{k=2}^{\infty}\{b_k \sin(k\omega t) + c_k \cos(k\omega t)\}. \tag{6.71}$$

By taking $d_0 = 0$ in (4.44), we find that the polynomial $D(\mu s)$ has the form

$$D(\mu s) = D_0(\mu s)\mu s \tag{6.72}$$

where

$$D_0(\mu s) = \mu^{q-1}s^{q-1} + d_{q-1}\mu^{q-2}s^{q-2} + \cdots + d_2\mu s + d_1;$$

as a result, the integral action is incorporated in the system of Fig. 6.16. Then we have

$$\int_t^{t+2\pi/\omega} e^F(\bar{t})\, d\bar{t} = 0 \tag{6.73}$$

for the stationary oscillations in the FMS (6.68). So the average value of e^F corresponds to the insensitivity condition (4.33), and the desired behavior of the output $y(t)$ with assigned dynamics (4.30) is satisfied if sufficiently fast oscillations take place.

The expression (6.73) represents the insensitivity condition for the output behavior with respect to parameter variations and external disturbances of the plant model in the average sense. Note that the existence of the fast oscillations in the FMS (6.68) is the essential requirement that allows us to reach the desired output behavior.

In accordance with the describing function method, let us replace the nonlinear element in Fig. 6.16 by its quasi-linear approximation as shown in Fig. 6.17. Here, for simplicity, the amplitude A and frequency ω of the stationary oscillations will be estimated to a first approximation on the condition that

$$F = f = 0 \tag{6.74}$$

and the nonlinearity is odd. Hence, we have that $u_0 = v_0 = 0$ and the

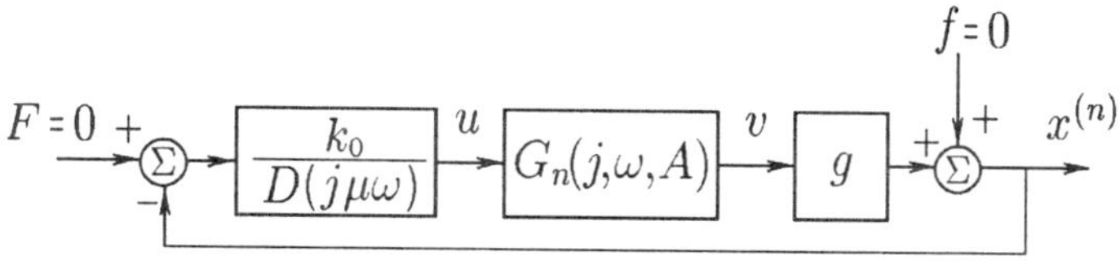

Fig. 6.17 Quasi-linear approximation of the FMS (6.68).

describing function representation of the nonlinear element is given by

$$G_n(j,\omega,A) = \frac{b_1(\omega,A)}{A} + j\frac{c_1(\omega,A)}{A}, \tag{6.75}$$

which may be complex and dependent on both ω and A in general.

Then the solutions of the first-order harmonic balance equation

$$1 + G_n(j,\omega,A)G_l(j\mu\omega) = 0 \tag{6.76}$$

represented in the form

$$G_l(j\mu\omega) = -\frac{1}{G_n(j,\omega,A)} \tag{6.77}$$

correspond to the points of intersection of the Nyquist plot of $G_l(j\mu\omega)$ and the $G_n^{-1}(j,\omega,A)$ locus. From this, the existence of the stable limit cycle in the FMS (6.68) and its parameters can be determined.

For instance, a limit cycle in the FMS (6.68) does not exist if the Nyquist plot of $G_l(j\mu\omega)$ and the $G_n^{-1}(j,\omega,A)$ locus are as shown in Fig. 6.18(a). Accordingly, there are two limit cycles as shown in Fig. 6.18(b). Here point 1 represents an unstable periodic solution and point 2 represents the stable periodic solution. Note that the arrows on $G_l(j\mu\omega)$ and $G_n^{-1}(j,\omega,A)$ indicate increases in ω and A, respectively. This example corresponds to a system preceded by a backlash hysteresis.

As another example, we have a system preceded by a relay with dead zone or hysteresis where the corresponding describing functions do not depend on the frequency ω. The qualitative Nyquist plots of $G_l(j\mu\omega)$ and $G_n^{-1}(j,A)$ loci for these two types of systems are shown in Fig. 6.19.

6.4.3 *Effect of chattering on control accuracy*

The oscillations in the FMS (6.68) induce the ripple in the output $y(t)$ of the system (6.62)–(6.63) and have an influence on the error of the output

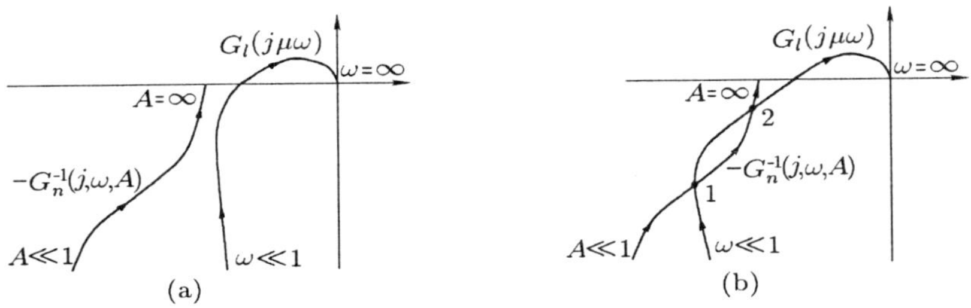

Fig. 6.18 (a) Limit cycle in the FMS (6.68) does not exist. (b) Two limit cycles in the FMS (6.68).

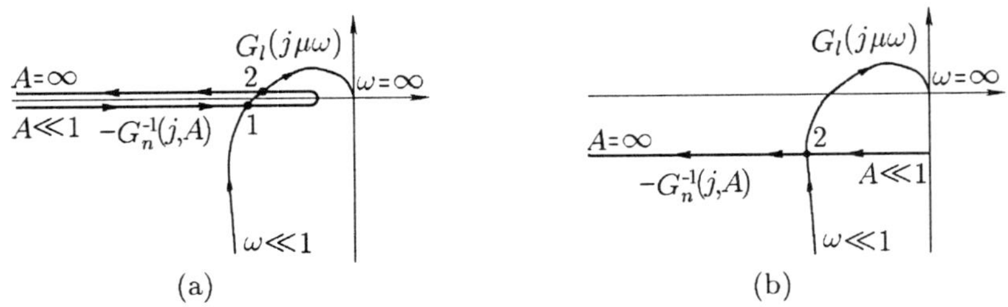

Fig. 6.19 Qualitative Nyquist plot of $G_l(j\mu\omega)$ and $G_n^{-1}(j,A)$ for a system preceded by a relay with dead zone (a), or hysteresis (b).

stabilization as shown in Fig. 6.20. Let e_{osc} be the amplitude of the stationary oscillations with frequency ω of the output $y(t)$. In order to estimate this when a stable limit cycle occurs in the FMS (6.68), let us assume that $r = \text{const}$ and denote

$$\bar{y} = \lim_{T\to\infty} \frac{1}{T}\int_0^T y(t)\,dt.$$

Then from $d_0 = 0$, because of the integral action incorporated in the system in Fig. 6.16, we have $r = \bar{y}$ in the stationary mode.

In accordance with the block diagram shown in Fig. 6.17, we have to a first approximation

$$\begin{aligned} e_{osc} &\approx \frac{|G_n(j,\omega,A)g|}{\omega^n}A \\ &= \frac{|D(j\mu\omega)|}{k_0\omega^n}A \end{aligned} \tag{6.78}$$

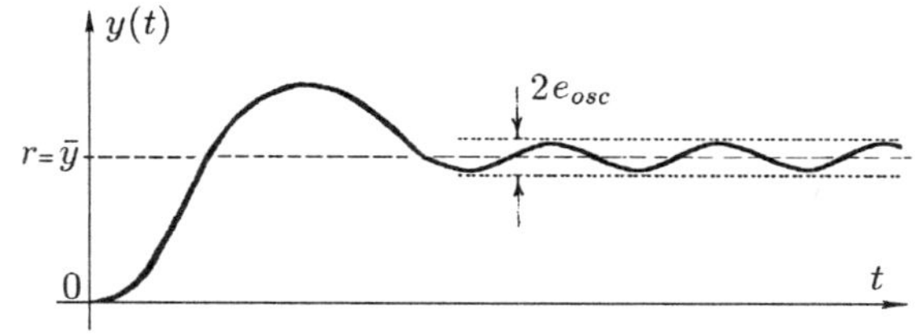

Fig. 6.20 Influence of oscillations in the FMS (6.68) on output behavior.

given that ω is sufficiently large. The effect of the fast oscillations in the FMS on the accuracy of the output stabilization can be reduced by proper choice of the controller parameters.

Note that the variations of the parameter g and nonzero values of F, f have an influence on the limit cycle parameters. Robustness of the prescribed output behavior is maintained within the region of limit cycle stability, given that the oscillation frequency ω is large enough.

6.4.4 *Example*

Let us consider the SISO nonlinear continuous-time system (5.80) (see p. 109)

$$x^{(2)} = [1 + \sin(x^{(1)})]x + [1 - 0.5\sin(x)]v + w, \quad y = x, \tag{6.79}$$

which is preceded by hysteresis $v = \varphi(u, u^{(1)})$ with parameters $\Delta = 0.1$ and $M = 5$ as shown in Fig. 6.14.

Assume $y(t) = x(t)$, and that the desired dynamics of $y(t)$ are assigned by the equation (5.81)

$$y^{(2)} = -\frac{a_1^d}{T}y^{(1)} - \frac{1}{T^2}y + \frac{b_1^d\tau}{T^2}r^{(1)} + \frac{1}{T^2}r.$$

Let $q = 2$. Then in accordance with (4.38) we have the control law structure given by (5.82):

$$\mu^2 u^{(2)} + d_1\mu u^{(1)} + d_0 u = k_0\left\{-y^{(2)} - \frac{a_1^d}{T}y^{(1)} + \frac{b_1^d\tau}{T^2}r^{(1)} + \frac{1}{T^2}[r - y]\right\},$$

where we assume that $\mu = 0.3$ s, $k_0 = 10$, $d_0 = 0$, and $d_1 = 2$. The state space representation (5.83) of (5.82) can be used as above in order to perform computer simulation.

In accordance with the given parameters we have

$$G_l(j\mu\omega) = \frac{k_0 g}{-\mu^2\omega^2 + jd_1\mu\omega} \tag{6.80}$$

and

$$G_n(j, A) = \frac{4M}{\pi A}\sqrt{1 - \left[\frac{\Delta}{A}\right]^2} - j\frac{4M}{\pi A^2}\Delta. \tag{6.81}$$

The harmonic balance equation (6.76) presented in the form

$$G_n(j, A) = -\frac{1}{G_l(j\mu\omega)}$$

yields

$$\frac{4M}{\pi A}\sqrt{1 - \left[\frac{\Delta}{A}\right]^2} - j\frac{4M}{\pi A^2}\Delta = \frac{\mu^2\omega^2}{k_0 g} - j\frac{d_1\mu\omega}{k_0 g}. \tag{6.82}$$

From (6.82) we get

$$\frac{\mu^3\pi\Delta}{d_1}\omega^3 + \Delta\pi d_1\mu\omega - 4Mk_0 g = 0. \tag{6.83}$$

The real positive solution of (6.83) is the frequency ω of the stationary oscillations, with amplitude A given by

$$A = \sqrt{\frac{4M\Delta k_0 g}{\pi d_1\mu\omega}}. \tag{6.84}$$

From (6.78) and (6.80) it follows that

$$e_{osc} \approx \sqrt{\frac{4Mg\mu\Delta(\mu^2\omega^2 + d_1^2)}{\pi d_1 k_0\omega^3}}. \tag{6.85}$$

In accordance with the above expressions and given values of the controller parameters, we can obtain $\omega \approx 28$ rad/s, $A \approx 0.43$, and $e_{osc} \approx 0.004$ for $g = 0.5$, and $\omega \approx 41$ rad/s, $A \approx 0.62$, and $e_{osc} \approx 0.006$ for $g = 1.5$.

The simulation results of the transients in the system (6.79) controlled by the algorithm (5.82) for a step reference input $r(t)$ and a step disturbance $w(t)$ are displayed in Figs. 6.21–6.22, where the initial conditions are zero and $T = \tau = 1$ s, $a_1^d = 1.4$, $b_1^d = 0$, $t \in [0, 8]$ s. We see that the simulation results confirm the analytical calculations.

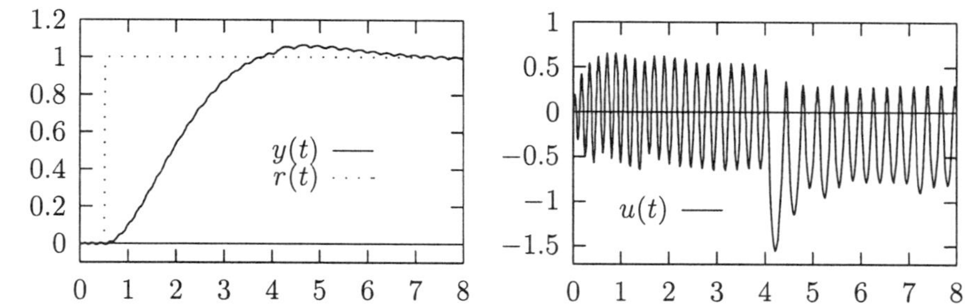

Fig. 6.21 Simulation results for the system (6.79) controlled by the algorithm (5.82).

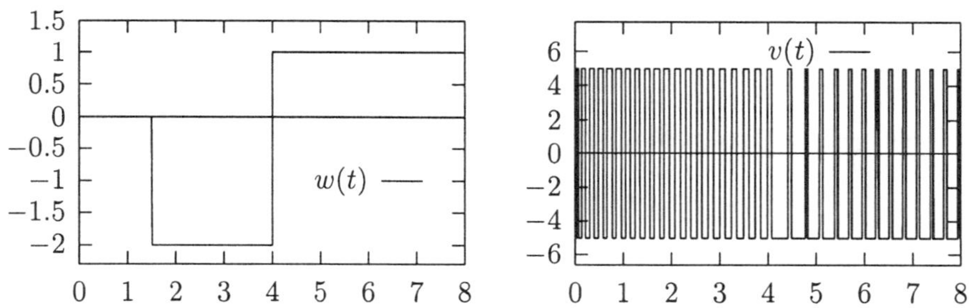

Fig. 6.22 Simulation results for the system (6.79) controlled by the algorithm (5.82).

6.5 Notes

The limit properties of the control systems with the highest derivative of the output signal in the feedback loop were discussed in Chapters 3 and 4 on the assumptions that the high gain $|k_0| \to \infty$ and the small parameter $\mu \to 0$. Obviously, such requirements are unrealized in practice. Therefore, in Chapter 5, problems of implementation were considered, and relationships were obtained from which we can choose finite values of the control law parameters in accordance with requirements on time-scale separation degree between the fast and slow modes, control accuracy, and the requirement placed on high-frequency sensor noise attenuation. In particular, it has been shown that the latter attenuation can be provided if and only if k_0 and μ are finite.

In this chapter, the restrictions caused by the unstructured uncertainties of the plant model in the form of a pure time-delay in control, regularly and singularly perturbed systems, and nonsmooth nonlinearities in control were investigated. In particular, it is assumed that the controller has been designed based on a simplified (approximate) model and in accordance with the methodology given in Chapter 5. Then the influence of unmodeled dynamics is considered when this controller is applied to the true system. As a result, the effect of the neglected parameters was investigated, and the additional restrictions on the controller parameters were obtained, which are caused by unmodeled dynamics.

6.6 Exercises

6.1 The plant model is given by

$$x^{(2)}(t) = 0.5x(t)x^{(1)}(t) + 0.1x^{(1)}(t) \\ +0.5x(t) + 0.5\sin(0.5t) + gu(t-\tau), \qquad (6.86)$$

where $g = 1$ and the reference model $x^{(2)} = F(x^{(1)}, x, r)$ is assigned by

$$x^{(2)} = T^{-2}\{-a_1 T x^{(1)} - x + r\}. \qquad (6.87)$$

The control law has the form

$$\mu^2 u^{(2)} + d_1 \mu u^{(1)} + d_0 u = k_0\{F(x^{(1)}, x, r) - x^{(2)}\}, \qquad (6.88)$$

where $T = 1$ s, $a_1 = 2$, $k_0 = 10$, $\mu = 0.1$ s, $d_0 = 0$, $d_1 = 4$. Determine the region of stability for τ of the FMS. Compare with simulation results of the closed-loop system.

6.2 Consider the closed-loop system with the input data of Exercise 6.1, where

$$g = 1 + 0.5\sin(0.2t).$$

Determine the region of stability for τ of the FMS. By computer simulation, determine the region of stability for τ of the closed-loop system in presence of the delay compensation.

6.3 Consider the plant model

$$x^{(2)}(t) = |x^{(1)}(t)|x(t) + x(t) + [1.5 + \sin(t)]u(t-\tau) \qquad (6.89)$$

together with the reference model (6.87) and the control law

$$\mu^3 u^{(3)} + d_2 \mu^2 u^{(2)} + d_1 \mu u^{(1)} + d_0 u = k_0\{F(x^{(1)}, x, r) - x^{(2)}\}, \quad (6.90)$$

where $T = 0.4$ s, $a_1 = 1.4$, $k_0 = 5$, $\mu = 0.05$ s, $d_0 = 0$, $d_1 = 17$, and $d_2 = 6$. Determine the region of stability for τ of the FMS. Compare with simulation results of the closed-loop system.

6.4 Determine the phase margin and gain margin of the FMS based on the input data of Exercise 6.1 for the time delay $\tau = 0.3\tau_m$, where $\tau = \tau_m$ corresponds to the marginally stable FMS.

6.5 Determine the phase margin and gain margin of the FMS based on the input data of Exercise 6.3 for the time delay $\tau = 0.5\tau_m$.

6.6 Let $\tau = 0.05$ s. Determine μ_m based on the input data of Exercise 6.1, where $\mu = \mu_m$ corresponds to the marginally stable FMS.

6.7 Consider the plant model

$$\begin{aligned} \dot{x} &= x\{1 + \cos(x)\} + 2u, \\ y &= x + \epsilon_0 u, \end{aligned} \tag{6.91}$$

where $|\epsilon_0| \leq 0.05$, and determine the parameters of the control law

$$\mu^2 u^{(2)} + d_1 \mu u^{(1)} + d_0 u = k_0\{[r - y]/T - y^{(1)}\} \tag{6.92}$$

to meet the following specifications: $\overline{\varepsilon}_r = 0$; $T = 1$ s; $\zeta_{FMS} \geq 0.2$; $\eta_3 \geq 10$. Note that η_3 is defined by (1.65).

6.8 Find the root loci of the FMS based on the input data of Exercise 6.7 for parameter the ϵ_0 where $k_0 = 5$, $\mu = 0.1$ s, $d_0 = 0$.

6.9 Consider the plant model

$$\begin{aligned} \dot{x}_1 &= x_2 + \epsilon_1 u, \\ \dot{x}_2 &= x_1(1 + x_1) + [1 + 0.5\cos(0.5t)]u, \\ y &= x_1, \end{aligned} \tag{6.93}$$

where $|x_1(t)| \leq 2$, $|x_2(t)| \leq 2$, and $|r(t)| \leq 1$. Take $\epsilon_1 = 0$ and determine the parameters of the control law (6.30) to meet the following specifications: $\overline{\varepsilon}_F = 0.1$; $\overline{\varepsilon}_r = 0.05$; $t_s^d \approx 3$ s; $\sigma^d \approx 10\,\%$; $\zeta_{FMS} \geq 0.8$; $\eta_3 \geq 15$. Determine the allowable region for ϵ_1, where the conditions $\zeta_{FMS} \geq 0.2$ and $\eta_3 \geq 10$ are satisfied. Run a computer simulation of the closed-loop system for $d_0 = 1$ and $d_0 = 0$.

6.10 Determine the parameters of the control law

$$\mu^3 u^{(3)} + d_2 \mu^2 u^{(2)} + d_1 \mu u^{(1)} + d_0 u = k_0\{F(y^{(1)}, y, r) - y^{(2)}\} \tag{6.94}$$

based on the input data of Exercise 6.9.

6.11 Consider the plant model

$$\dot{x}_1 = x_2, \quad \dot{x}_2 = x_1^2 + x_2 \sin(t) + 2x_3,$$
$$\nu \dot{x}_3 = -x_3 + u, \quad y = x_1,$$

where $|x_1(t)| \leq 1$, $|x_2(t)| \leq 1$, $|x_3(t)| \leq 1$, and $|r(t)| \leq 1$. Take $\nu = 0$ and determine the parameters of the control law (6.30) to meet the following specifications: $\bar{\varepsilon}_F = 0.1$; $\bar{\varepsilon}_r = 0.05$; $t_s^d \approx 6$ s; $\sigma^d \approx 10\,\%$; $\zeta_{FMS} \geq 0.8$; $\eta_3 \geq 15$. Determine the allowable region for ν, where the conditions $\zeta_{FMS} \geq 0.2$ and $\eta_3 \geq 10$ are satisfied. Run a computer simulation of the closed-loop system for $d_0 = 1$ and $d_0 = 0$.

6.12 Solve Exercise 6.11 where the control law is given by (6.94).

6.13 Consider the plant model

$$\dot{x}_1 = x_1|\sin(t)| + x_2, \quad \nu \dot{x}_2 = x_3, \tag{6.95}$$
$$\nu \dot{x}_3 = -x_2 - 3x_3 + u, \quad y = x_1,$$

where the control law is given by

$$\mu u^{(1)} + d_0 u = k_0\{[r - y]/T - y^{(1)}\} \tag{6.96}$$

and $T = 2$ s, $k_0 = 10$, $\mu = 0.4$ s, $d_0 = 0$. Determine the region of stability for ν of the FMS and the phase margin and gain margin if $\nu = 0.5\nu_m$, where $\nu = \nu_m$ corresponds to marginally stable FMS.

6.14 Consider the system (6.89) preceded by relay with dead zone (see Fig. 6.14) together with the reference model (6.87) and the control law given by (6.90) where $\tau = 0$, $\Delta = 0.4$, $M = 10$, $T = 1$ s, $a_1 = 1.4$, $k_0 = 10$, $\mu = 0.1$ s, $d_0 = 0$, $d_1 = 15$, $d_2 = 5$. Using the describing function method, determine the frequency of the oscillations in the FMS and estimate the amplitude of the oscillations in the output variable $x(t)$. Compare with simulation results.

6.15 Consider the system

$$\dot{x} = x|\sin(t)| + [1 + 0.2\sin(t)]u(t - \tau)$$

preceded by relay with dead zone (see Fig. 6.14) together with the control law given by (6.96) where $\tau = 0.03$ s, $\Delta = 0.5$, $M = 2$, $T = 1$ s, $k_0 = 10$, $\mu = 0.1$ s, $d_0 = 0$. Using the describing function method, determine the frequency of the oscillations in the FMS and estimate the amplitude of the oscillations in the output variable $x(t)$. Compare with simulation results.

Chapter 7

Realizability of desired output behavior

Before carrying out a design, we must analyze the realizability of the desired output behavior. In the preceding chapters attention was devoted to the problem of control system design with the highest derivative in the feedback loop for the SISO plant model given by (4.27), where output regulation with prescribed dynamics may be provided if the condition (4.35) holds. This chapter is devoted to consideration of conditions that allow us to provide desired output behavior for more general dynamic systems. It will be shown that, in general, the analysis of the realizability of the desired output behavior is a much more complicated problem, and involves such concepts as invertibility of a dynamic system, nonlinear inverse dynamics, and internal behavior analysis of the system. In this chapter, concepts such as invertibility index (relative degree), normal form of nonlinear systems, internal stability analysis, degenerated system on the condition of output stabilization, and zero-dynamics are discussed. Finally, the design procedure for SISO nonlinear control systems is discussed in the presence of internal dynamics.

7.1 Control problem statement for MIMO control system

7.1.1 *MIMO plant model*

Let us consider a nonlinear time-varying system in the following form:

$$\dot{X} = f(t, X) + G(t, X)u, \quad X(0) = X^0, \tag{7.1}$$

$$y = h(t, X), \tag{7.2}$$

where

t denotes time, $t \in [0, \infty)$;
X is the state vector, $X = \{x_1, x_2, \ldots, x_n\}^T$;
$X(0) = X_0$ is the initial state, $X_0 \in \Omega_X$, Ω_X is a bounded set of $\mathbb{R}^n$;
y is the output of the system (7.1)–(7.2), available for measurement, $y = \{y_1, y_2, \ldots, y_p\}^T$;
u is the control signal, $u = \{u_1, u_2, \ldots, u_m\}^T$, $u \in \Omega_u \subset \mathbb{R}^m$;
$p \leq m \leq n$.

It is assumed that the vector functions $f(t, X), h(t, X)$ and the elements of the matrix $G(t, X)$ are smooth and bounded for all $(t, X) \in \Omega_{t,X}$.

The model extension is given by

$$\dot{X} = f(t, w, X) + G(t, w, X)u, \quad X(0) = X^0, \tag{7.3}$$

$$y = h(t, w, X), \tag{7.4}$$

where $w(t)$ is a vector of external disturbances and varying parameters (unavailable for measurement). We assume that $w(t)$ is smooth and bounded for all $t \in [0, \infty)$. Then the system (7.3)–(7.4) can be represented in the form (7.1)–(7.2) because $w = w(t)$. As a result, the influence of all external disturbances and varying parameters of the system (7.1)–(7.2) is represented implicitly by the dependence of $f(t, X), h(t, X)$ and $G(t, X)$ on the time variable t.

In some references (e.g., [Isidori and Byrnes (1990); Isidori (1995); Marconi (1998)]) the external disturbance model $w(t)$, called the exogenous system, is included in the system description. For systems with sliding mode or high gain in feedback, and for the discussed approach to control system design, the varying parameters and external disturbances, and their manner of entering into the system, need not be known. Then explicit reference to $w(t)$ may be omitted if certain additional conditions for disturbance rejection are satisfied. These conditions are the main subject matter of this chapter and will be presented below.

7.1.2 *Control problem*

We seek a control system for which

$$\lim_{t \to \infty} e(t) = 0, \tag{7.5}$$

where $e(t)$ is the error of the reference input realization (tracking error), $e = \{e_1, e_2, \ldots, e_p\}^T$, $e(t) = r(t) - y(t)$; $r(t)$ is the reference input, $r = \{r_1, r_2, \ldots, r_p\}^T$.

Moreover, the controlled transients of the ith component $y_i(t)$ of the output vector $y(t)$ should have desired performance indices that are separately assigned, such as overshoot σ_i^d, settling time t_i^d, and system type. These transients should not depend on external disturbance or varying parameters of the system (7.1)–(7.2).

7.2 Invertibility of dynamical systems

7.2.1 *Role of invertibility of dynamical systems*

The invertibility of dynamical systems was first widely investigated in mechanics, in order to find the forces that cause the observable behavior of mechanical systems [Santilli (1978)]. In the general case, this meaning of the term leads to the concept of left invertibility. A system is said to be left invertible if a unique control function exists and can be found for the given system model, initial state, and output function.

So, on one hand, the left invertibility condition for a dynamical control system is the condition for uniqueness of the control function that provides the desired output behavior [Zadeh and Desoer (1963)]. On the other hand, right invertibility is the necessary condition for the existence of a control function such that the output behavior is an arbitrarily assigned smooth function [Brockett and Mesarovic (1965); Porter (1970)].

From a theoretical viewpoint, the desired input-controlled output map can be provided by a controller in the form of a serial system of the reference model and the right inverse system. Control of nonlinear systems through the use of their inverse dynamics is a topic that has received much attention [Boychuk (1966); Silverman (1969); Porter (1970); Popov and Krutko (1979); Petrov and Krutko (1980); Singh (1980); Slotine and Li (1991)]. The application of the inversion method in the discrete-time nonlinear control system synthesis problem was discussed in [Kotta (1995)]. Note that the well known input-output linearization technique for nonlinear systems is based on the inclusion of the right inverse system into the control law structure [Isidori and Byrnes (1990); Nijmeijer and Schaft (1990); Isidori (1995); Sastry (1999); Khalil (2002)]. Obviously, linearization may be used only if complete information is available about the disturbances, model parameters, and system state. That technique is useless for nonlinear control system design on the condition of incomplete information about varying system parameters and external disturbances.

Note that control laws based on nonlinear inversion theoretically allow us to determine the potential for improving the system output behavior. An examination of these laws is essential for analysis.

Control system design can efficiently be done under uncertainty via the design methodologies for control systems with sliding motions [Utkin (1977); Utkin (1992)], control systems with high gain in feedback [Meerov (1965); Young *et al.* (1977)], and control systems with the highest-order derivative and high gain in feedback [Vostrikov (1977a); Utkin and Vostrikov (1978); Vostrikov and Sarycheva (1982); Vostrikov *et al.* (1982); Vostrikov (1988b); Vostrikov (1990)]. Note that if output regulation is based on the above design methodologies, then invertibility of the system and stability of the inverse system (more precisely, uniform ultimate boundedness of that system) are the conditions under which the solution of the output regulation problem exists.

So, invertibility is a fundamental characteristic of dynamical control systems, along with controllability, observability, and stability.

7.2.2 *Definition of invertibility of dynamic control system*

Let us consider a LTI control system in the form

$$\dot{X} = AX + Bu, \quad X(0) = X^0, \tag{7.6}$$

$$y = CX + Du, \tag{7.7}$$

where $X \in \mathbb{R}^n$; $u \in \mathbb{R}^m$; $y \in \mathbb{R}^p$; A, B, C, D are real constant matrices with appropriate dimensions. We assume that $u(t) \in \mathbf{U} \subseteq (\mathbf{C}[0,T], \mathbb{R}^m)$.

The system (7.6)–(7.7) with a given initial state $X(0) = X^0$ may be considered as a mapping (operator)

$$\mathcal{A}(X^0, u)\colon\ \mathbf{U} \to \mathbf{Y},$$

where $\mathbf{U}$ is the set of all control functions $u(t)$ and $\mathbf{Y}$ is the set of corresponding output functions $y(t)$.

Let us consider the definition of inverse operator given by [Zadeh and Desoer (1963)], which may be clarified by the block diagram of Fig. 7.1.

Definition 7.1 An operator $\mathcal{A}_r^{-1}(\mathcal{A}_l^{-1})$ is said to be a right (left) inverse of the operator $\mathcal{A}$ if

$$\mathcal{A}\mathcal{A}_r^{-1} = I_y \quad (\mathcal{A}_l^{-1}\mathcal{A} = I_u) \tag{7.8}$$

holds, where $I_y\colon \mathbf{Y} \to \mathbf{Y}$ ($I_u\colon \mathbf{U} \to \mathbf{U}$) is the identity operator.

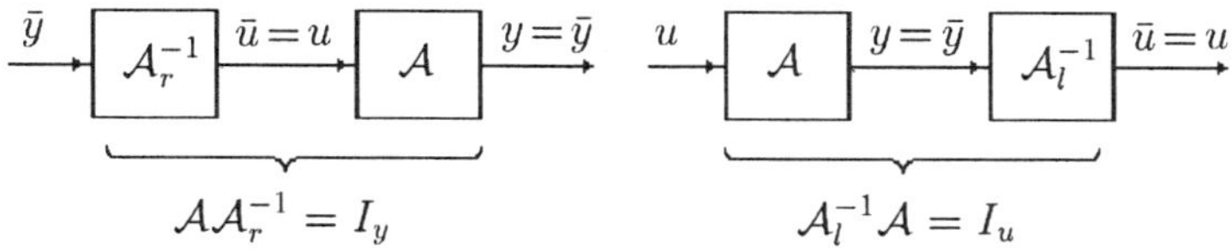

Fig. 7.1 Serial connection of inverse systems.

In particular, if $p = m$ and both the right and left inverses of the system (7.6)–(7.7) exist, then these inverses are identical and the system is said to be invertible [Wang and Davison (1973)].

Note that in Definition 7.1, zero initial conditions are implicitly assumed. In the general case, matching of the initial states of the inverse and original systems should be provided [Zadeh and Desoer (1963)].

Let us consider a set of definitions for the concept of invertibility as given by [Hirschorn (1979a); Hirschorn (1979b)] for a nonlinear system of the form

$$\dot{X} = f(X) + G(X)u, \quad X(0) = X^0, \tag{7.9}$$

$$y = h(X). \tag{7.10}$$

Here $X \in \mathbf{M}$ where $\mathbf{M}$ is an n-dimensional smooth manifold (e.g., a subset of $\mathbb{R}^n$), $y \in \mathbb{R}^p$, and $u \in \mathbb{R}^p$. It is assumed that $f(X)$ and $h(X)$ are analytic vector functions and that $G(X)$ is a matrix all of whose elements are analytic functions. Moreover, $u(t) \in \mathbf{U}$ where $\mathbf{U}$ is a set of analytic vector functions of t. It is assumed that the solution of (7.9)–(7.10) exists and is unique for any given $X^0 \in \mathbf{M}$.

Definition 7.2 The system (7.9)–(7.10) is said to be invertible in $X^0 \in \mathbf{M}$ if for any $u_1, u_2 \in \mathbf{U}$ where $u_1 \neq u_2$ it follows that

$$y(t, X^0, u_1) \neq y(t, X^0, u_2).$$

Definition 7.3 The system (7.9)–(7.10) is said to be strictly invertible in $X^0 \in \mathbf{M}$ if there exists some open neighborhood $O(X^0) \subseteq \mathbf{M}$ of X^0 such that the system is invertible $\forall\ X \in O(X^0)$.

Definition 7.4 The system (7.9)–(7.10) is said to be strictly invertible if there exists some open dense manifold $\mathbf{M} \subseteq \mathbb{R}^n$ such that $\forall\ X^0 \in \mathbf{M}$ the system is strictly invertible in X^0.

There is a broad set of invertibility conditions for LTI systems, expressed in terms of rank conditions on the matrix transfer function

$$H(s) = C(sI_n - A)^{-1}B + D$$

or the system matrix

$$P(s) = \begin{bmatrix} sI_n - A & B \\ -C & D \end{bmatrix}.$$

The interested reader can also find various criteria for invertibility of the system (7.6)–(7.7), expressed as rank conditions for real matrices constructed through the matrices A, B, C, D [Brockett and Mesarovic (1965); Dorato (1969); Emre and Hüseyin (1974); Sain and Massey (1969); Wang and Davison (1973)].

Analysis of the invertibility of nonlinear control systems is often approached through sequential differentiation of the output variables of $y(t)$ in order to construct some special subsystem from which the vector of control variables $u(t)$ may be explicitly derived. This approach was developed and used by [Silverman (1969); Porter (1970); Hirschorn (1979a); Hirschorn (1979b); Singh (1980)].

7.2.3 *Invertibility condition for nonlinear systems*

Let us consider the nonlinear time-varying control system given by (7.1), (7.2):

$$\dot{X} = f(t, X) + G(t, X)u, \quad X(0) = X^0,$$
$$y = h(t, X),$$

where $X \in \mathbb{R}^n$, $u \in \mathbb{R}^m$, $y \in \mathbb{R}^p$, and $p \leq m \leq n$.

Assume that the invertibility condition [Porter (1970)] is satisfied: there exists a system of equations for the output derivatives which makes it possible to derive the input $u(t)$ of the system (7.1)–(7.2).

In accordance with [Porter (1970)], let the expression

$$y_1(t) = h_1(t, X) \tag{7.11}$$

be the first component of the output vector function $y(t)$, where $h = \{h_1, h_2, \ldots, h_p\}^T$. For convenience, denote

$$h_{10}(t, X) = h_1(t, X).$$

Differentiating (7.11) along the solutions of the system (7.1), we obtain

$$y_1^{(1)} = \partial h_{10}/\partial t + \{\partial h_{10}/\partial X\}\{f + Gu\},$$

where

$$\partial y_1^{(1)}/\partial u = \{\partial h_{10}/\partial X\}G(t, X).$$

Assume that the condition

$$\{\partial h_{10}/\partial X\}G(t, X) = 0, \quad \forall\ (t, X) \in \Omega_{t,X}$$

holds, where $\Omega_{t,X} = [0, \infty) \times \Omega_X$ and Ω_X is a connected domain in $\mathbb{R}^n$. Denote

$$h_{11} = \partial h_{10}/\partial t + \{\partial h_{10}/\partial X\}f$$

and then derive the expression for $y_1^{(2)}$. Repetition of this procedural cycle will continue until the control variable $u(t)$ appears explicitly. Let us assume that the α_1th step of this procedure yields the following expression:

$$y_1^{(\alpha_1)} = h_{1,\alpha_1}(t, X) + \{\partial h_{1,\alpha_1-1}/\partial X\}G(t, X)u,$$

where

$$\partial y_1^{(\alpha_1)}/\partial u = \{\partial h_{1,\alpha_1-1}/\partial X\}G(t, X).$$

Assume that the conditions

$$\{\partial h_{1,j}/\partial X\}G(t, X) = 0, \quad \forall\ (t, X) \in \Omega_{t,x}, \quad \forall j = 1, \dots, \alpha_1 - 2$$

and

$$\{\partial h_{1,\alpha_1-1}/\partial X\}G(t, X) \neq 0, \quad \forall\ (t, X) \in \Omega_{t,X}$$

are satisfied.

As a result of providing the above procedure for each component of the output vector $y(t)$, the equations

$$y_i^{(\alpha_i)} = h_{i,\alpha_i}(t, X) + \{\partial h_{i,\alpha_i-1}/\partial X\}G(t, X)u, \quad i = 1, \dots, p \tag{7.12}$$

are obtained. Let us rewrite (7.12) as

$$y_* = H^*(t, X) + G^*(t, X)u, \tag{7.13}$$

where

$$y_* = \{y_1^{(\alpha_1)}, y_2^{(\alpha_2)}, \dots, y_p^{(\alpha_p)}\}^T, \tag{7.14}$$

$$H^*(t,X) = \begin{bmatrix} h_1^*(t,X) \\ h_2^*(t,X) \\ \cdots \\ h_p^*(t,X) \end{bmatrix}, \quad G^*(t,X) = \begin{bmatrix} g_1^*(t,X) \\ g_2^*(t,X) \\ \cdots \\ g_p^*(t,X) \end{bmatrix}, \tag{7.15}$$

and $h_i^*(t,X) = h_{i,\alpha_i}(t,X)$, $g_i^*(t,X) = \{\partial h_{i,\alpha_i-1}/\partial X\}G(t,X)$. By definition, $h_i^*(t,X)$ is the ith element of the column vector $H^*(t,X)$ and $g_i^*(t,X)$ is the ith row of the matrix $G^*(t,X)$, where $H^* \in \mathbb{R}^{p\times 1}$, $G^* \in \mathbb{R}^{p\times m}$.

In accordance with the discussed algorithm scheme, the recurrence relationship

$$h_{i,j}(t,X) = \partial h_{i,j-1}/\partial t + \{\partial h_{i,j-1}/\partial X\}f(t,X),\ \forall\, i=1,\ldots,p,\ \forall\, j=1,\ldots,\alpha_i$$

holds, where $h_{i,0}(t,X) = h_i(t,X)$ is the ith element of the column vector $h(t,X)$. Moreover, the conditions

$$\{\partial h_{i,j}/\partial X\}G(t,X) = 0,\ \forall\,(t,X) \in \Omega_{t,X},\ \forall\, i = 1,\ldots,p,\ \forall\, j = 0,\ldots,\alpha_i - 2$$

are satisfied in accordance with the algorithmic scheme. So we have

$$y_i^{(j)} = h_{i,j}(t,X), \quad \forall\, j = 0,\ldots,\alpha_i - 1 \quad \text{and} \quad \forall\, i = 1,\ldots,p. \tag{7.16}$$

Remark 7.1 *The positive integer value α_i is a structural invariant of the system (7.1)–(7.2) and is known as the invertibility index (or recently more often as the relative degree) of the system with respect to the output y_i [Silverman (1969)]. Throughout the text below, we assume a uniform relative degree of the system (7.1)–(7.2), i.e., that $\alpha_i = const$ for all $(t,X) \in \Omega_{t,X}$.*

Assumption 7.1 The condition

$$\operatorname{rank} G^*(t,X) = p, \quad \forall\,(t,X) \in \Omega_{t,X} \tag{7.17}$$

is satisfied.

Theorem 7.1 *From (7.17) it follows that the right inverse of the system (7.1)–(7.2) for all $(t,X) \in \Omega_{t,X}$ has the following form:*

$$\begin{aligned} \bar{X}^{(1)} &= f(t,\bar{X}) + G(t,\bar{X})[G^*(t,\bar{X})]^T\{G^*(t,\bar{X})[G^*(t,\bar{X})]^T\}^{-1} \\ &\qquad \times\{\bar{y}_* - H^*(t,\bar{X})\}, \quad \bar{X}(0) = X^0, \end{aligned} \tag{7.18}$$

$$\bar{u} = [G^*(t,\bar{X})]^T\{G^*(t,\bar{X})[G^*(t,\bar{X})]^T\}^{-1}\{\bar{y}_* - H^*(t,\bar{X})\}. \tag{7.19}$$

Proof. A proof may be based on Definition 7.1, by direct substitution of (7.19) into (7.13), in view of the condition $X(t) = \bar{X}(t)$, $\forall\ t$ which follows from $u(t) = \bar{u}(t)$, $\forall\ t$ and $\bar{X}(0) = X(0)$. ∎

Remark 7.2 *From (7.17) it follows that $m \geq p$ is a necessary condition for right inversion of the system (7.1)-(7.2).*

Remark 7.3 *If $p = m$, then from (7.17) it follows that*

$$\det G^*(t, X) \neq 0, \quad \forall\ (t, X) \in \Omega_{t,X}. \tag{7.20}$$

As a result, from (7.18)-(7.19) we can obtain

$$\bar{X}^{(1)} = f(t, \bar{X}) + G(t, \bar{X})\{G^*(t, \bar{X})\}^{-1}\{\bar{y}_* - H^*(t, \bar{X})\}, \tag{7.21}$$

$$\bar{u}(t) = \{G^*(t, \bar{X})\}^{-1}\{\bar{y}_* - H^*(t, \bar{X})\}, \quad \bar{X}(0) = X^0. \tag{7.22}$$

This is the inverse system (both right and left) of the system (7.1)-(7.2).

7.3 Insensitivity condition for MIMO control system

7.3.1 *Desired dynamics equations*

From (7.13) and (7.17), it follows that for $i = 1, \ldots, p$ the relative highest derivative $y_i^{(\alpha_i)}$ depends explicitly on the control vector $u(t)$. Therefore an arbitrary behavior of the relative highest derivative vector $y_*(t)$ may be provided by appropriate selection of the control function $u(t)$. Let us construct the reference model of the desired behavior of the relative highest derivative of $y_i(t)$ for each $i = 1, \ldots, p$ in the form of the stable differential equation

$$y_i^{(\alpha_i)} = F_i(Y_i, R_i). \tag{7.23}$$

This is called the desired dynamics equation of $y_i(t)$, where

$$Y_i = \{y_i, y_i^{(1)}, y_i^{(2)}, \ldots, y_i^{(\alpha_i - 1)}\}^T, \quad R_i = \{r_i, r_i^{(1)}, \ldots, r_i^{(\rho_i)}\}^T, \quad \rho_i < \alpha_i.$$

Equation (7.23) is the counterpart of (4.31), and can be selected in the form of the linear differential equation (2.8). The parameters of (7.23) for each ith output component are assigned in accordance with the time-domain specifications on the desired output behavior of $y_i(t)$ and the requirement

$$\lim_{t \to \infty} e_i(t) = 0, \tag{7.24}$$

where $e_i = r_i - y_i$ and $i = 1, \ldots, p$. Hence we have $y_i = r_i$ at the equilibrium of (7.23) for r_i = const.

As a result, we have the system of stable differential equations

$$y_* = F(Y, R) \tag{7.25}$$

composed of the equations (7.23), where $F = \{F_1, F_2, \ldots, F_p\}^T$.

7.3.2 *Insensitivity condition*

Let us denote

$$e^F = F(Y, R) - y_*, \tag{7.26}$$

where e^F is the realization error of the desired dynamics $F(Y, R)$, assigned by (7.25). Accordingly, if

$$e^F = 0, \tag{7.27}$$

then the desired behavior of $y(t)$ with prescribed dynamics (7.25) is fulfilled. Expression (7.27) is the insensitivity condition for the output transient performance with respect to the external disturbances and varying parameters of the plant model (7.1)–(7.2). In other words, the control design problem (7.5) has been reformulated as the requirement (7.27).

So if (7.27) is satisfied, then the following hold simultaneously:

(i) The behavior of each ith output component is insensitive to parameter variations and external disturbances.
(ii) The output behavior of each ith output component is assigned by the parameters of the desired dynamics equation (7.23).
(iii) The output behavior of each ith output component does not depend on the behavior of the other output components.

From (7.13) it follows that (7.27) can be rewritten as

$$F - H^*(t, X) - G^*(t, X)u = 0. \tag{7.28}$$

So the control problem (7.5) for the MIMO system (7.1)–(7.2) has been reformulated as a problem to provide the requirement (7.28), despite the presence of unknown parameters in (7.28).

7.4 Internal stability

7.4.1 *Boundedness of NID-control function*

Let us consider the behavior of the control function $u(t)$ and the state vector $X(t)$ of the system (7.1)–(7.2), given that the desired reference input-controlled output map assigned by (7.25) is satisfied.

Assume henceforth that equidimensional input and output vectors of the system (7.1)–(7.2) are considered, i.e., that $m = p$.

From (7.28) it follows that the function

$$u^{NID}(t) = \{G^*(t,X)\}^{-1}\{F(Y,R) - H^*(t,X)\} \tag{7.29}$$

uniquely satisfies (7.27) and is the solution of the nonlinear inverse dynamics. By substituting (7.29) into the state equation (7.1), we obtain

$$\dot{X} = f(t,X) + G(t,X)\{G^*(t,X)\}^{-1}\{F - H^*(t,X)\},\ X(0) = X^0. \tag{7.30}$$

The system (7.30) describes the behavior of the state vector $X(t)$ of the system (7.1)–(7.2), given that the desired output behavior assigned by (7.25) is fulfilled.

Remark 7.4 *It is easy to see that (7.29)–(7.30) and the inverse system equations (7.21)–(7.22) are the same. So, the boundedness of the NID-control function (7.29) corresponds to bounded-input-bounded-output (BIBO) stability of the inverse system (7.21)–(7.22). Related remarks can be found in [Silverman (1969); Porter (1970); Fomin et al. (1981)].*

If the nonlinear inverse dynamics solution $u^{NID}(t)$ of (7.29) is an unbounded function, i.e.,

$$\lim_{t\to\infty} \sup_{t>0} \|u^{NID}(t)\| = \infty,$$

then the desired output behavior assigned by the reference model (7.25)

$$y_* = F(Y,R)$$

is unrealizable in the system (7.1)–(7.2) where $t \in [t_0, \infty)$. This is because the increase in the control function $u^{NID}(t)$ leads inevitably to saturation of the control variables in practice.

So, the desired output behavior assigned by (7.25) is realizable for all $t \in [0,\infty)$, that is, some bounded set Ω_u exists such that the inclusion $u(t) \in \Omega_u$ holds, if the following two requirements are satisfied:

(i) The inverse system of (7.1)–(7.2) exists.
(ii) The NID-control function $u^{NID}(t)$ is bounded, i.e.,

$$\limsup_{t\to\infty\ t>0} \|u^{NID}(t)\| < \infty. \tag{7.31}$$

Definition 7.5 The property expressed by (7.31) is called the boundedness of the NID-control function in the system (7.1)–(7.2), given that the desired output behavior assigned by the stable system (7.25) is satisfied.

Note that (7.31) corresponds to (3.14) discussed above for the SISO control systems given by (3.1).

7.4.2 *Concept of internal stability*

The boundedness (7.31) of the NID-control function in the system (7.1)–(7.2) does not exclude the existence of unstable motions in the state space of (7.1)–(7.2), where the unstable motions have no effect on the behavior of the output $y(t)$ and the NID-control function $u^{NID}(t)$, given that the desired output behavior assigned by the stable system (7.25) is satisfied. From a theoretical viewpoint, we can obtain the solution of the output regulation problem despite the fact that there are unobserved unstable motions in the state space of (7.1)–(7.2). From a practical standpoint, the model of the form (7.1)–(7.2) is usually valid for only some bounded subset Ω_X of the state space $\mathbb{R}^n$. The reason is that the range of permissible variations of any technical plant variable is bounded in practice by a set of limitations on thermal stability, electric strength, mechanical strength, and other physical characteristics.

In accordance with these technical limitations, we must further restrict the trajectories of (7.30) so that

$$\limsup_{t\to\infty\ t>0} \|X(t)\| < \infty. \tag{7.32}$$

By this we mean that internal unbounded motions in (7.1)–(7.2) do not exist when the desired reference input-controlled output map assigned by (7.25) is satisfied.

Definition 7.6 Let us say that the internal stability of the system (7.1)–(7.2) is satisfied if the inequality (7.32) holds for all trajectories of (7.30) in the specified region of the state space (or, in other words, for all trajectories of (7.1)–(7.2) when the desired reference input-controlled output map assigned by (7.25) is fulfilled).

Remark 7.5 *From the internal stability of (7.1)–(7.2) and the condition (7.20), the boundedness of the NID-control function (7.31) results.*

Remark 7.6 *Let the allowable subset* Ω_u *be specified, and suppose* $m = p$ *where* $\Omega_u \subset R^m$. *Then the desired output behavior assigned by (7.25) is realizable if the inclusion* $u^{NID}(t) \in \Omega_u, \ \forall\, t \in [0, \infty)$ *holds.*

Remark 7.7 *By taking into account Remark 7.4 and Definition 7.6, we find that the internal stability of (7.1)–(7.2) corresponds to bounded-input-bounded-state (BIBS) stability of the inverse system (7.21)–(7.22).*

7.4.3 *Normal form of the plant model*

The internal stability of the system (7.1)–(7.2) can be easily verified by transformation to a special canonical form known as the normal form. This form of state space representation for dynamical systems was discussed in a broad set of references, e.g., [Brunovský (1970); Luenberger (1967); O'Reilly (1983); Sira-Ramirez (1989); Byrnes and Isidori (1991); Isidori (1995); Fradkov *et al.* (1999); Khalil (2002); Vostrikov and Yurkevich (1991)].

Remark 7.8 *In the normal form, new state variables are introduced which include each ith component* y_i *of the output vector* y *and its first* $\alpha_i - 1$ *derivatives.*

Let us consider hereafter the system (7.1)–(7.2), where Assumption 7.1 is satisfied, i.e., the system (7.1)–(7.2) is right invertible, and $n \geq m \geq p$.

By taking into account (7.16), let us introduce the vector function

$$Y = Q_1(t, X), \tag{7.33}$$

where

$$\begin{aligned} Y &= \{y_{10}, y_{11}, \ldots, y_{1,\alpha_1-1}, y_{20}, \ldots, y_{p,\alpha_p-1}\}^T \\ &= \{y_1, y_1^{(1)}, \ldots, y_1^{(\alpha_1-1)}, y_2, \ldots, y_p^{(\alpha_p-1)}\}^T, \\ y_{i0} &= y_i, \ y_{ij} = y_i^{(j)}, \quad \forall\, i = 1, \ldots, p, \quad \forall\, j = 0, \ldots, \alpha_i - 1, \\ Q_1(t, X) &= \{h_{10}(t, X), h_{11}(t, X), \ldots, h_{1,\alpha_1-1}(t, X), h_{20}(t, X), \ldots \\ &\quad \ldots, h_{p,0}(t, X), \ldots, h_{p,\alpha_p-1}(t, X)\}^T. \end{aligned} \tag{7.34}$$

Theorem 7.2 *From the given system (7.1)–(7.2) where the conditions (7.16)–(7.17) hold, we find that the vector function* $Y = Q_1(t, X)$ *is such* *that*

$$\mathit{rank}\ \frac{\partial Q_1(t,X)}{\partial X} = l, \quad \forall\ (t,X) \in \Omega_{t,X} \quad \mathit{where}\ \ l = \sum_{i=1}^{p} \alpha_i.$$

Proof. First, let us construct a control function of the form

$$u = -[G^*(t,X)]^T \{G^*(t,X)[G^*(t,X)]^T\}^{-1}\tilde{u}, \tag{7.35}$$

where

$$\tilde{u} = \{\tilde{u}_1, \tilde{u}_2, \ldots, \tilde{u}_p\}^T$$

and

$$\tilde{u}_i = h_i^*(t,X) + \sum_{j=0}^{\alpha_i - 1} \xi_{ij} y_i^{(j)}, \quad i = 1, \ldots, p.$$

By substituting (7.35) into (7.13), we find that for all $(t,X) \in \Omega_{t,X}$ the behavior of the ith component y_i of the output vector y is described by the differential equation

$$y_i^{(\alpha_i)} + \sum_{j=0}^{\alpha_i - 1} \xi_{ij} y_i^{(j)} = 0, \tag{7.36}$$

where the parameters ξ_{ij} may be assigned arbitrarily.

Next, the proof proceeds by contradiction. Let us assume the existence of a point $(\hat{t}, \hat{X}) \in \Omega_{t,X}$ where

$$\text{rank}\ \frac{\partial Q_1(t,X)}{\partial X} < l \ \ \text{for} \ \ (t,X) = (\hat{t}, \hat{X}). \tag{7.37}$$

In accordance with (7.37), a column vector

$$\xi(\hat{t},\hat{X}) = \left\{\xi_1(\hat{t},\hat{X}), \xi_2(\hat{t},\hat{X}), \ldots, \xi_l(\hat{t},\hat{X})\right\}^T$$

exists such that the following conditions are satisfied:

$$\|\xi(\hat{t},\hat{X})\| > 0 \text{ and } \xi^T(t,X)\frac{\partial Q_1(t,X)}{\partial X} = \bar{0} \text{ for } \ (t,X) = (\hat{t},\hat{X}). \tag{7.38}$$

Here $\bar{0} = \{0, 0, \ldots, 0\}$ is the null row vector, $\bar{0} \in \mathbb{R}^n$.

From (7.33) it follows that

$$\dot{Y}(t) = \frac{\partial Q_1(t,X)}{\partial X}\dot{X} + \frac{\partial Q_1(t,X)}{\partial t}.$$

Then, in accordance with (7.38), we obtain

$$\xi^T(t,X)\dot{Y}(t) = \partial Q_1(t,X)/\partial t \quad \text{for} \quad (t,X) = (\hat{t},\hat{X}). \tag{7.39}$$

So, from Assumption (7.37) it follows that at the point $(t,X) = (\hat{t},\hat{X})$ the condition (7.39) is satisfied, where the vector $\xi(\hat{t},\hat{X})$ depends on the parameters of the system (7.1)–(7.2). This contradicts the condition (7.36), which proves Theorem 7.2. ■

In accordance with Theorem 7.2, the vector function (7.33) can be supplemented by another vector function $z = Q_2(t,X)$ in such a way that the transformation

$$\begin{bmatrix} Y \\ z \end{bmatrix} = \begin{bmatrix} Q_1(t,X) \\ Q_2(t,X) \end{bmatrix} = Q(t,X) \tag{7.40}$$

is a one-one mapping (bijection) where $z \in \mathbb{R}^{n-l}$. Consequently, the choice of the vector function $Q_2(t,X)$ should be provided in accordance with the requirement

$$\operatorname{rank} \frac{\partial Q(t,X)}{\partial X} = n, \quad \forall\ (t,X) \in \Omega_{t,X}. \tag{7.41}$$

From the implicit function theorem it follows that an inverse transformation $X = Q^{-1}(t,Y,z)$ exists for all $(t,X) \in \Omega_{t,X}$. Then, by the change of variables (7.40), from the system (7.1)–(7.2) the normal form of the state space representation follows, that is

$$\begin{aligned} \frac{d}{dt}y_{i0} &= y_{i1}, \\ \cdots\cdots & \quad \cdots\cdots \\ \frac{d}{dt}y_{i,\alpha_i-2} &= y_{i,\alpha_i-1}, \\ \frac{d}{dt}y_{i,\alpha_i-1} &= h_i^*(t,Y,z) + g_i^*(t,Y,z)u, \quad i = 1,\ldots,p, \end{aligned} \tag{7.42}$$

$$\frac{d}{dt}z = \tilde{f}(t,Y,z) + \tilde{G}(t,Y,z)u. \tag{7.43}$$

In accordance with (7.14), the system (7.42)–(7.43) may be rewritten concisely as

$$y_* = H^*(t,Y,z) + G^*(t,Y,z)u, \tag{7.44}$$

$$\dot{z} = \tilde{f}(t,Y,z) + \tilde{G}(t,Y,z)u, \tag{7.45}$$

where the new state vector (Y,z) includes the ith component y_i of the output vector y and its first $\alpha_i - 1$ derivatives. Here we have $Y \in \mathbb{R}^l$,

$z \in \mathbb{R}^{n-l}$, $p \leq l \leq n$, and

$$H^*(t,Y,z) = \begin{bmatrix} h_1^*(t,Y,z) \\ h_2^*(t,Y,z) \\ \cdots \\ h_p^*(t,Y,z) \end{bmatrix}, \quad G^*(t,Y,z) = \begin{bmatrix} g_1^*(t,Y,z) \\ g_2^*(t,Y,z) \\ \cdots \\ g_p^*(t,Y,z) \end{bmatrix}. \tag{7.46}$$

Note that the system (7.1)–(7.2) presented in the normal form (7.42)–(7.43) consists of two subsystems. The first one, (7.42), is the subsystem for the output variables $y(t)$ and their derivatives; it is called an external subsystem. The second one, (7.43), is the subsystem for the internal variables $z(t)$. Accordingly, it is known as an internal subsystem.

7.4.4 *Internal stability of linear systems*

First, let us clarify the core of the approach to internal stability analysis by considering the linear system given by

$$\dot{X} = AX + Bu, \quad X(0) = X^0, \tag{7.47}$$

$$y = CX, \tag{7.48}$$

where $X \in \mathbb{R}^n$, $y \in \mathbb{R}^p$, $u \in \mathbb{R}^p$, and $p < n$. Also assume that

$$\det CB \neq 0. \tag{7.49}$$

So the system (7.47)–(7.48) is invertible and the vector of the relative degree equals $\alpha = \{1, \ldots, 1\}$.

For internal stability analysis of the system (7.47)–(7.48), let us transform it to the normal form (7.42)–(7.43). Note that if (7.49) is satisfied, then the normal form corresponds to the one discussed in [O'Reilly (1983)].

In accordance with (7.49), the transformation (7.40) has the form

$$\begin{bmatrix} y \\ z \end{bmatrix} = QX, \tag{7.50}$$

where

$$Q = \begin{bmatrix} Q_1 \\ Q_2 \end{bmatrix} \quad \text{and} \quad Q_1 = C. \tag{7.51}$$

The matrix Q_2 is chosen so that $\det Q \neq 0$. For example, Q_1 may be supplemented with appropriate identity matrix rows. Note that Q_2 may be assigned in the form $Q_2 = Q_{1R}^T$, where Q_{1R} is a right annihilating matrix of the maximal rank for Q_1. This follows from the next theorem.

Theorem 7.3 *Let $Q_1 \in \mathbb{R}^{p\times(n-p)}$ where* $\operatorname{rank} Q_1 = p$. *Suppose Q_{1R} is a right annihilating matrix of the maximal rank for Q_1. Then*

$$\det \begin{bmatrix} Q_1 \\ Q_{1R}^T \end{bmatrix} \neq 0.$$

Proof. Since Q_{1R} is a right annihilating matrix, we have $Q_1 Q_{1R} = 0$ where 0 is a null matrix. Furthermore, Q_{1R} is the right annihilating matrix of maximal rank for Q_1. This and the fact that $\operatorname{rank} Q_1 = p$ imply $\operatorname{rank} Q_{1R} = n - p$.

As a result we have

$$\begin{bmatrix} Q_1 \\ Q_{1R}^T \end{bmatrix} \begin{bmatrix} Q_1 \\ Q_{1R}^T \end{bmatrix}^T = \begin{bmatrix} Q_1 Q_1^T & 0 \\ 0^T & Q_{1R}^T Q_{1R} \end{bmatrix},$$

where $\det Q_1 Q_1^T \neq 0$ and $\det Q_{1R}^T Q_{1R} \neq 0$. The proof is complete. ∎

Remark 7.9 *If Q_{1R} is the right annihilating matrix of maximal rank for Q_1, then Q_{1R}^T is the left annihilating matrix of maximal rank for Q_1^T.*

As a result of the transformation (7.50), we have the normal form of the system (7.47)–(7.48):

$$\dot{y} = A_{11} y + A_{12} z + B_1 u, \tag{7.52}$$

$$\dot{z} = A_{21} y + A_{22} z + B_2 u. \tag{7.53}$$

Here (7.52) is the external subsystem and (7.53) is the internal subsystem. Note that $B_1 = CB$.

Since the relative degree of each output component equals unity, the reference model of the desired behavior of the output $y(t)$ in the system (7.47)–(7.48) can be constructed such that

$$\dot{y} = F(y, r) \tag{7.54}$$

is a system of stable differential equations, where $y = r$ at the equilibrium point for $r = \text{const}$. For instance, we may assume that (7.54) consists of p separate differential equations of the form (4.6) (see p. 58).

As a result, the control problem (7.5) for the system (7.47)–(7.48) corresponds to the insensitivity condition given by

$$e^F = 0, \tag{7.55}$$

where

$$e^F = F(y, r) - \dot{y}. \tag{7.56}$$

From (7.52)–(7.54) it follows that the solution of (7.55) is given by

$$u^{LID}(t) = B_1^{-1}\{F(y,r) - A_{11}y - A_{12}z\}. \tag{7.57}$$

This is called the solution of the linear inverse dynamics (or LID-control function by analogy with the NID-control function). Note that it corresponds to the desired input-controlled output map assigned by (7.54).

Substitution of (7.57) into (7.52)–(7.53) yields

$$\dot{y} = F(y,r), \tag{7.58}$$

$$\begin{aligned}\dot{z} &= \{A_{22} - B_2B_1^{-1}A_{12}\}z \\ &\quad + \{A_{21} - B_2B_1^{-1}A_{11}\}y + B_2B_1^{-1}F(y,r).\end{aligned} \tag{7.59}$$

These describe the behavior of the state variables $y(t), z(t)$ on the condition that the desired output behavior assigned by (7.54) occurs.

In the closed-loop system (7.58)–(7.59) the external subsystem equation (7.58) equals the reference model equation (7.54), and the stability of the internal subsystem equation (7.59) depends only on the inherent properties of the system (7.47)–(7.48).

Since the reference model equation (7.54) is stable (meaning that (7.54) has an unique asymptotically stable equilibrium at the point $r = y$ for $r = \text{const}$), the internal stability (7.32) of the system (7.47)–(7.48) occurs if the trajectories of the internal subsystem equation (7.59) are asymptotically stable. This is the case if all roots of the internal subsystem characteristic polynomial

$$\det\{sI_{n-p} - A_{22} + B_2B_1^{-1}A_{12}\} \tag{7.60}$$

lie in the left half of the s-plane. Note that these roots correspond to zeros of the transfer matrix of the system (7.47)–(7.48).

Consequently, the analysis of the internal stability (7.32) of the system (7.47)–(7.48) reduces to the stability analysis of the internal subsystem of the form (7.59).

In compliance with the characteristic polynomial (7.60), it is easy to see that internal stability of (7.47)–(7.48) depends only on the inherent properties of the system (7.47)–(7.48). Such an inference corresponds to the invariance of the transfer matrix zeros with respect to feedback in the system with equidimensional input and output vectors [Rosenbrock (1970); Rosenbrock (1973)].

7.4.5 *Internal stability of nonlinear systems*

Returning to the nonlinear control system given by (7.44)–(7.45), we see that the desired output behavior assigned by (7.25) is reproduced with the NID-control function (7.29), which may be presented in the form

$$u^{NID}(t) = \{G^*(t, Y, z)\}^{-1}\{F(Y, R) - H^*(t, Y, z)\}. \tag{7.61}$$

By substituting (7.61) into (7.44)–(7.45), we get

$$y_* = F(Y, R), \tag{7.62}$$

$$\dot{z} = \tilde{f}(t, Y, z) + \tilde{G}(t, Y, z)\{G^*(t, Y, z)\}^{-1} \times\{F(Y, R) - H^*(t, Y, z)\}. \tag{7.63}$$

So, in the closed-loop system equations (7.62)–(7.63) the external subsystem equation (7.62) equals the stable reference model equation (7.25). Hence

$$\lim_{t\to\infty} \sup_{t>0} \|Y(t)\| < \infty. \tag{7.64}$$

Assume that for all trajectories of the internal subsystem (7.63), the condition

$$\lim_{t\to\infty} \sup_{t>0} \|z(t)\| < \infty \tag{7.65}$$

is satisfied. Then, by (7.40) and (7.41), we can conclude that the internal stability property (7.32) holds in the specified region Ω_X of the state space. The result can be formulated as the following theorem.

Theorem 7.4 *If the property (7.65) of the internal subsystem (7.63) trajectories holds in the specified region Ω_X of the state space, then a bounded set Ω_u exists such that the desired output behavior assigned by the reference model of the form (7.25) is realizable in the system (7.1)–(7.2) for all $X \in \Omega_X$ and $t \in [0, \infty)$.*

Remark 7.10 *If $l = n$ then the closed-loop system equations (7.62)–(7.63) reduce to the external subsystem equation (7.62) without the internal subsystem (7.63). In this case, from the invertibility condition (7.20) it follows that the internal stability (7.32) and the boundedness of the NID-control function (7.31) in the system (7.1)–(7.2) with the desired stable output dynamics assigned by (7.25) are maintained.*

Remark 7.11 *In the context of the discussed output regulation problem, the effect of the internal subsystem (7.63) on the output behavior can be rejected if the solutions of (7.63) are smooth and bounded. The existence of the equilibrium point of (7.63) and its stability in some sense are not required. For instance, the solutions of (7.63) may have a stable cycle or reveal chaotic behavior as shown in the concluding example of this chapter.*

So we may conclude that internal stability of the system (7.1)–(7.2) corresponds to the uniform ultimate boundedness of the internal subsystem (7.63).

7.4.6 *Degenerated motions and zero-dynamics*

Let us consider the case when a controller keeps the output $y(t)$ the same as a constant reference input $r = \text{const}$, i.e.,

$$y(t) = r = \text{const}, \quad \forall\, t \in [0, \infty). \tag{7.66}$$

Then from (7.33), (7.34), and (7.66) it follows that the state vector $X(t)$ of (7.62)–(7.63) belongs to the manifold given by

$$y_i(t) \overset{t}{\equiv} r_i, \;\; y_i^{(j)} \overset{t}{\equiv} 0, \;\; i = 1, \dots, p, \;\; j = 1, \dots, \alpha_i - 1, \;\; t \in [0, \infty). \tag{7.67}$$

In accordance with (7.16), this manifold can be represented as the system of l algebraic equations

$$h_{i,0}(t, X) \overset{t}{\equiv} r_i, \;\; h_{i,j}(t, X) \overset{t}{\equiv} 0, \;\; j = 1, \dots, \alpha_i - 1, \;\; i = 1, \dots, p. \tag{7.68}$$

Motions along this manifold are called degenerated motions. Then the system (7.62)–(7.63) of dimension n degenerates into a system of dimension $n - l$. So, the internal stability analysis of the system (7.1)–(7.2) on the additional condition (7.66) leads to analysis of the stability of the trajectories on the $n - l$ dimensional manifold (7.68). Here we recall that l is the sum of the relative degrees α_i where $i = 1, \dots, p$.

From (7.67) it follows that the NID-control function (7.61) is given by

$$u^{NID}(t) = -\{G^*(t, Y_r, z)\}^{-1} H^*(t, Y_r, z), \tag{7.69}$$

where $r = \text{const}$ and

$$Y_r = \{r_1, 0, \dots, 0, r_2, \dots, 0, r_p, \dots, 0\}^T. \tag{7.70}$$

The vector Y_r follows from the vector Y (7.34) by taking into account the additional restrictions given by (7.67).

From (7.67) and the closed-loop system equations (7.62)–(7.63), it follows also that the behavior of the internal variables $z(t)$ on the manifold (7.67) in the state space (Y, z) is described by the system of $n-l$ differential equations

$$\dot{z} = \tilde{f}(t, Y_r, z) - \tilde{G}(t, Y_r, z)\{G^*(t, Y_r, z)\}^{-1}H^*(t, Y_r, z). \tag{7.71}$$

System (7.71) is called the degenerated system [Vostrikov and Yurkevich (1991)].

In a particular case, let us assume that $r \overset{t}{\equiv} 0$ and, accordingly, obtain $Y_r = Y_{r=0} \overset{t}{\equiv} 0$, $\forall\ t \in [0, \infty)$. Then from the degenerated system (7.71) the equation of zero-dynamics

$$\dot{z} = \tilde{f}(t, Y_{r=0}, z) - \tilde{G}(t, Y_{r=0}, z)\{G^*(t, Y_{r=0}, z)\}^{-1}H^*(t, Y_{r=0}, z) \tag{7.72}$$

follows, where the stability of transients in (7.72) corresponds to the concept of the minimum phase of nonlinear system discussed in [Byrnes and Isidori (1991)].

The main result of this section is given in the following theorem.

Theorem 7.5 *If in the system (7.1)–(7.2) the conditions (7.67) hold, then in the normal form representation (7.44)–(7.45) the external subsystem (7.62) reduces to the system of l algebraic equations (7.67), and the transient process of the internal variables $z(t)$ occurs along the manifold (7.67), where the transients of $z(t)$ are described by the degenerated system (7.71) of the $(n-l)$th order.*

It is clear that verification of the uniform ultimate boundedness of the degenerated system (7.71) trajectories on the manifold (7.66) is often much simpler than the analysis of the internal subsystem (7.63).

Remark 7.12 *Note that the relations between zeros of the transfer function and the zero-dynamics system were discussed in [Levy and Sivan (1966)]. Canonical representation of nonlinear systems where the internal subsystem gets rid of the control variable $u(t)$ is discussed in [Byrnes and Isidori (1991); Isidori (1995)].*

7.4.7 Example

Let us consider the nonlinear time-varying system

$$\begin{aligned}
\dot{x}_1 &= x_2 + x_3(x_1 - x_3)(x_3 + x_4 - x_1) + [2 + \sin(x_4)]u_1 + u_2,\\
\dot{x}_2 &= -(x_1 - x_3)(x_3 + x_4 - x_1) + w_2(t) - u_1 + [1 + 0.5\sin(x_3)]u_2,\\
\dot{x}_3 &= x_3(x_1 - x_3)(x_3 + x_4 - x_1) + [2 + \sin(x_4)]u_1 + u_2,\\
\dot{x}_4 &= x_2 - 12(x_3 + x_4 - x_1) + w_4(t) + u_1 + u_2,\\
y_1 &= x_1 - x_3,\\
y_2 &= x_3,
\end{aligned} \tag{7.73}$$

where

$x_1(t), \ldots, x_4(t)$ are the state variables;
$y_1(t), y_2(t)$ are the output variables;
$u_1(t), u_2(t)$ are the input variables;
$w_2(t)$, $w_4(t)$ are the external disturbances;
$|w_2(t)| \leq w_{\max}$ and $|w_4(t)| \leq w_{\max}$.

From (7.73) we can find the relative degrees given by $\alpha_1 = 2$, $\alpha_2 = 1$. The matrix G^* has the form

$$G^* = \begin{bmatrix} -1 & 1 + 0.5\sin(x_5) \\ 2 + \sin(x_4) & 1 \end{bmatrix}. \tag{7.74}$$

From this it is easy to verify that $\det G^* \neq 0$, i.e., (7.73) is an invertible system since (7.20) holds. Denote

$$Y = \{y_{10}, y_{11}, y_{20}\}^T = \{y_1, y_1^{(1)}, y_2\}^T. \tag{7.75}$$

From (7.73) we find that

$$y_{10} = x_1 - x_3, \quad y_{11} = x_2, \quad y_{20} = x_3. \tag{7.76}$$

Let us introduce the internal variable

$$z = x_3 + x_4 - x_1. \tag{7.77}$$

Expressions (7.76) and (7.77) give the desired transformation (7.40) of the form

$$\begin{bmatrix} y_{10} \\ y_{11} \\ y_{20} \\ z \end{bmatrix} = \begin{bmatrix} 1 & 0 & -1 & 0 \\ 0 & 1 & 0 & 0 \\ 0 & 0 & 1 & 0 \\ -1 & 0 & 1 & 1 \end{bmatrix} \begin{bmatrix} x_1 \\ x_2 \\ x_3 \\ x_4 \end{bmatrix}. \tag{7.78}$$

By the change of variables (7.78) we find that the normal form of (7.73) is given by

$$\begin{aligned}
\dot{y}_{10} &= y_{11}, \\
\dot{y}_{11} &= -y_{10}z + w_2(t) - u_1 + [1 + 0.5\sin(y_{20})]u_2, \\
\dot{y}_{20} &= y_{10}y_{20}z + [2 + \sin(z + y_{10})]u_1 + u_2, \\
\dot{z} &= -12z + w_4(t) + u_1 + u_2.
\end{aligned} \tag{7.79}$$

Next, the degenerated system (7.71) can be obtained by assuming that for all $t \in [0, \infty)$ the conditions

$$y_{10} = r_1 = 1, \quad y_{10}^{(1)} = 0, \quad y_{10}^{(2)} = 0, \quad y_{20} = r_2 = 1, \quad y_{20}^{(1)} = 0 \tag{7.80}$$

hold. On these conditions the system (7.79) reduces to the system

$$\begin{aligned}
0 &= y_{11}, \\
0 &= -y_{10}z + w_2(t) - u_1 + [1 + 0.5\sin(y_{20})]u_2, \\
0 &= y_{10}y_{20}z + [2 + \sin(z + y_{10})]u_1 + u_2, \\
\dot{z} &= -12z + w_4(t) + u_1 + u_2.
\end{aligned}$$

Accordingly, the degenerated system is given by

$$\dot{z} = \tilde{a}z + \phi(w_2, w_4, r_1, r_2), \tag{7.81}$$

where

$$\begin{aligned}
\tilde{a} = -12 + r_1[3 + \sin(r_2) + \sin(z + r_1) \\
+0.5\sin(r_2)\sin(z + r_1)]^{-1}[3 + \sin(z + r_1) + 0.5r_2\sin(r_2)].
\end{aligned} \tag{7.82}$$

For instance, let us assume that $|r_1(t)| \leq 1$ and $|r_2(t)| \leq 1$. Then $\tilde{a} < 0$. Hence, the boundedness of the trajectories of the degenerated system (7.81) follows on the given bounded sets of $r_1(t), r_2(t)$, and $w_2(t), w_4(t)$. Therefore, the bounded subset Ω_u exists such that the inclusion $u^{NID}(t) \in \Omega_u$, $\forall\, t \in [0, \infty)$ holds on the condition (7.66).

7.5 Output regulation of SISO systems

7.5.1 *Realizability of desired output behavior*

Before addressing MIMO control system design, let us consider the problem of nonlinear time-varying SISO control system design in the presence of the internal subsystem. We will deal with the system given by

$$\dot{X} = f(t, X) + G(t, X)u, \quad X(0) = X^0, \tag{7.83}$$

$$y = h(t, X), \tag{7.84}$$

where $X \in \mathbb{R}^n$, $u \in \mathbb{R}^1$, and $y \in \mathbb{R}^1$.

Denote $h_0(t, X) = h(t, X)$. Differentiation of the output equation $y = h_0(t, X)$ yields

$$y^{(1)} = \frac{\partial h_0}{\partial t} + \frac{\partial h_0}{\partial X}\{f(t, X) + G(t, X)u\}.$$

Assume that

$$\frac{\partial h_0(t, X)}{\partial X} G(t, X) = 0, \quad \forall\ (t, X) \in \Omega_{t,X},$$

where $\Omega_{t,X} = [0, \infty) \times \Omega_X$. Then, by denoting

$$h_1(t, X) = \frac{\partial h_0(t, X)}{\partial t} + \frac{\partial h_0(t, X)}{\partial X} f(t, X),$$

we can find the expression for $y^{(2)}$.

Let us assume that there exists an $\alpha \leq n$ such that

$$y^{(j)} = h_j(t, X), \qquad \forall\ j = 0, \ldots, \alpha - 1 \tag{7.85}$$

and

$$y^{(\alpha)} = h_\alpha(t, X) + \frac{\partial h_{\alpha-1}(t, X)}{\partial X} G(t, X)u, \tag{7.86}$$

where the conditions

$$\frac{\partial h_j(t, X)}{\partial X} G(t, X) = 0, \quad \forall j = 0, \ldots, \alpha - 2, \quad \forall\ (t, X) \in \Omega_{t,X},$$

$$\frac{\partial h_{\alpha-1}(t, X)}{\partial X} G(t, X) \neq 0, \quad \forall\ (t, X) \in \Omega_{t,X}$$

are satisfied. That is, the system (7.83)–(7.84) satisfies the well known sufficient condition for invertibility [Porter (1970)].

The value α is the relative degree (or invertibility index) of the system (7.83)–(7.84). Then $y^{(\alpha)}$ is the relative highest output derivative of that system.

For convenience, let us denote

$$h^*(t, X) = h_\alpha(t, X), \qquad g^*(t, X) = \frac{\partial h_{\alpha-1}(t, X)}{\partial X} G(t, X),$$

and rewrite (7.86) as

$$y^{(\alpha)} = h^*(t, X) + g^*(t, X)u. \tag{7.87}$$

We wish to design a control system to provide the condition

$$\lim_{t \to \infty} e(t) = 0, \tag{7.88}$$

where $e(t)$ is the error of the reference input realization (tracking error); $e(t) = r(t) - y(t)$; $r(t)$ being the reference input.

Assume that, in accordance with the time-domain specifications of the desired output behavior, the reference model of the form

$$y^{(\alpha)} = F(y^{(\alpha-1)}, \dots, y^{(1)}, y, r^{(\rho)}, \dots, r^{(1)}, r) \tag{7.89}$$

has been constructed as a stable differential equation where $\rho < \alpha$ and $y = r$ at the equilibrium point for $r =$ const. Let us rewrite (7.89) in the concise form

$$y^{(\alpha)} = F(Y, R), \tag{7.90}$$

where $Y = \{y, y^{(1)}, \dots, y^{(\alpha-1)}\}^T$ and $R = \{r, r^{(1)}, \dots, r^{(\rho)}\}^T$.

The deviation between the desired dynamics $F(Y, R)$ assigned by (7.90) and the actual value of the relative highest output derivative $y^{(\alpha)}$ is denoted by

$$e^F = F(Y, R) - y^{(\alpha)},$$

where e^F is the error of the desired dynamics realization. Then the discussed control problem for the SISO system (7.83)–(7.84) corresponds to the insensitivity condition given by

$$e^F = 0.$$

Assume that the internal stability of the system (7.83)–(7.84) holds, that is, condition (7.32) is satisfied in a specified region of the state space of (7.83)–(7.84) given that $e^F = 0$.

7.5.2 Closed-loop system analysis

In accordance with (4.38), let us consider the control law with the relative highest output derivative $y^{(\alpha)}$ in feedback:

$$\begin{aligned}&\mu^q u^{(q)} + d_{q-1}\mu^{q-1}u^{(q-1)} + \cdots + d_1\mu u^{(1)} + d_0 u \\ &\quad = k_0\{F(Y,R) - y^{(\alpha)}\},\ U(0) = U^0,\end{aligned} \tag{7.91}$$

where $\mu > 0$, $U = \{u, u^{(1)}, \ldots, u^{(q-1)}\}^T$, and $q \geq \alpha$. Assume that $g^*(t,X)k_0 > 0$ for all $(t,X) \in \Omega_{t,X}$.

The closed-loop system equations become

$$\dot{X} = f(t,X) + G(t,X)u,\ y = h(t,X),\ X(0) = X^0, \tag{7.92}$$

$$\begin{aligned}&\mu^q u^{(q)} + d_{q-1}\mu^{q-1}u^{(q-1)} + \cdots + d_1\mu u^{(1)} + d_0 u \\ &\quad = k_0\{F(Y,R) - y^{(\alpha)}\},\ U(0) = U^0.\end{aligned} \tag{7.93}$$

where the two-time-scale motions are induced as $\mu \to 0$.

From (7.92)–(7.93) and (7.87), by the above procedure for two-time-scale motion separation, the FMS

$$\begin{aligned}&\mu^q u^{(q)} + d_{q-1}\mu^{q-1}u^{(q-1)} + \cdots + d_1\mu u^{(1)} + \{d_0 + k_0 g^*(t,X)\}u \\ &\quad = k_0\{F(Y,R) - h^*(t,X)\},\ U(0) = U^0\end{aligned} \tag{7.94}$$

results, where $g^*(t,X)$ is the frozen parameter during the transients in (7.94).

Assume that the FMS (7.94) is stable. Then, letting $\mu \to 0$ in (7.94), we obtain the steady state (more precisely, quasi-steady state) of the FMS (7.94), where $u(t) = u^s(t)$ and

$$u^s(t) = [d_0 + k_0 g^*(t,X)]^{-1} k_0\{F(Y,R) - h^*(t,X)\}. \tag{7.95}$$

Substitution of (7.95) into (7.86) yields

$$\begin{aligned}y^{(\alpha)} &= F(Y,R) \\ &\quad + d_0[d_0 + k_0 g^*(t,X)]^{-1}\{h^*(t,X) - F(Y,R)\},\end{aligned} \tag{7.96}$$

where (7.96) is the SMS of the output behavior. So, the desired dynamical properties of $y(t)$ are provided in a specified region of the state space of the system (7.83)–(7.84) if $d_0 = 0$ or for $d_0 = 1$ as $g^*(t,X)k_0 \to \infty$, despite the existence of unknown varying system parameters and external disturbances.

7.5.3 *Example*

Let us consider a nonlinear system

$$\begin{aligned}\dot{y} &= y^2 + z_1 z_2 + u,\\ \dot{z}_1 &= y^2 + z_1 z_2 + z_2 + u,\\ \dot{z}_2 &= y^2 - z_1 + z_1 z_2 + (1 - z_1^2) z_2 + u,\end{aligned} \tag{7.97}$$

which is represented in the normal form. From (7.97) it follows that the relative degree $\alpha = 1$.

Let the desired output behavior be assigned by

$$y^{(1)} = F(y, r),$$

where

$$y^{(1)} = \frac{1}{T}[r - y] \tag{7.98}$$

and $T = 1$ s.

The counterpart of the system (7.62)–(7.63) has the following form:

$$\begin{aligned}\dot{y} &= \frac{1}{T}[r - y],\\ \dot{z}_1 &= z_2 + \frac{1}{T}[r - y],\\ \dot{z}_2 &= -z_1 + (1 - z_1^2) z_2 + \frac{1}{T}[r - y].\end{aligned} \tag{7.99}$$

Similar to (7.66), assume that

$$y(t) = r = \text{const}, \quad \forall\, t \in [0, \infty).$$

Then the system (7.99), having dimension 3, degenerates into the system

$$\begin{aligned}y &= r = \text{const},\\ \dot{z}_1 &= z_2,\\ \dot{z}_2 &= -z_1 + (1 - z_1^2) z_2,\end{aligned} \tag{7.100}$$

having dimension 2. This Van der Pol oscillator equation possesses a stable limit cycle [Khalil (2002)].

Since the solutions are bounded, by Theorem 7.4 and Remark 7.11 we know that the desired output behavior assigned by the reference model of the form (7.98) is realizable in the system (7.97).

Therefore, consider the control law given by

$$\mu u^{(1)} + d_0 u = k_0\{T^{-1}[r - y] - y^{(1)}\}, \tag{7.101}$$

which is the counterpart of the control law (4.16). In order to perform numerical simulation, let us rewrite (7.101) in the state-space form

$$\frac{du_1}{dt} = -\frac{d_0}{\mu}\{u_1 - k_0 y\} + \frac{k_0}{T}[r - y], \tag{7.102}$$

$$u = \frac{1}{\mu}[u_1 - k_0 y]. \tag{7.103}$$

Simulation results for the system (7.97) controlled by the algorithm (7.102)–(7.103) are displayed in Fig. 7.2. Here the initial conditions are zero, and $k_0 = 10$, $d_0 = 0$, $\mu = 0.1$ s, $t \in [0, 10]$ s.

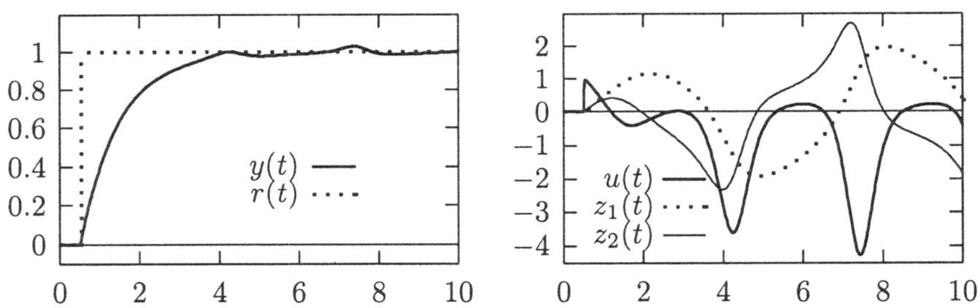

Fig. 7.2 Simulation results for the system (7.97) and (7.102)–(7.103).

7.6 Switching regulator for boost DC-to-DC converter

7.6.1 *Boost DC-to-DC converter circuit model*

Let us consider the application of the above design methodology to the problem of switching controller design for a boost DC-to-DC converter circuit as shown in Fig. 7.3.

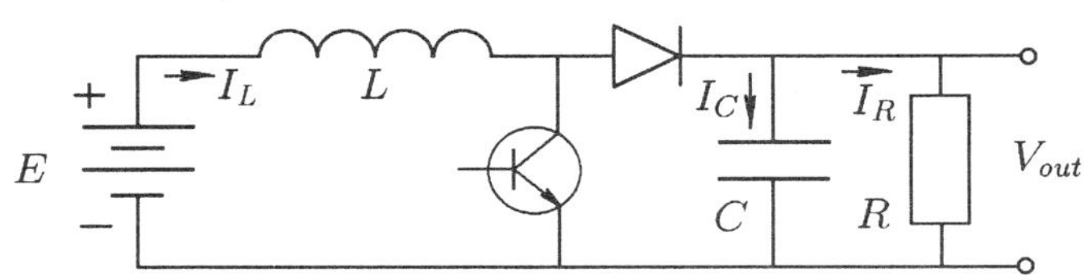

Fig. 7.3 Boost DC-to-DC converter circuit.

By Kirchhoff's voltage law we have $E = V_L$ when the transistor is ON and $E = V_L + V_{out}$ when the transistor is OFF; hence, we get

$$E = V_L + uV_{out},$$

where the state of the ideal switches (transistor and diode) is represented by the control variable u and $u \in \{0, 1\}$.

The voltage drop V_L across an inductor is given by $V_L = L\dot{I}_L$, where I_L is the current through the inductor with inductance L. We have $C\dot{V}_C = I_C$ for the voltage drop V_C across a capacitor, where I_C is the current through the capacitor with capacitance C. By Ohm's law, the voltage drop V_R across a resistor is given by $V_R = RI_R$, where I_R is the current through the resistor with resistance R. By Kirchhoff's current law we get $I_L u = I_C + I_R$. Denote $x_1 = I_L$, $x_2 = V_C$. By taking into account that $V_R = V_C = V_{out}$, we obtain the bilinear switched model given by (see, e.g., [Escobar *et al.* (1999); Sira-Ramirez (2002)])

$$\dot{x}_1 = \frac{E}{L} - \frac{x_2}{L}u, \tag{7.104}$$

$$\dot{x}_2 = -\frac{1}{RC}x_2 + \frac{x_1}{C}u, \tag{7.105}$$

where E is the value of the DC voltage source, $E > 0$.

7.6.2 *Model with continuous control variable*

Assume that $y = x_1$ is the measurable output of the system (7.104)–(7.105) and consider the model with continuous control variable given by

$$\dot{x}_1 = \frac{E}{L} - \frac{x_2}{L}\bar{u}, \tag{7.106}$$

$$\dot{x}_2 = -\frac{1}{RC}x_2 + \frac{x_1}{C}\bar{u}, \tag{7.107}$$

$$y = x_1, \tag{7.108}$$

where $\bar{u}$ is the continuous control variable and $\bar{u} \in [0, 1)$.

Similar to (7.66), assume that

$$y(t) = x_1(t) = r = \text{const}, \quad \forall\, t \in [0, \infty). \tag{7.109}$$

Then from (7.106) and (7.109), we get

$$\bar{u}^{NID} = E/x_2, \tag{7.110}$$

where (7.110) is the counterpart of (7.69). Since $\bar{u} \in [0, 1)$, we obtain $x_2 \in [E, \infty)$. Then the system (7.106)–(7.108), having dimension 2, degenerates into the system

$$\begin{aligned} x_1 &= r = \text{const}, \\ \dot{x}_2 &= -\frac{1}{RC}x_2 + \frac{Er}{Cx_2}, \end{aligned} \tag{7.111}$$

having dimension 1. The degenerated system (7.111) has the unique asymptotically stable equilibrium point x_2^s given by

$$x_2^s = \sqrt{ERr}. \tag{7.112}$$

Therefore, the internal stability of the system (7.106)–(7.108) is satisfied in the neighborhood of the point $\{x_1, x_2\} = \{r, x_2^s\}$ (or, by [Byrnes and Isidori (1991)], the system (7.106)–(7.108) is the minimum phase system).

Since $\bar{u} \in [0, 1)$ and $x_2 \in [E, \infty)$, we obtain from (7.112) the allowable interval for r, which is $r \in [E/R, \infty)$.

Assume that [Sira-Ramirez (2002)]

$$E = 15\,\text{V}, \;\; L = 0.02\,\text{H}, \;\; C = 0.001\,\text{F}, \;\; R = 200\,\Omega.$$

Let $r \in [1, 4]$. By (7.112), we get $x_2 \in [54, 110]$. Then, by linearization of (7.111) at the equilibrium point x_2^s we obtain

$$a_{int} = \frac{\partial}{\partial x_2}\left[\frac{Er}{Cx_2} - \frac{x_2}{RC}\right]\bigg|_{x_2=\sqrt{ERr}} = -\frac{2}{RC},$$

where $T_{int} = -1/a_{int}$ is the time constant of the linearized internal subsystem and for the above parameters we obtain $T_{int} = 0.1$ s.

The relative degree of the discussed system equals 1 and the internal stability is satisfied. Therefore, the control law

$$\mu_1 \bar{u}^{(1)} + d_{10}\bar{u} = k_1\{T_1^{-1}[r - y] - y^{(1)}\} \tag{7.113}$$

can be applied in order to obtain the desired output behavior of the system (7.106)–(7.108). Hence, in the closed-loop system given by (7.106)–(7.108) and (7.113), we have that the two-time-scale motions are induced as $\mu_1 \to 0$. Hence, we obtain the FMS given by

$$\mu_1 \bar{u}^{(1)} + \left[d_{10} - \frac{k_1 x_2}{L}\right]\bar{u} = k_1\left\{\frac{1}{T_1}[r - y] - \frac{E}{L}\right\}, \tag{7.114}$$

and the SMS given by

$$\dot{y} = \frac{1}{T_1}[r - y] + d_{10}k_1^{-1}\left[d_{10}k_1^{-1} - \frac{x_2}{L}\right]^{-1}\left[\frac{E}{L} - \frac{1}{T_1}[r - y]\right], \quad (7.115)$$

$$\dot{x}_2 = -\frac{1}{RC}x_2 + \frac{y}{C}\left[d_{10}k_1^{-1} - \frac{x_2}{L}\right]^{-1}\left[\frac{1}{T_1}[r - y] - \frac{E}{L}\right] \quad (7.116)$$

where x_1 is replaced by y.

Take $T_1 = 0.02$ s, $\mu_1 = 0.002$ s, $k_1 = -0.001$, $d_{10} = 0$ (as a result, we obtain that $\gamma_{\min} = d_{10} - k_1 x_2^{\min} L^{-1} = 2.7$). Simulation results for the model (7.106)–(7.108) controlled by the algorithm (7.113) are displayed in Figs. 7.4–7.5, where the initial conditions are $x_1(0) = 0$, $x_2(0) = 15$, $\bar{u}(0) = 0$, and $t \in [0, 0.4]$ s. The external disturbance is represented by the varying resistance $R = R(t)$.

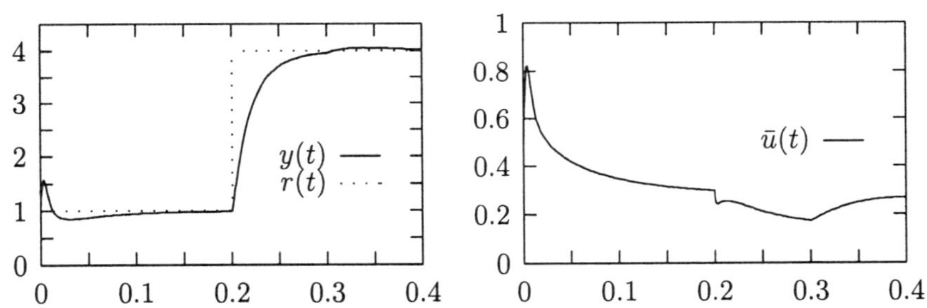

Fig. 7.4 Simulation results for the model (7.106)–(7.108) controlled by the algorithm (7.113).

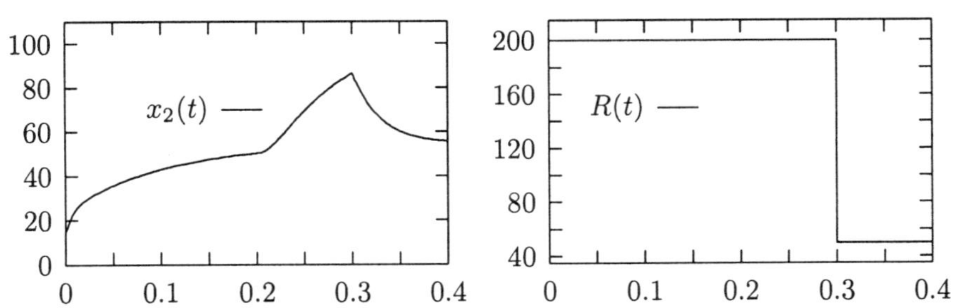

Fig. 7.5 Simulation results for the model (7.106)–(7.108) controlled by the algorithm (7.113).

7.6.3 *Switching regulator*

As the next step, let us consider the switching regulator given by

$$\mu_1 u_1^{(1)} + d_{10} u_1 = k_1\{T_1^{-1}[r - y] - y^{(1)}\}, \tag{7.117}$$

$$u_2(t) = u_1(t - \tau), \tag{7.118}$$

$$u = u_{\max}\frac{1 + \operatorname{sgn}(u_2)}{2}, \tag{7.119}$$

where $u_{\max} = 1$ and the time delay τ is included in order to satisfy the conditions for limit cycle existence in the FMS of the closed-loop system (7.104)–(7.105) controlled by the algorithm (7.117)–(7.119).

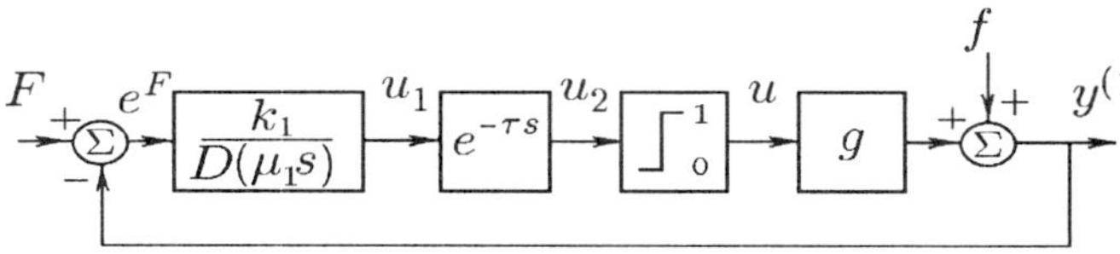

Fig. 7.6 Block diagram of the FMS in the closed-loop system (7.104)–(7.105) controlled by the algorithm (7.117)–(7.119).

Let us consider the block diagram of the FMS shown in Fig. 7.6, where

$$g = -\frac{x_2}{L} \quad \text{and} \quad f = \frac{E}{L}.$$

Assume that there are fast oscillations in this FMS. Then, in accordance with the describing function method, let us replace the relay switch by its quasi-linear approximation. Let the transfer function $k_1/D(\mu_1 s)$ display a low-pass filtering property and assume that the nonlinearity input is $u_2(t)$, where

$$u_2(t) = u_2^0 + A\sin(\omega t)$$

with u_2^0 the constant bias signal. Consider the output $u(t)$ of the nonlinearity represented by its Fourier series

$$u(t) = u_0 + b_1\sin(\omega t) + c_1\cos(\omega t) + \sum_{k=2}^{\infty}\{b_k\sin(k\omega t) + c_k\cos(k\omega t)\}. \tag{7.120}$$

Note that the particular feature of the discussed system is the nonsymmetrical limits of the nonlinearity. Hence, $u_2^0 \neq 0$, $u_0 \neq 0$, and it is known that for the given nonlinearity we have[1] (see, e.g., [Paltov (1975)])

$$u_0 = \frac{u_{\max}}{2} + \frac{u_{\max}}{\pi} \sin^{-1}\left(\frac{u_2^0}{A}\right), \tag{7.121}$$

$$b_1 = \frac{2u_{\max}}{\pi} \sqrt{1 - \left[\frac{u_2^0}{A}\right]^2}, \tag{7.122}$$

$$c_1 = 0, \tag{7.123}$$

where $A \geq |u_2^0|$. Therefore, the sinusoidal plus bias describing function[2] of the discussed nonlinear element has the gain for the bias

$$u_0/u_2^0$$

and the gain for the sinusoid

$$G_n(j, A) = \frac{2u_{\max}}{\pi A} \sqrt{1 - \left[\frac{u_2^0}{A}\right]^2}. \tag{7.124}$$

By the block diagram of the FMS shown in Fig. 7.6, we have

$$f = \frac{E}{L} \neq 0$$

and assume that

$$F = 0.$$

Then, by taking into account that $d_{10} = 0$, we get the balance equation for the constant bias signal u_2^0 of the discussed FMS:

$$\frac{E}{L} - \frac{x_2}{L}\left[\frac{u_{\max}}{2} + \frac{u_{\max}}{\pi} \sin^{-1}\left(\frac{u_2^0}{A}\right)\right] = 0. \tag{7.125}$$

The 1st order harmonic balance equation (6.76) yields

$$1 - \frac{2x_2 u_{\max} k_1 e^{-j\tau\omega}}{j\,\mu_1\,\omega\,\pi\,A\,L} \sqrt{1 - \left[\frac{u_2^0}{A}\right]^2} = 0. \tag{7.126}$$

From (7.126) we obtain

$$m^2 A^4 - A^2 + (u_2^0)^2 = 0, \tag{7.127}$$

$$\omega = \frac{\pi}{2\tau}, \tag{7.128}$$

[1] $y = \sin^{-1}(x)$ denotes the inverse sine of x, i.e., $\sin(y) = x$.

[2] The notion of the sinusoidal plus bias describing function can be found, for instance, in [Atherton (1981)].

where

$$m = \frac{\mu_1 L \pi^2}{4|k_1|x_2 \tau u_{\max}}. \tag{7.129}$$

Take $T_1 = 0.02$ s, $\mu_1 = 0.002$ s, $k_1 = -0.001$, and $\tau = 0.001$ s. From (7.128) we get $\omega \approx 1570.8$ rad/s. The joint numerical resolution[3] of (7.125) and (7.127) yields $u_2^0 \approx -0.362$, $A \approx 0.556$ when $x_2 = 54.7$, and $u_2^0 \approx -0.977$, $A \approx 1.074$ when $x_2 = 110$.

Simulation results for the bilinear switched model (7.104)–(7.105) controlled by the algorithm (7.117)–(7.119) are displayed in Figs. 7.7–7.8, where the initial conditions are $x_1(0) = 0$, $x_2(0) = 15$, $u_1(0) = 0$ and $R = 200\ \Omega$ for all $t \in [0, 0.4]$ s. We see that the simulation results confirm the analytical calculations.

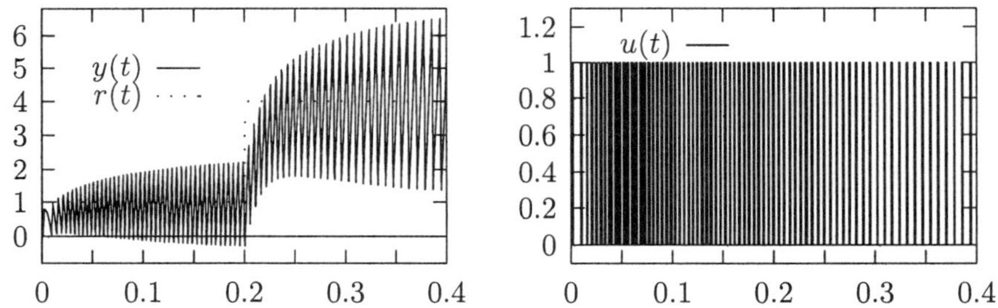

Fig. 7.7 Simulation results for the bilinear switched model (7.104)–(7.105) controlled by the algorithm (7.117)–(7.119).

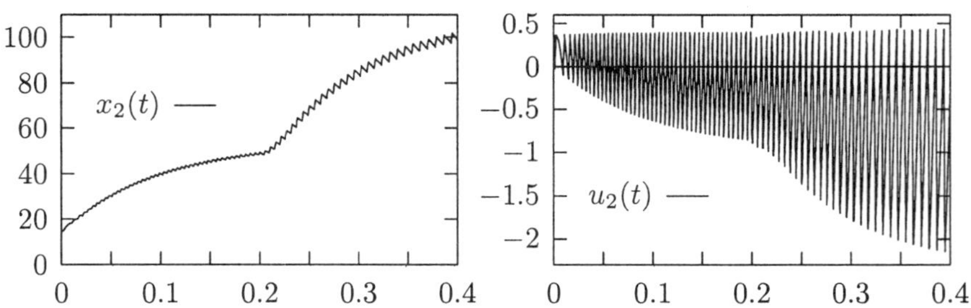

Fig. 7.8 Simulation results for the bilinear switched model (7.104)–(7.105) controlled by the algorithm (7.117)–(7.119).

[3]Such software as, for instance, "Mathcad" can be easily used for the joint numerical resolution of (7.125) and (7.127).

Let us, for simplicity, omit from our consideration the nonsymmetrical property of the discussed limit oscillations and determine the limit cycle parameters under the assumption that $F = 0$, $f = 0$.

Denote by $u_2(t) = A\sin(\omega t)$ the first harmonic of the Fourier series for the actual periodic signal $u_2(t)$ in the FMS. Consider the harmonic balance equation (6.76) where

$$G_l(j\mu_1\omega) = -\frac{x_2 k_1 e^{-\tau s}}{L(\mu_1 s + d_{10})}, \quad G_n(j, A) = \frac{4u_{\max}}{\pi A}, \text{ and } u_{\max} = \frac{1}{2}. \quad (7.130)$$

Let $k_1 < 0$ and $d_{10} = 0$. Hence, the solution of (6.76) yields the frequency ω and amplitude A of the stationary oscillations in the FMS shown in Fig. 7.6:

$$\omega = \frac{\pi}{2\tau} \quad \text{and} \quad A = \frac{8k_1\tau u_{\max} x_2}{\pi^2 \mu_1 L}.$$

Let e_{osc} be the amplitude of the stationary oscillations with frequency ω of the output $y(t)$. In accordance with the block diagram shown above in Fig. 7.6, we have to a first approximation

$$e_{osc} \approx \frac{8\tau u_{\max} x_2}{\pi^2 L}$$

given that ω is sufficiently large. The effect of the fast oscillations in the FMS on the accuracy of the output stabilization can be reduced by selection of τ. By taking into account (7.112), we obtain

$$A = A^s \approx \frac{8k_1\tau u_{\max}\sqrt{ERr}}{\pi^2\mu_1 L}, \qquad e_{osc} = e_{osc}^s \approx \frac{8\tau u_{\max}\sqrt{ERr}}{\pi^2 L}, \quad (7.131)$$

when the steady state x_2^s of the internal subsystem (7.111) holds.

Take

$$T_1 = 0.02 \text{ s}, \ \mu_1 = 0.002 \text{ s}, \ k_1 = -0.001, \ \tau = 0.001 \text{ s}. \quad (7.132)$$

From (7.131) we get $\omega \approx 1570.8$ rad/s, $A^s \approx 0.55$, $e_{osc}^s \approx 1.11$ when $r = 1$, and $A^s \approx 1.11$, $e_{osc}^s \approx 2.22$ when $r = 4$. There is perfect coincidence between the analytical calculations and the simulation results again.

7.6.4 *External disturbance attenuation*

Let us introduce the external disturbance $w(t)$ represented by the varying resistance $R = R(t)$ and consider the block diagram of the control system shown in Fig. 7.9 where there is an inner switching controller C_1 and an outer continuous controller C_2. Here, we denote $r_2 = V^d_{out}$, $x_2 = V_{out}$.

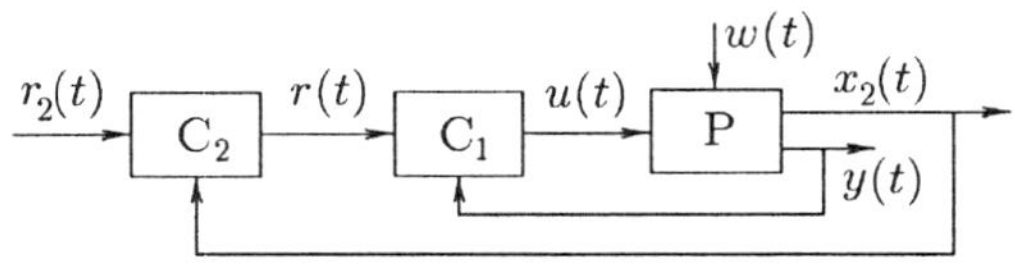

Fig. 7.9 Block diagram of the closed-loop system with an inner switching controller C_1 and an outer continuous controller C_2.

Since T_1 has been chosen so that the inequality $T_1 \ll T_{int}$ holds, we may assume that $y = r$ in the average sense for the stationary oscillations in the FMS. Then, in accordance with our usual two-time-scale design methodology, we have that the behavior of $x_2(t)$ can be approximately described by the degenerated system (7.111) of the 1st order, where r is the new control variable, $x_2(t)$ is the new output variable, and the relative degree equals 1. Therefore, the structure of C_2 can be selected in the form

$$\mu_2 r^{(1)} + d_{20} r = k_2\{T_2^{-1}[r_2 - r] - r^{(1)}\} \tag{7.133}$$

and designed similar to the control law (7.113). For instance, we can take

$$T_2 = 0.1 \text{ s}, \ \mu_2 = 0.01 \text{ s}, \ k_2 = 0.002, \ d_{20} = 0. \tag{7.134}$$

Finally, in order to perform numerical simulation, let us rewrite the switching regulator (7.117)–(7.119) and the outer continuous controller (7.133) in the form

$$\begin{aligned} \frac{du_{11}}{dt} &= -\frac{d_{10}}{\mu_1}\{u_{11} - k_1 y\} + \frac{k_1}{T_1}[r - y], \\ u_1 &= \frac{1}{\mu_1}[u_{11} - k_1 y], \\ u_2(t) &= u_1(t - \tau), \\ u &= u_{\max}\frac{1 + \operatorname{sgn}(u_2)}{2}, \\ \frac{du_{21}}{dt} &= -\frac{d_{20}}{\mu_2}\{u_{21} - k_2 x_2\} + \frac{k_2}{T_2}[r_2 - x_2], \\ r &= \frac{1}{\mu_2}[u_{21} - k_2 x_2]. \end{aligned} \tag{7.135}$$

Simulation results for the bilinear switched model (7.104)–(7.105) controlled by the algorithm (7.135) with parameters given by (7.132) and (7.134) are displayed in Figs. 7.10–7.11, where $r_2 = 54$ and the initial conditions are the following:

$$x_1(0) = 1,\ x_2(0) = 15,\ u_{11}(0) = -0.001,\ u_{21}(0) = 0.116.$$

Note that we need to match the initial conditions here in order to avoid undesirable transients in the closed-loop system.

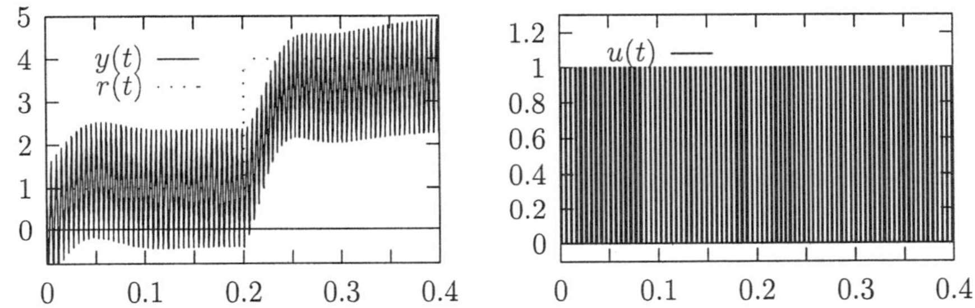

Fig. 7.10 Simulation results for the bilinear switched model (7.104)–(7.105) controlled by the algorithm (7.135).

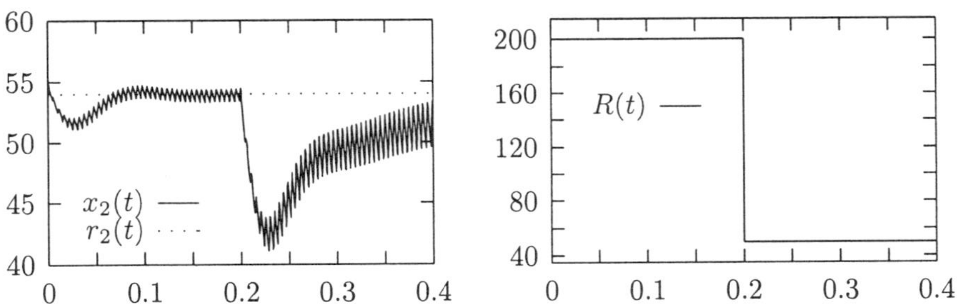

Fig. 7.11 Simulation results for the bilinear switched model (7.104)–(7.105) controlled by the algorithm (7.135).

Remark 7.13 *Note that various applications of the describing function method in periodically switched circuits can be found in a broad set of references, for instance, [Sanders (1993)].*

7.7 Notes

As shown above, the analysis of the realizability of the desired output behavior for the dynamic system given by (7.1)–(7.2) requires a more careful investigation of the intrinsic properties of the plant model. In particular, the invertibility of a dynamic system and the internal behavior of its state variables on the condition of the assigned output behavior should be examined. Such investigation leads to consideration of the invertibility indices (relative degrees), normal form of a nonlinear system, internal behavior analysis of the state variables, degenerated system, and zero-dynamics.

Finally, in this chapter the design procedure for SISO nonlinear control systems has been discussed in the presence of internal dynamics. In the next chapter, this procedure will be extended to MIMO nonlinear control systems, provided that the realizability of the desired output behavior presented in this chapter holds.

The concluding example of the boost DC-to-DC converter paves the way for extension of the book's design methodology to the problem of switching controller design.

7.8 Exercises

7.1 The differential equations of a plant model are given by

$$\begin{aligned} \dot{x}_1 &= x_1 + x_2, \quad \dot{x}_2 = x_1 + x_2 + x_3 + u, \\ \dot{x}_3 &= 2x_1 - x_2 + 2x_3 + a\,u, \quad y = x_1. \end{aligned} \tag{7.136}$$

Verify the invertibility and internal stability of the given system (7.136), where (a) $a = 1$, and (b) $a = 3$. Find the degenerated system.

7.2 The plant model is given by

$$\begin{aligned} \dot{x}_1 &= x_2, \quad \dot{x}_2 = x_1^2 + x_2 + x_3^3 + u, \\ \dot{x}_3 &= x_1^2 + x_2 + x_3 + a\,u, \quad y = x_1. \end{aligned} \tag{7.137}$$

Verify the invertibility and internal stability of the given system (7.137), where (a) $a = 1$, and (b) $a = -1$. Find the degenerated system.

7.3 The differential equations of a plant model are given by

$$\begin{aligned} \dot{x}_1 &= x_2, \quad \dot{x}_2 = |x_1| + x_3^3 + u, \\ \dot{x}_3 &= x_1 x_2^2 - x_3 + u, \quad y = x_1. \end{aligned} \tag{7.138}$$

Verify the invertibility and internal stability of the given system (7.138). Find the degenerated system.

7.4 Verify the invertibility and internal stability of the system

$$\begin{aligned} \dot{x}_1 &= 0.5(x_1 - x_2) - 0.5(x_1 - x_2)^3 + 0.5u + 2w_1, \\ \dot{x}_2 &= 0.5(x_1 - x_2) - 0.5(x_1 - x_2)^3 + 1.5u - 0.5w_2, \\ y &= x_1 + x_2. \end{aligned} \tag{7.139}$$

7.5 Verify the invertibility and internal stability of the system

$$\begin{aligned} \dot{x}_1 &= 0.5x_1 - 0.5x_2 - 0.5x_3 - 0.5(x_1 - x_2 - x_3)^3 + 0.5u, \\ \dot{x}_2 &= 0.5x_1 + 1.5x_2 + 1.5x_3 + 0.5(x_1 - x_2 - x_3)^3 - 0.5u, \\ \dot{x}_3 &= -2x_2 - 2x_3 + 2u + w, \quad y = x_1 + x_2. \end{aligned} \tag{7.140}$$

7.6 Verify the invertibility and internal stability of the system

$$\begin{aligned} \dot{x}_1 &= x_1 + x_2 + u_1 + u_2, \\ \dot{x}_2 &= 2x_1 + x_2 + 2u_1 + u_2, \\ y_1 &= x_1 - x_2, \quad y_2 = x_1 + x_2. \end{aligned} \tag{7.141}$$

7.7 Verify the invertibility and internal stability of the system

$$\begin{aligned} \dot{x}_1 &= x_1 + x_2 + u_1 + 2u_2, \\ \dot{x}_2 &= -x_1 + x_2 - x_3 + 2u_1 + u_2, \\ \dot{x}_3 &= x_1 + x_2 + x_3 + u_1 + 6u_2, \\ y_1 &= x_1 - x_2, \; y_2 = x_1 + x_2. \end{aligned} \tag{7.142}$$

7.8 Verify the invertibility and internal stability of the system

$$\begin{aligned} \dot{x}_1 &= x_1^2 + |x_2| + u_1 + u_2, \\ \dot{x}_2 &= x_1 x_2 + x_3^3 + 2u_1 + u_2, \\ \dot{x}_3 &= x_3 - u_2, \\ y_1 &= x_1, \quad y_2 = x_2. \end{aligned} \tag{7.143}$$

7.9 Verify the invertibility and internal stability of the system

$$\begin{aligned} \dot{x}_1 &= x_2 + x_3 + u, \\ \dot{x}_2 &= x_1 - x_2 - x_3 - u, \\ \dot{x}_3 &= x_3 + 2u + w, \\ y &= x_1 + x_2. \end{aligned} \tag{7.144}$$

Assume that the inequalities $|x_i(t)| \leq 1 \; \forall \; i$, $|r(t)| \leq 0.5$ hold for all $t \in [0, \infty)$. Find the control law of the form (7.91) such that $\overline{\varepsilon}_r = 0$,

$t_s^d \approx 2$ s, $\sigma^d \approx 10\%$. Run a computer simulation of the closed-loop system with zero initial conditions. Compare simulation results of the output response with the assignment for $r(t) = 0.5,\ \forall\, t > 0$.

7.10 Verify the invertibility and internal stability of the system

$$\dot{x}_1 = x_1^2 - x_2^3 + u, \quad \dot{x}_2 = |x_2| - u, \quad y = x_1 \tag{7.145}$$

Assume that the inequalities $|x_1(t)| \leq 1.5$, $|x_2(t)| \leq 1.5$, $|r(t)| \leq 1$ hold for all $t \in [0, \infty)$. Find the control law of the form (7.91) such that $\bar{\varepsilon}_r = 0$, $t_s^d \approx 3$ s, $\sigma^d \approx 0\%$. Run a computer simulation of the closed-loop system with zero initial conditions. Compare simulation results of the output response with the assignment for $r(t) = 1,\ \forall\, t > 0$.

7.11 Verify the invertibility and internal stability of the system

$$\begin{aligned}
\dot{x}_1 &= x_3 - x_1^7(x_1^4 + 2x_3^2 - 1), \\
\dot{x}_2 &= x_2^2 + x_3^2 - (x_1^2 + x_3^2 + 0.2)u, \\
\dot{x}_3 &= -x_1^3 - 3x_3^5(x_1^4 + 2x_3^2 - 1) + x_2, \\
\dot{x}_4 &= x_2, \qquad y = x_4.
\end{aligned} \tag{7.146}$$

Assume that the inequalities $|x_i(t)| \leq 1\ \forall\, i$, $|r(t)| \leq 0.5$ hold for all $t \in [0, \infty)$. Find the control law of the form (7.91) such that $\bar{\varepsilon}_r = 0$, $t_s^d \approx 3$ s, $\sigma^d \approx 10\%$. Run a computer simulation of the closed-loop system with zero initial conditions. Compare simulation results of the output response with the assignment for $r(t) = 0.5,\ \forall\, t > 0$.

7.12 Solve Exercise 7.10 such that $e_{osc} \leq 0.1$, where the switching controller is given by

$$\begin{aligned}
\mu u_1^{(1)} + d_{10}u &= k_0\{T^{-1}[r - y] - y^{(1)}\}, \\
u_2(t) &= u_1(t - \tau), \\
u &= u_{\max}\,\mathrm{sgn}(u_2).
\end{aligned}$$

7.13 Solve Exercise 7.11 such that $e_{osc} \leq 0.01$, where the switching controller is given by

$$\begin{aligned}
\mu^2 u_1^{(2)} &+ d_1\mu u_1^{(1)} + d_0 u_1 \\
&= k_0\left\{-y^{(2)} - \frac{a_1^d}{T}y^{(1)} + \frac{b_1^d \tau}{T^2}r^{(1)} + \frac{1}{T^2}[r - y]\right\}, \\
u_2(t) &= u_1(t - \tau), \\
u &= u_{\max}\,\mathrm{sgn}(u_2).
\end{aligned}$$

Chapter 8

Design of MIMO continuous-time control systems

This chapter is devoted to extending the design methodology for SISO control systems with the highest derivative in feedback to the multi-input multi-output (MIMO) nonlinear time-varying case. In particular, the problem of output regulation is discussed, where the goals of MIMO control system design are to provide (1) output decoupling and disturbance rejection, i.e., each output should be independently controlled by a single input, and (2) the desired output transient performance indices on the condition of incomplete information about varying parameters of the plant model and unknown external disturbances. The control law structure with the relative highest derivatives in feedback is used in order to provide the desired dynamical properties globally in a specified region of the state space. A systematic design procedure for the control laws with the relative highest output derivatives in feedback is presented.

Note that throughout this chapter the problem of output regulation is discussed on the assumption that the previously presented realizability of the desired output behavior is maintained.

8.1 MIMO system without internal dynamics

8.1.1 *MIMO system with identical relative degrees*

The final purpose of this chapter is a controller design methodology for MIMO nonlinear time-varying systems given by

$$\dot{X} = f(t, X) + G(t, X)u, \quad X(0) = X^0, \tag{8.1}$$

$$y = h(t, X), \tag{8.2}$$

where t is time, $t \in [0, \infty)$; X is the state vector, $X = \{x_1, x_2, \dots, x_n\}^T$; y is the output, $y = \{y_1, y_2, \dots, y_p\}^T$; u is the input, $u = \{u_1, u_2, \dots, u_p\}^T$; $u \subseteq \Omega_u \subset \mathbb{R}^p$, Ω_u is an allowable bounded set for the control variable u; and $p \leq n$.

Assume the vector functions $f(t, X), h(t, X)$ and the matrix $G(t, X)$ are smooth with respect to t and X (or are at least n-times differentiable) in a specified bounded region of the state space Ω_X. The initial state X^0 of the system (8.1) is within the specified bounded region Ω_X^0, where $\Omega_X^0 \subseteq \Omega_X \subset \mathbb{R}^n$.

The influences of all external disturbances and varying parameters of the system (8.1)–(8.2) are represented implicitly by the time dependences of $f(t, X), h(t, X)$, and $G(t, X)$.

Note that throughout this chapter we assume identical dimensions for the vectors u and y, while the other case will be discussed in the next chapter.

First, for the sake of simplicity, let us consider the nonlinear time-varying system given by

$$y^{(n_0)} = f(t, Y_0) + G(t, Y_0)u, \quad Y_0(0) = Y_0^0, \tag{8.3}$$

where y is the measurable output of the system, $y \in \mathbb{R}^p$; Y_0 is the state vector, $Y_0 = \{y^T, [y^T]^{(1)}, [y^T]^{(2)}, \dots, [y^T]^{(n_0-1)}\}^T$; u is the control, $u \in \mathbb{R}^p$; Y_0^0 is the initial state, $Y_0^0 \in \Omega_{Y_0}^0$; $\Omega_{Y_0}^0$ is the specified bounded region, $\Omega_{Y_0}^0 \subseteq \Omega_{Y_0} \subset \mathbb{R}^{n_0 p}$; and $n_0 p = n$.

The system (8.3) has the following special features:

(i) It has identical relative degrees for each output component $y_i(t)$ of the vector $y(t)$, i.e., $\alpha_i = n_0, \ \forall \ i = 1, \dots, p$.
(ii) There is no internal subsystem (7.43) in the normal form representation, since

$$\sum_{i=1}^{p} \alpha_i = n_0 p = n.$$

Assumption 8.1 The functions $f(t, Y_0)$ and $G(t, Y_0)$ are bounded in Ω_{Y_0}, and

$$\det G(t, Y_0) \neq 0, \quad \forall \ (t, Y_0) \in \Omega_{t, Y_0}, \tag{8.4}$$

where $\Omega_{t, Y_0} = [0, \infty) \times \Omega_{Y_0}$. That is, the condition (7.20) (see p. 157) is satisfied and the system (8.3) is invertible.

Remark 8.1 *Let us consider, for instance, a Lagrangian system given by*

$$M(\boldsymbol{q})\boldsymbol{q}^{(2)} + C(\boldsymbol{q}, \boldsymbol{q}^{(1)}) = \boldsymbol{u}.$$

This is the typical model for a rigid link manipulator, where $\boldsymbol{q}$ is the vector of joint variables; $M(\boldsymbol{q})$ is the nonsingular inertia matrix; $C(\boldsymbol{q}, \boldsymbol{q}^{(1)})$ accounts for centripetal, Coriolis, and gravitational torques; $\boldsymbol{u}$ is the vector of input torques [Spong and Vidyasagar (1989); Slotine and Li (1991)]. The model presented is the counterpart of (8.3) for $n_0 = 2$.

8.1.2 MIMO system with different relative degrees

Let us consider the next particular case of the nonlinear time-varying control systems (8.1),(8.2) given by

$$\begin{aligned} \frac{d}{dt}y_{io} &= y_{i1}, \\ \frac{d}{dt}y_{i1} &= y_{i2}, \\ \cdots &\quad \cdots \\ \frac{d}{dt}y_{i,\alpha_i-2} &= y_{i,\alpha_i-1}, \\ \frac{d}{dt}y_{i,\alpha_i-1} &= h_i^*(t,Y) + g_i^*(t,Y)u, \end{aligned} \tag{8.5}$$

where

$$\sum_{i=1}^{p} \alpha_i = n \tag{8.6}$$

and

$$Y = \{y_{10}, y_{11}, \ldots, y_{1,\alpha_1-1}, y_{20}, y_{21}, \ldots, y_{2,\alpha_2-1}, \ldots, y_{p,\alpha_p-1}\}^T, \tag{8.7}$$

$$y_i = y_{io}, \quad \forall\ i = 1, \ldots, p.$$

The system (8.5) has the following special features:

(i) It may have different relative degrees.
(ii) There is no internal subsystem (7.43) in the normal form representation. This follows from (8.6).

Let us rewrite (8.5), for convenience, as

$$y_i^{(\alpha_i)} = h_i^*(t,Y) + g_i^*(t,Y)u, \quad i = 1, \ldots, p. \tag{8.8}$$

The representation

$$y_* = H^*(t,Y) + G^*(t,Y)u \tag{8.9}$$

of (8.8) will be used hereafter, where y_* is the vector (7.14),

$$y_* = \{y_1^{(\alpha_1)}, y_2^{(\alpha_2)}, \ldots, y_p^{(\alpha_p)}\}^T,$$

and

$$H^*(t,Y) = \begin{bmatrix} h_1^*(t,Y) \\ h_2^*(t,Y) \\ \cdots \\ h_p^*(t,Y) \end{bmatrix}, \quad G^*(t,Y) = \begin{bmatrix} g_1^*(t,Y) \\ g_2^*(t,Y) \\ \cdots \\ g_p^*(t,Y) \end{bmatrix}. \tag{8.10}$$

Assumption 8.2 We have

$$\det G^*(t,Y) \neq 0, \quad \forall\ (t,Y) \in \Omega_{t,Y}, \tag{8.11}$$

i.e., the system (8.9) is invertible [Porter (1970)].

8.2 MIMO control system design (identical relative degrees)

8.2.1 *Insensitivity condition*

We wish to design a control system for which

$$\lim_{t\to\infty} e(t) = 0, \tag{8.12}$$

where $e(t)$ is the tracking error, $e(t) = r(t) - y(t)$, $r(t)$ is the reference input, $r = \{r_1, r_2, \ldots, r_p\}^T$. Moreover, the controlled transients of all components of the output vector $y(t)$ should have desired performance indices. These transients should not depend on the external disturbances and varying parameters of the system (8.9).

First, let us consider the system given by (8.3) where the vector $y^{(n_0)}$ depends explicitly on the control $u(t)$. From (8.4) it follows that the desired behavior of the highest derivative vector $y^{(n_0)}$ can be provided by the control $u(t)$. Assume that the reference model of the desired output behavior is given by

$$y^{(n_0)} = F(Y_0, R), \tag{8.13}$$

where $R = \{r^T, [r^{(1)}]^T, \ldots, [r^{(\rho)}]^T\}^T$, $\rho < n_0$, and the parameters of (8.13) are chosen in accordance with the requirement (8.12) and the desired time-domain specifications of the stable output behavior of $y(t)$.

In order to ensure decoupling of the controlled transients of y_i for all components of the output vector $y(t)$, the reference model (8.13) of the desired output behavior is constructed in the form

$$y_i^{(n_0)} = F_i(Y_{i0}, R_i), \quad i = 1, \ldots, p, \tag{8.14}$$

where

$$Y_{i0} = \{y_i, y_i^{(1)}, y_i^{(2)}, \ldots, y_i^{(n_0-1)}\}^T; \; R_i = \{r_i, r_i^{(1)}, \ldots, r_i^{(\rho_i)}\}^T; \; \rho_i < n_0.$$

It is more convenient to construct (8.14) in the form of (5.11) or (2.8).

Denote by

$$e_0^F = F(Y_0, R) - y^{(n_0)} \tag{8.15}$$

the realization error of the desired dynamics, where $e_0^F \in \mathbb{R}^p$.

By introducing the reference model (8.13), the control problem (8.12) is reformulated as the requirement

$$e_0^F = 0. \tag{8.16}$$

This is the insensitivity condition for the behavior of the output $y(t)$ with respect to parameter variations and external disturbances in (8.3). It is also the condition for decoupling of the output transients if (8.14) is satisfied.

In accordance with (8.3), (8.13), and (8.15), the expression (8.16) can be considered as the equation

$$F(Y_0, R) - f(t, Y_0) - G(t, Y_0)u = 0. \tag{8.17}$$

From (8.17) it follows that the function

$$u^{NID} = \{G(t, Y_0)\}^{-1}\{F(Y_0, R) - f(t, Y_0)\} \tag{8.18}$$

is the nonlinear inverse dynamics solution.

Remark 8.2 *Condition (8.4) is necessary and sufficient for existence of a solution of (8.16), without taking the boundedness of Ω_u into consideration, that is $\Omega_u = \mathbb{R}^p$.*

Remark 8.3 *Expression (8.18) cannot be used as the control law function in practice, because $G(t, Y_0)$ and $f(t, Y_0)$ are unknown. But from a theoretical viewpoint this type of control function is widely discussed [Silverman (1969); Slotine and Li (1991)].*

8.2.2 *Control system with the relative highest derivatives in feedback*

Just as for the SISO case (4.38), in order to fulfill the requirement of (8.16) let us consider the control law in the form of the following system of differential equations:

$$\begin{aligned} D_q\boldsymbol{\mu}^q u^{(q)} + D_{q-1}\boldsymbol{\mu}^{q-1}u^{(q-1)} + \cdots + D_1\boldsymbol{\mu}u^{(1)} \\ +D_0 u = Ke_0^F,\ U(0) = U^0. \end{aligned} \tag{8.19}$$

Here $q \geq n_0$ and $U^0 = \{u^T(0), [u^T]^{(1)}(0), \ldots, [u^T]^{(q-1)}(0)\}^T$ is the initial condition.

Assume, for simplicity, that $\boldsymbol{\mu}$ and D_j are diagonal matrices:

$$\boldsymbol{\mu} = \operatorname{diag}\{\mu_1, \ldots, \mu_p\},\ D_j = \operatorname{diag}\{d_{1j}, \ldots, d_{pj}\},\ j = 0, \ldots, q. \tag{8.20}$$

Note that the degrees of the control law differential equations may differ between control channels, for instance, $d_{iq} \in \{0, 1\}$, $\forall\ i = 1, \ldots, p$ in the diagonal matrix D_q.

Assume that the reference model (8.13) is given by a system of linear differential equations. Then (8.19) has the form

$$\begin{aligned} &D_q\boldsymbol{\mu}^q u^{(q)} + D_{q-1}\boldsymbol{\mu}^{q-1}u^{(q-1)} + \cdots + D_1\boldsymbol{\mu}u^{(1)} + D_0 u \\ &= K\{-y^{(n_0)} - A^d_{n_0-1}y^{(n_0-1)} - \cdots - A^d_1 y^{(1)} - A^d_0 y \\ &\quad +B^d_\rho r^{(\rho)} + B^d_{\rho-1}r^{(\rho-1)} + \cdots + B^d_1 r^{(1)} + B^d_0 r\}, \end{aligned} \tag{8.21}$$

where $A^d_0 = B^d_0$ and $q \geq n_0 \geq \rho$. The expression (8.21) reduces to (4.41) (see p. 67) for $p = 1$.

Remark 8.4 *Note that the particular feature of the control law (8.21) is that the vector of the highest derivative $y^{(n_0)}$ is used in feedback.*

Remark 8.5 *If $q \geq n_0$ and $q \geq \rho$, then the control law (8.21) corresponds to a proper matrix transfer function and, therefore, may be realized without an ideal differentiation of $y(t)$ or $r(t)$.*

8.2.3 Fast-motion subsystem

From (8.3), (8.13), and (8.19), we obtain the closed-loop system equations

$$y^{(n_0)} = f(t, Y_0) + G(t, Y_0)u, \; Y_0(0) = Y_0^0, \tag{8.22}$$

$$D_q \boldsymbol{\mu}^q u^{(q)} + D_{q-1}\boldsymbol{\mu}^{q-1}u^{(q-1)} + \cdots + D_1\boldsymbol{\mu}u^{(1)} + D_0 u = K e_0^F, \; U(0) = U^0. \tag{8.23}$$

In accordance with (8.3), (8.13), and (8.15), (8.22)–(8.23) can be rewritten in the form

$$y^{(n_0)} = f(t, Y_0) + G(t, Y_0)u, \; Y_0(0) = Y_0^0, \tag{8.24}$$

$$D_q \boldsymbol{\mu}^q u^{(q)} + D_{q-1}\boldsymbol{\mu}^{q-1}u^{(q-1)} + \cdots + D_1\boldsymbol{\mu}u^{(1)} + \{D_0 + KG(t, Y_0)\}u = K\{F(Y_0, R) - f(t, Y_0)\}, \; U(0) = U^0. \tag{8.25}$$

Let all elements of the diagonal matrix $\boldsymbol{\mu} = \text{diag}\{\mu_1, \mu_2, \ldots, \mu_p\}$ be small positive parameters. Then (8.24)–(8.25) is the system of singularly perturbed differential equations where fast and slow modes are induced as $\boldsymbol{\mu} \to 0$.

Remark 8.6 *Note that the particular feature of the system (8.24)–(8.25) is that instead of a single small parameter, there are p distinct small parameters μ_i in the general case. Therefore, in order to implement the standard procedure [Tikhonov (1952)] of two-time-scale separation, let us provide the normalization $\boldsymbol{\mu} = \mu\, diag\{\bar{\mu}_1, \ldots, \bar{\mu}_p\}$ by introducing the single small parameter μ where*

$$\mu = \max_{i=1,\ldots,p} \{\mu_1, \ldots, \mu_p\}. \tag{8.26}$$

Further, let us introduce the new fast time scale $t_0 = t/\mu$ into the closed-loop system (8.24)–(8.25) and find the limit as $\mu \to 0$. Then, by returning to the primary time scale $t = \mu t_0$, we obtain the following FMS:

$$D_q \boldsymbol{\mu}^q u^{(q)} + D_{q-1}\boldsymbol{\mu}^{q-1}u^{(q-1)} + \cdots + D_1\boldsymbol{\mu}u^{(1)} + \{D_0 + KG(t, Y_0)\}u = K\{F(Y_0, R) - f(t, Y_0)\}, \; U(0) = U^0, \tag{8.27}$$

where Y_0 is the vector of the frozen variables during the transients in (8.27), i.e., $Y_0(t) = \text{const}$.

Remark 8.7 *Note that if $\boldsymbol{\mu} \to 0$, then during the fast transients in (8.27) the condition $\{D_0 + KG(t, Y_0)\} \approx const$ holds. This occurs since the rate of external disturbance and parameter variations in the plant model is bounded due to the smoothness assumption on $G(t, Y_0)$, and $\boldsymbol{\mu}$ can be selected as an* *arbitrary small value. Hence, the FMS (8.27) may be examined as the linear system with the frozen parameters of the matrix* $\{D_0 + KG(t, Y_0)\}$, *where the vectors* $F(Y_0, R)$ *and* $f(t, Y_0)$ *play the role of external disturbances.*

Assumption 8.3 The stability and sufficiently rapid decay of the FMS transients in (8.27) are provided for all $(t, Y_0) \in \Omega_{t,Y_0}$ by selection of the control law parameters $D_j, K, \boldsymbol{\mu}$.

In accordance with Assumption 8.3, let us consider the steady state (more precisely, quasi-steady state) of the FMS (8.27). As $\boldsymbol{\mu} \to 0$ in (8.27) we obtain

$$u(t) = u^s(t),$$

where

$$u^s = \{D_0 + KG(t, Y_0)\}^{-1} K \{F(Y_0, R) - f(t, Y_0)\}. \tag{8.28}$$

From (8.18) and (8.28) it follows that

$$u^s = u^{NID} + \{D_0 + KG(t, Y_0)\}^{-1} D_0 G^{-1}(t, Y_0) \{f(t, Y_0) - F(Y_0, R)\}, \tag{8.29}$$

where $u^s(t)$ is the value of the control in the closed-loop system which corresponds to the quasi-steady state of the FMS (8.27). Proof of (8.29) see in Appendix A.1.

For simplicity, let us assume that the matrix K has the form

$$K = k_0 K_0, \tag{8.30}$$

where k_0 is the gain, $k_0 > 0$. Note that the matching matrix K_0 is defined in accordance with the requirement of FMS stability; for instance, let $K_0 \approx G^{-1}$. Then, by taking into account (8.30), let us rewrite (8.29) in the form

$$u^s = u^{NID}$$
$$+ k_0^{-1} \{k_0^{-1} D_0 + K_0 G(t, Y_0)\}^{-1} D_0 \{G(t, Y_0)\}^{-1} \{f(t, Y_0) - F(Y_0, R)\}, \tag{8.31}$$

where we have $u^s(t) - u^{NID}(t) \to 0$ as $k_0 \to \infty$.

In particular, from (8.29) the equality $u^s(t) - u^{NID}(t) = 0$ follows if $D_0 = 0$.

8.2.4 *Slow-motion subsystem*

In order to obtain the SMS, assume that the FMS (8.27) is stable and let $\mu \to 0$ in (8.24)–(8.25). As a result, from (8.24) and (8.25) we find that

$$y^{(n_0)} = F(Y_0, R) + G(t, Y_0)\{D_0 + KG(t, Y_0)\}^{-1} \times D_0\{G(t, Y_0)\}^{-1}\{f(t, Y_0) - F(Y_0, R)\} \tag{8.32}$$

is the SMS, where $Y_0(0) = Y_0^0$ and (8.32) corresponds to the steady state of the FMS (8.27).

Remark 8.8 *Note that the SMS (8.32) can also be obtained by direct substitution of (8.31) into (8.3).*

Assume that the condition (8.30) for the matrix K holds. Then the SMS (8.32) can be rewritten in the form

$$y^{(n_0)} = F(Y_0, R) + k_0^{-1}G(t, Y_0)\{k_0^{-1}D_0 + K_0G(t, Y_0)\}^{-1} \times D_0\{G(t, Y_0)\}^{-1}\{f(t, Y_0) - F(Y_0, R)\}, \tag{8.33}$$

where $Y_0(0) = Y_0^0$. From (8.33) it follows that the error of the desired dynamics realization is given by

$$e_0^F = k_0^{-1}G\{k_0^{-1}D_0 + K_0G\}^{-1}D_0G^{-1}\{F - f\}$$

if the steady state of the FMS (8.27) takes place.

The main results of this examination can be expressed by the following theorems.

Theorem 8.1 *The fast and slow motions are induced in the system (8.3) with control law (8.19) as $\mu \to 0$, where the equation of the FMS subsystem is given by (8.27).*

Theorem 8.2 *If the FMS (8.27) is stable and $\mu \to 0$, then the SMS is given by (8.32).*

Theorem 8.3 *If $\det D_0 \neq 0$, then the SMS equation (8.32) tends to the reference equation (8.13) as $k_0 \to \infty$. If $D_0 = 0$, then (8.32) is identical to (8.13) and, accordingly, the integral action is incorporated in the control loop; that is, robust zero steady-state error of the output regulation is provided.*

So, the discussed dynamical controller with the relative highest derivatives in feedback induces the two-time-scale separation of the fast and slow modes in the closed-loop system, where after damping of the stabilized fast transients for $u(t)$, the behavior of the output $\boldsymbol{y}(t)$ corresponds to the reference model (8.14) and is insensitive to external disturbances and variations in the parameters of the system (8.3).

8.2.5 *Control system design with zero steady-state error*

From (8.32) it follows that $e_0^F = 0$ if $D_0 = 0$; that is, the SMS has the form

$$y^{(n_0)} = F(Y_0, R)$$

for any $k_0 > 0$. So, the integral action is incorporated in the control loop for all components of the output vector $y(t)$ simultaneously.

Let us consider a slight modification of the above control law structure, which allows us to include the integral action in the control loop of an arbitrarily chosen subset of the output variables $y_i(t)$, where

$$i = i_1, i_2, \ldots, i_\zeta$$

and $\zeta \leq p$.

Let us introduce an auxiliary control vector $\tilde{u}(t)$ and a control law structure of the form

$$D_q \boldsymbol{\mu}^q \tilde{u}^{(q)} + D_{q-1} \boldsymbol{\mu}^{q-1} \tilde{u}^{(q-1)} + \cdots + D_1 \boldsymbol{\mu} \tilde{u}^{(1)} + D_0 \tilde{u} = K_1 e_0^F, \qquad (8.34)$$

where

$$K_1 = \operatorname{diag}\{k_1, k_2, \ldots, k_p\} \qquad (8.35)$$

and the actual control vector $u(t)$ depends on $\tilde{u}(t)$ through

$$u = K_0 \tilde{u}. \qquad (8.36)$$

The closed-loop system equations become

$$y^{(n_0)} = f(t, Y_0) + G(t, Y_0) K_0 \tilde{u}, \quad Y_0(0) = Y_0^0, \qquad (8.37)$$

$$D_q \boldsymbol{\mu}^q \tilde{u}^{(q)} + D_{q-1} \boldsymbol{\mu}^{q-1} \tilde{u}^{(q-1)} + \cdots + D_1 \boldsymbol{\mu} \tilde{u}^{(1)} + D_0 \tilde{u} = K_1 e_0^F, \; \tilde{U}(0) = \tilde{U}^0. \qquad (8.38)$$

In accordance with (8.3), (8.13), and (8.15), the closed-loop system equations (8.37), (8.38) can be rewritten as

$$y^{(n_0)} = f(t, Y_0) + G(t, Y_0) K_0 \tilde{u}, \quad Y_0(0) = Y_0^0, \qquad (8.39)$$

$$D_q \boldsymbol{\mu}^q \tilde{u}^{(q)} + D_{q-1} \boldsymbol{\mu}^{q-1} \tilde{u}^{(q-1)} + \cdots + D_1 \boldsymbol{\mu} \tilde{u}^{(1)} + \{D_0 + K_1 G(t, Y_0) K_0\} \tilde{u} = K_1 \{F(Y_0, R) - f(t, Y_0)\}, \; \tilde{U}(0) = \tilde{U}^0. \qquad (8.40)$$

From (8.39), (8.40) the FMS

$$D_q\boldsymbol{\mu}^q\tilde{u}^{(q)} + D_{q-1}\boldsymbol{\mu}^{q-1}\tilde{u}^{(q-1)} + \cdots + D_1\boldsymbol{\mu}\tilde{u}^{(1)}$$
$$+\{D_0 + K_1G(t, Y_0)K_0\}\tilde{u} = K_1\{F(Y_0, R) - f(t, Y_0)\}, \; \tilde{U}(0) = \tilde{U}^0 \quad (8.41)$$

follows, where it is assumed that $G(t, Y_0) = \text{const}$ during the transients in the FMS (8.41).

Remark 8.9 *The matching matrix K_0 is selected in accordance with the requirement for stability of the FMS (8.41). For instance, $K_0 \approx \{G(t, Y_0)\}^T$ or, preferably, $K_0 \approx \{G(t, Y_0)\}^{-1}$.*

Remark 8.10 *If $K_0 = \{G(t, Y_0)\}^{-1}$ and all matrices $\boldsymbol{\mu}, D_0, \ldots, D_q$ are diagonal in form, then the decoupling of the FMS (8.41) into p mutually independent fast-motion subsystems for the control variables $u_i(t)$ is provided. As a result, the design of controller parameters for each control loop can be done in succession using the above design procedures for SISO systems.*

Assumption 8.4 The stability and sufficiently rapid decay of the FMS transients in (8.41) are provided for all $(t, Y) \in \Omega_{t,Y_0}$ by selection of the parameters D_j, K_1, K_0, and $\boldsymbol{\mu}$ in the control law (8.41).

In accordance with Assumption 8.4, let us consider the steady state (more precisely, quasi-steady state) of the FMS (8.41). To that end, by taking $\boldsymbol{\mu} \to 0$ in (8.41), we obtain

$$\tilde{u}(t) = \tilde{u}^s(t),$$

where

$$\tilde{u}^s = \tilde{u}^{NID}$$
$$+\{GK_0\}^{-1}K_1^{-1}D_0\{K_1^{-1}D_0 + GK_0\}^{-1}\{f(t, Y_0) - F(Y_0, R)\} \quad (8.42)$$

and

$$\tilde{u}^{NID} = \{G(t, Y_0(t))K_0\}^{-1}\{F(Y_0, R) - f(t, Y_0)\}. \quad (8.43)$$

Proof of (8.42) see in Appendix A.2.

Assume that the steady state of the FMS (8.41) takes place. Substitution of (8.42) into (8.39) yields the SMS of the form

$$y^{(n_0)} = F(Y_0, R)$$
$$+K_1^{-1}D_0\{K_1^{-1}D_0 + G(t, Y_0)K_0\}^{-1}\{f(t, Y_0) - F(Y_0, R)\}, \quad (8.44)$$

where $Y_0(0) = Y_0^0$. Finally, from (8.44) it follows that the error of the desired dynamic realization e_0^F is given by

$$e_0^F = K_1^{-1}D_0\{K_1^{-1}D_0 + G(t, Y_0)K_0\}^{-1}\{F(Y_0, R) - f(t, Y_0)\}. \quad (8.45)$$

In accordance with (8.20) and (8.35), the matrices K_1 and D_0 are diagonal. By (8.45) we have $e_{i0}^F = 0$ for all $i = i_1, i_2, \ldots, i_\varsigma$ if $d_{i0} = 0$ where $e_0^F = \{e_{10}^F, e_{20}^F, \ldots, e_{p0}^F\}^T$.

It follows that the modified control law (8.34) and (8.36) allows us to incorporate the integral action into the control loop for any arbitrarily chosen output variable $y_i(t)$ without increasing the controller's order, so that the robust zero steady-state error of the reference input realization is provided in the closed-loop system.

8.2.6 *Example*

Let us consider a 3-dimensional, articulated manipulator with the kinematic scheme shown in Fig. 8.1.

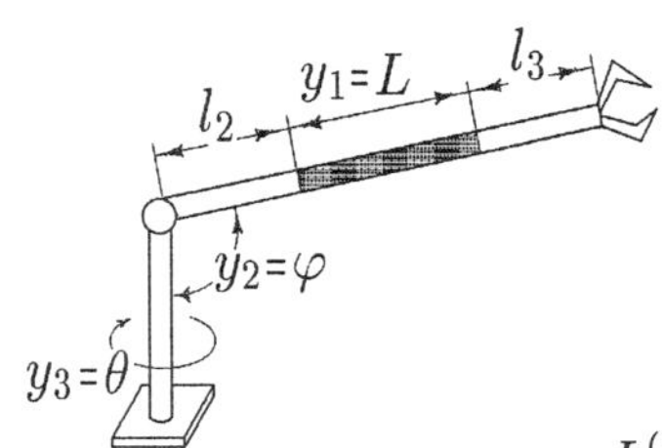

Fig. 8.1 Kinematic scheme of the robot manipulator.

$$\begin{aligned} m_3L^{(2)} - m_3\, g\, \sin(\theta) &= u_1, \\ \hat{q}\varphi^{(2)} + 0.5\, m_3\, L^{(1)}\, \varphi^{(1)} &= u_2, \\ \hat{q}^2\, \cos^2(\theta)\, \theta^{(2)} - \hat{q}^2\, \sin(\theta)\, \theta^{-2} & \\ +\hat{q}\, m_3\, \cos(\theta)\, \theta^{(1)}\, L^{(1)} - g\, \hat{q}\, \cos(\theta) &= u_3, \end{aligned} \quad (8.46)$$

where $\{y_1, y_2, y_3\} = \{L, \varphi, \theta\}$ is the output vector, $\{u_1, u_2, u_3\}$ is the control vector (torques), and $\hat{q} = l_1m_2 + (l_2 + L + l_3)m_3$.

Assume that the parameters of (8.46) are as follows: $l_1 = 0.25$ m; $l_2 = l_3 = 0.5$ m; $m_2 = m_3 = 5$ kg; $g = 9.8$ m/s^2.

In accordance with (8.34) and (8.36), let us consider the control law given by

$$\mu_i^2 u_i^{(2)} + d_{i1}\mu_i u_i^{(1)} + d_{i0}u_i$$
$$= k_i\left\{-y_i^{(2)} - \frac{a_{i1}^d}{T_i}y_i^{(1)} + \frac{b_{i1}^d}{T_i}r_i^{(1)} + \frac{1}{T_i^2}[r_i - y_i]\right\}, \quad (8.47)$$

where $i = 1, 2, 3$ and $K_0 = \text{diag}\{m_3,\ \hat{q},\ \hat{q}^2 \cos^2(\theta)\}$. Assume the parameters of (8.47) are as follows:

$$T_i = 1 \text{ s},\ a_{i1}^d = b_{i1}^d = 2,\ d_{i0} = 0, \quad \forall\, i = 1, 2, 3,$$
$$\mu_1 = \mu_2 = 0.5 \text{ s},\ \mu_3 = 0.4 \text{ s},\ d_{11} = 7,\ d_{21} = 8,\ d_{31} = 16,$$
$$k_1 = k_2 = 20,\ k_3 = 40.$$

Simulation results for the system (8.46) controlled by the algorithm (8.47) for a ramp reference input $r(t)$ are displayed in Figs. 8.2–8.4, where the initial conditions are all zero and $dr_1/dt = dr_2/dt = dr_3/dt = 0.02$, $t \in [0, 8]$ s.

Note that the allowable error of the reference input realization can be provided by decreasing the parameters μ_1, μ_2, μ_3.

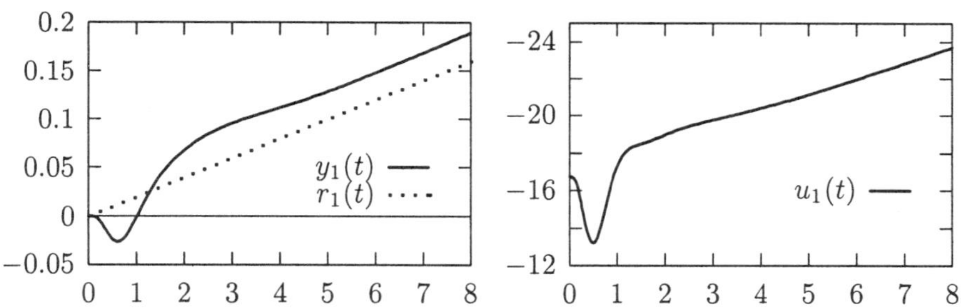

Fig. 8.2 Simulation results for $r_1(t), y_1(t), u_1(t)$ in the system (8.46)–(8.47).

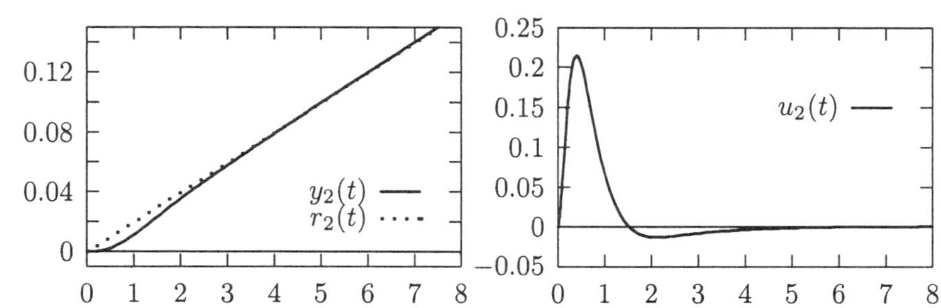

Fig. 8.3 Simulation results for $r_2(t), y_2(t), u_2(t)$ in the system (8.46)–(8.47).

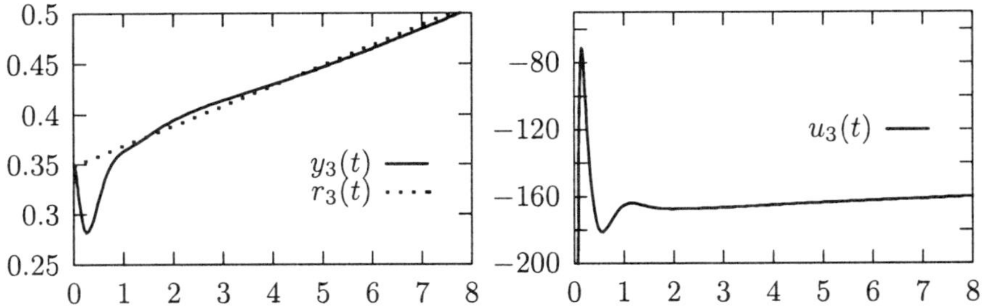

Fig. 8.4 Simulation results for $r_3(t), y_3(t), u_3(t)$ in the system (8.46)–(8.47).

8.3 MIMO control system design (different relative degrees)

8.3.1 *Insensitivity condition and control law structure*

Next, let us consider controller design with the relative highest output derivatives in feedback for nonlinear time-varying systems given by (8.5), where there is no internal subsystem in the normal form of the state space representation since (8.6) is satisfied.

From (8.9) and (8.11) we know that the vector y_* is given by (7.13), and that it depends explicitly on the control $u(t)$. Consequently, the desired behavior of the relative highest derivative vector y_* can be provided by selection of the control $u(t)$. Let us construct the reference model for this desired behavior in the form (7.25):

$$y_* = F(Y, R),$$

where the reference model for each variable $y_i(t)$ has the form (7.23).

In accordance with (7.26) we know that

$$\mathbf{e}^F = F(Y, R) - y_*$$

measures the output behavior deviation of (8.5) from the desired behavior assigned by (7.25). Hence the solution of the output regulation problem corresponds to the condition (7.27),

$$\mathbf{e}^F = 0,$$

which is the insensitivity condition for the behavior of the output $y(t)$ with respect to parameter variations and external disturbances in (8.5).

In order to fulfill the requirement of (7.27), let us consider a control law in the form of the system of differential equations

$$D_q\boldsymbol{\mu}^q\tilde{u}^{(q)} + D_{q-1}\boldsymbol{\mu}^{q-1}\tilde{u}^{(q-1)} + \cdots + D_1\boldsymbol{\mu}\tilde{u}^{(1)} + D_0\tilde{u} = K_1\boldsymbol{e}^F, \qquad (8.48)$$

where

$$u = K_0\tilde{u}, \quad K_1 = \operatorname{diag}\{k_1, k_2, \ldots, k_p\}. \qquad (8.49)$$

8.3.2 *Closed-loop system analysis*

From (8.9) and (8.48), we get the closed-loop system equations

$$y_* = H^*(t,Y) + G^*(t,Y)K_0\tilde{u}, \ Y(0) = Y^0, \qquad (8.50)$$

$$D_q\boldsymbol{\mu}^q\tilde{u}^{(q)} + D_{q-1}\boldsymbol{\mu}^{q-1}\tilde{u}^{(q-1)} + \cdots + D_1\boldsymbol{\mu}\tilde{u}^{(1)} + D_0\tilde{u} = K_1\boldsymbol{e}^F, \ \tilde{U}(0) = \tilde{U}^0. \qquad (8.51)$$

In accordance with (7.25), (7.26), and (8.9), the system (8.50)–(8.51) can be rewritten in the form

$$y_* = H^*(t,Y) + G^*(t,Y)K_0\tilde{u}, \ Y(0) = Y^0, \qquad (8.52)$$

$$D_q\boldsymbol{\mu}^q\tilde{u}^{(q)} + D_{q-1}\boldsymbol{\mu}^{q-1}\tilde{u}^{(q-1)} + \cdots + D_1\boldsymbol{\mu}\tilde{u}^{(1)} + \{D_0 + K_1G^*(t,Y)K_0\}\tilde{u} = K_1\{F(Y,R) - H^*(t,Y)\}, \ \tilde{U}(0) = \tilde{U}^0. \qquad (8.53)$$

By (8.52)–(8.53) and Remark 8.6, the FMS is given by

$$D_q\boldsymbol{\mu}^q\tilde{u}^{(q)} + D_{q-1}\boldsymbol{\mu}^{q-1}\tilde{u}^{(q-1)} + \cdots + D_1\boldsymbol{\mu}\tilde{u}^{(1)} + \{D_0 + K_1G^*(t,Y)K_0\}\tilde{u} = K_1\{F(Y,R) - H^*(t,Y)\}, \ \tilde{U}(0) = \tilde{U}^0. \qquad (8.54)$$

Here, by Remark 8.7, we have $G^*(t,Y) = \text{const}$ during the transients in the FMS (8.54).

Similar to what was done above, assume that the stability and sufficiently rapid decay of the FMS transients in (8.54) are provided for all $(t,Y) \in \Omega_{t,Y}$ by selection of the control law parameters D_i, K_1, K_0, $\boldsymbol{\mu}$. Then consider the quasi-steady state of the FMS (8.54). To that end, by taking $\boldsymbol{\mu} \to 0$ in (8.54), we obtain

$$\tilde{u}(t) = \tilde{u}^s(t),$$

where

$$\tilde{u}^s = \tilde{u}^{NID} + \{G^*K_0\}^{-1}K_1^{-1}D_0\{K_1^{-1}D_0 + G^*K_0\}^{-1}\{H^*(t,Y) - F(Y,R)\} \qquad (8.55)$$

and

$$\tilde{u}^{NID} = \{G^*(t,Y)K_0\}^{-1}\{F(Y,R) - H^*(t,Y)\}.$$

From (8.55) it follows that $\tilde{u}^s(t) - \tilde{u}^{NID}(t) \to 0$ as $k_i \to \infty$, $\forall\ i = 1,\dots,p$. Moreover $\tilde{u}^s(t) - \tilde{u}^{NID}(t) = 0$ if $D_0 = 0$.

Let the quasi-steady state of the FMS (8.54) occur. Then substitution of (8.55) into (8.52) yields the SMS of the form

$$\begin{aligned} y_* = F(Y,R) + K_1^{-1}D_0\{K_1^{-1}D_0 + G^*(t,Y)K_0\}^{-1} \\ \times\{H^*(t,Y) - F(Y,R)\}, \quad Y(0) = Y^0. \end{aligned} \tag{8.56}$$

From (8.56) we have

$$e^F = K_1^{-1}D_0\{K_1^{-1}D_0 + G^*(t,Y)K_0\}^{-1}\{F(Y,R) - H^*(t,Y)\}, \tag{8.57}$$

where

$$K_1 = \text{diag}\{k_1, k_2, \dots, k_p\}, \quad D_0 = \text{diag}\{d_{10}, d_{20}, \dots, d_{p0}\}. \tag{8.58}$$

In accordance with (8.57) and (8.58), the condition $e_i^F = 0$ holds for all $i = i_1, i_2, \dots, i_\varsigma$ if $d_{i0} = 0$, where $\mathbf{e}^F = \{e_1^F, e_2^F, \dots, e_p^F\}^T$. Since the property (7.24) of the reference model equation (7.23) is maintained, the discussed control law allows us to incorporate the integral action into the control loop without increasing the controller's order, that is, the robust zero steady-state error of the reference input realization is fulfilled.

So, the dynamical controller (8.48) with the relative highest output derivatives in feedback induces the two-time-scale separation of the fast and slow modes in the closed-loop system (8.52)–(8.53) as $\mu \to 0$ and, after damping of the stabilized fast transients in (8.54), the behavior of the output $\mathbf{y}(t)$ in the SMS (8.56) corresponds to that of the reference model (7.25) and is insensitive to external disturbances and parameter variations in the system (8.5).

The main results of the above consideration may be expressed by the following theorems.

Theorem 8.4 *The fast and slow motions are induced in the system (8.5) with control law (8.48) as $\mu \to 0$, where the FMS equation is given by (8.54).*

Theorem 8.5 *If the FMS (8.54) is stable and $\mu \to 0$, then the SMS is given by (8.56).*

Theorem 8.6 *Let (8.58) hold and $d_{i0} \neq 0$. Then the equation for $y_i(t)$ in (8.56) tends to the reference equation (7.23) as $k_i \to \infty$. If $d_{i0} = 0$, then the equation for $y_i(t)$ in (8.56) is the same as the reference equation (7.23) and, accordingly, the integral action is incorporated in the control loop; that is, robust zero steady-state error is provided.*

8.3.3 *Control accuracy*

First, let us consider the realization error of the desired dynamics assigned by (7.25) in the closed-loop system (8.50), (8.51) given that the steady state of the FMS (8.54) is reached. Then from the expression for the SMS (8.56), in particular, by (8.57), we can select matrices K_1 and D_0 for which

$$\|e^F\| \leq e^F_{\max}, \quad \forall\ (t, Y) \in \Omega_{t,Y},$$

where $e^F_{\max} > 0$. In particular, for the sake of simplicity, let us assume that

$$K_0 = \{G^*\}^{-1}. \tag{8.59}$$

By this, the decomposition of (8.57) to the expressions

$$e_i^F = \frac{d_{i0}}{d_{i0} + k_i}\{F_i - h_i^\star\}, \quad i = 1, \ldots, p \tag{8.60}$$

is provided, where (8.60) is the counterpart of (5.2).

Second, let us consider the steady-state error in the closed-loop system (8.50)–(8.51), that is, on the assumption of steady state of the SMS (8.56).

Let

$$r = \text{const} \quad \text{and} \quad y(t) = y^s,$$

where y^s is the steady state of the output $y(t)$, i.e.,

$$y^s = \lim_{t \to \infty} y(t).$$

Denote

$$e^s = r - y^s,$$

where e^s is the steady-state error under the condition that the steady state of the SMS (8.56) takes place. Let the reference model of the desired output behavior of $y_i(t)$ be given by a linear differential equation of the form

$$\begin{aligned} &T_i^{\alpha_i} y_i^{(\alpha_i)} + a^d_{i,\alpha_i-1} T_i^{\alpha_i-1} y_i^{(\alpha_i-1)} + \cdots + a^d_{i,1} T_i y_i^{(1)} + y_i \\ &= b^d_{i,\rho_i} \tau_i^{\rho_i} r_i^{(\rho_i)} + b^d_{i,\rho_i-1} \tau_i^{\rho_i-1} r_i^{(\rho_i-1)} + \cdots + b^d_{i,1} \tau_i r_i^{(1)} + r_i, \end{aligned} \tag{8.61}$$

where $i = 1, \ldots, p$. Then the steady state of (8.61) for all $i = 1, \ldots, p$ corresponds to the following relationships:

$$y_i^{(j)} \overset{\forall\, t}{\equiv} 0,\ \forall\, j = 1, \ldots, \alpha_i,\ \forall\, i = 1, \ldots, p \tag{8.62}$$

and

$$r_i^{(\chi)} \overset{\forall\, t}{\equiv} 0,\ \forall\, \chi = 1, \ldots, \rho_i,\ \forall\, i = 1, \ldots, p. \tag{8.63}$$

From (8.61), (8.62), and (8.63), we obtain

$$F = \mathcal{T}^{-1} e^s, \tag{8.64}$$

where

$$\mathcal{T} = \mathrm{diag}\{T_1^{\alpha_1}, T_2^{\alpha_2}, \ldots, T_p^{\alpha_p}\}.$$

From (8.57), (8.62), (8.63), and (8.64), we can obtain

$$\begin{aligned} e^s = & -\mathcal{T}\,\{K_1^{-1}D_0 + G^* K_0\}\{G^* K_0\}^{-1} K_1^{-1} D_0 \\ & \times \{K_1^{-1} D_0 + G^* K_0\}^{-1} H_s^*. \end{aligned} \tag{8.65}$$

Proof of (8.65) see in Appendix A.3.

Expression (8.65) allows us to evaluate the error e^s of the reference input realization for the steady state of the SMS (8.56), and to select K_1 such that the assigned restriction

$$\|e^s\| \leq e^s_{\mathrm{max}}, \quad \forall\, (t, Y) \in \Omega_{t,Y}$$

is fulfilled where $e^s_{\mathrm{max}} > 0$.

For instance, from (8.59) and (8.65) we obtain

$$e_i^s = -\frac{d_{i0}\ T_i^{\alpha_i}}{k_i} h_{i\,s}^*, \quad \forall\, j = 1, \ldots, p, \tag{8.66}$$

where (8.66) is the counterpart of (5.5).

Remark 8.11 *From (8.57) and (8.65) it follows that under the condition (8.59) each gain k_i can be evaluated independently for all $i = 1, \ldots, p$ in order to reach the desired control accuracy.*

Note that the expressions (8.60) and (8.66) can be easily used for selection of K_1 given that the domain $\Omega_{t,Y}$ is assigned.

8.4 MIMO control system in presence of internal dynamics

In the previous sections the design methodology for MIMO control systems was discussed for nonlinear time-varying systems given by (8.3) or (8.5), where there is no internal subsystem (7.43) in the normal form representation. This section is devoted to design of MIMO control systems with the relative highest output derivatives in feedback for generalized systems of the form (8.1)–(8.2):

$$\dot{X} = f(t, X) + G(t, X)u, \quad X(0) = X^0,$$
$$y = h(t, X),$$

where $X \in \mathbb{R}^n$, $u \in \mathbb{R}^m$, $y \in \mathbb{R}^p$, $p = m \leq n$. The problem of output regulation for $y(t)$ with prescribed performance indices (such as overshoot σ_i^d, settling time t_i^d, and system type) is the subject of this section.

It is assumed that $p = m$; the effects of having $p < m$ will be discussed in the next chapter.

Let us assume that by procedure [Porter (1970)], the expression (7.13)

$$y_* = H^*(t, X) + G^*(t, X)u$$

is derived, where y_* is the vector of the form (7.14).

Assumption 8.5 The sufficient invertibility condition (7.20),

$$\det G^*(t, X) \neq 0, \quad \forall\ (t, X) \in \Omega_{t,X}$$

is fulfilled for the given system (8.1)–(8.2).

Let the reference model of the form (7.25) be constructed in accordance with the time-domain specifications on the desired output behavior. As a result, the problem of output regulation has been reformulated as the insensitivity condition (7.27):

$$\boldsymbol{e}^F = 0.$$

Assumption 8.6 The internal stability of the system (8.1)–(8.2) is maintained; that is, condition (7.32) is satisfied in a specified region of the state space of the system (8.1)–(8.2), given that the condition $e^F = 0$ is fulfilled.

Similar to (8.48) and (8.49), let us introduce an auxiliary vector $\tilde{u}(t)$

and consider the control law given by

$$\begin{aligned}\mu_i^{q_i}\tilde{u}_i^{(q_i)} &+ d_{i,q_i-1}\mu_i^{q_i-1}\tilde{u}_i^{(q_i-1)} + \cdots + d_{i,1}\mu\tilde{u}_i^{(1)} + d_{i,0}\tilde{u}_i \\ &= k_i e_i^F, \quad \tilde{U}_i(0) = \tilde{U}_i^0, \quad i = 1, \ldots, p, \end{aligned} \tag{8.67}$$

where

$$\mu_i > 0, \quad k_i > 0, \quad \tilde{U}_i = \{\tilde{u}_i, \tilde{u}_i^{(1)}, \ldots, \tilde{u}_i^{(q_i-1)}\}^T, \quad q_i \geq \alpha_i.$$

We can see that the control law (8.67) is the perfect counterpart of (4.38). Let us rewrite (8.67) in the vector form

$$\begin{gathered} D_q\boldsymbol{\mu}^q\tilde{u}^{(q)} + D_{q-1}\boldsymbol{\mu}^{q-1}\tilde{u}^{(q-1)} + \cdots + D_1\boldsymbol{\mu}\tilde{u}^{(1)} + D_0\tilde{u} = K_1\mathbf{e}^F, \\ \tilde{U}(0) = \tilde{U}^0, \end{gathered} \tag{8.68}$$

where $\boldsymbol{\mu}$, K_1, and D_j are the diagonal matrices given by (8.20)–(8.35).

As a result, the closed-loop system equations become

$$\dot{X} = f(t, X) + G(t, X)K_0\tilde{u}, \quad X(0) = X^0, \tag{8.69}$$

$$D_q\boldsymbol{\mu}^q\tilde{u}^{(q)} + D_{q-1}\boldsymbol{\mu}^{q-1}\tilde{u}^{(q-1)} + \cdots + D_1\boldsymbol{\mu}\tilde{u}^{(1)} + D_0\tilde{u}$$

$$= K_1\mathbf{e}^F, \quad \tilde{U}(0) = \tilde{U}^0. \tag{8.70}$$

The general block diagram representation of the closed-loop MIMO control system, described by (8.69)–(8.70), is shown in Fig. 8.5. We can see that the controller of the MIMO system consists of p linear controllers $\mathrm{C}_1, \ldots, \mathrm{C}_p$ generating the auxiliary control vector $\tilde{u}(t)$ and accompanied by the matching matrix K_0. The linear controllers are described by (8.67).

Note that information exchange between the outputs of the MIMO system is used in order to generate the actual control vector $u(t)$ due to the presence of the matching matrix K_0. Therefore, this controller is called the centralized output feedback controller with the relative highest derivatives in feedback.

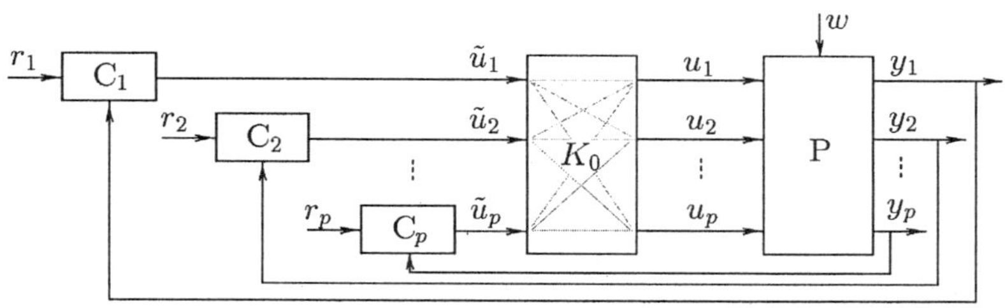

Fig. 8.5 General block diagram representation of the closed-loop MIMO control system (8.69)–(8.70).

8.4.1 *Fast-motion subsystem*

By taking into account the expressions (7.13), (7.25), and (7.26), we can rewrite the closed-loop system equations (8.69)–(8.70) in the form

$$\dot{X} = f(t, X) + G(t, X)K_0\tilde{u}, \quad X(0) = X^0, \tag{8.71}$$

$$\begin{aligned} D_q\boldsymbol{\mu}^q\tilde{u}^{(q)} &+ D_{q-1}\boldsymbol{\mu}^{q-1}\tilde{u}^{(q-1)} + \cdots + D_1\boldsymbol{\mu}\tilde{u}^{(1)} + \Gamma(t, X)\tilde{u} \\ &= K_1\{F - H^*(t, X)\}, \quad \tilde{U}(0) = \tilde{U}^0, \end{aligned} \tag{8.72}$$

where

$$\Gamma(t, X) = D_0 + K_1G^*(t, X)K_0.$$

We already know that, by Remark 8.6 (see p. 195), the standard procedure can be applied to derive the FMS from the closed-loop system (8.71)–(8.72). That is, the new fast time scale $t_0 = t/\mu$ is introduced into (8.71)–(8.72) and the limit $\mu \to 0$ is taken. Then, by returning to the primary time scale $t = \mu t_0$, we obtain the FMS

$$\begin{aligned} D_q\boldsymbol{\mu}^q\tilde{u}^{(q)} &+ D_{q-1}\boldsymbol{\mu}^{q-1}\tilde{u}^{(q-1)} + \cdots + D_1\boldsymbol{\mu}\tilde{u}^{(1)} + \Gamma(t, X)\tilde{u} \\ &= K_1\{F - H^*(t, X)\}, \quad \tilde{U}(0) = \tilde{U}^0. \end{aligned} \tag{8.73}$$

Here, by Remark 8.7, we have $\Gamma(t, X) = \text{const}$ during the transients in (8.73). So the FMS (8.73) may be considered as the linear system with the frozen elements of the matrix $\Gamma(t, X)$, where the vectors F and $H^*(t, X)$ play the role of external disturbances.

Assumption 8.7 Stability and sufficiently rapid decay of the FMS transients in (8.73) are provided for all $(t, X) \in \Omega_{t,X}$ by selection of the control law parameters D_j, K, and $\boldsymbol{\mu}$.

Under the condition of Assumption 8.7, let us consider the steady state (more precisely, quasi-steady state) of the FMS (8.73). By taking $\mu \to 0$ in (8.73), we obtain the control function

$$u(t) = u^s(t)$$

in the closed-loop system for the quasi-steady state of the FMS (8.27), where

$$\tilde{u}^s = \Gamma^{-1} K_1 \{F - H^*\}. \tag{8.74}$$

From (7.29), (8.36), and (8.74), it follows that

$$\tilde{u}^s = \tilde{u}^{NID} + \{G^* K_0\}^{-1} K_1^{-1} D_0 \{K_1^{-1} D_0 + G^* K_0\}^{-1} \{H^*(t, X) - F\}, \tag{8.75}$$

where $\tilde{u}^{NID}(t) = K_0^{-1} u^{NID}(t)$ and $u^{NID}(t)$ is given by (7.29) (see p. 159).

If (8.59) holds, then

$$\tilde{u}_i^{NID} = F_i - h_i^*, \quad \forall\, i = 1, \ldots, p \tag{8.76}$$

follows from (7.29).

The result can be expressed by the following theorem.

Theorem 8.7 *If Assumptions 8.5 and 8.6 are satisfied, and the closed-loop system is given by (8.71)–(8.72), then two-time-scale motions are induced in the closed-loop system as $\mu \to 0$ and the behavior of the FMS is described by (8.73).*

8.4.2 *Slow-motion subsystem*

Let Assumption 8.7 holds and consider the steady state of the FMS (8.73). Passing to the limit $\mu \to 0$ in (8.71)–(8.72), we obtain the SMS given by

$$\begin{aligned} \dot{X} &= f + G\{G^*\}^{-1}\{F - H^*\} \\ &+ G\{G^*\}^{-1} K_1^{-1} D_0 \{K_1^{-1} D_0 + G^* K_0\}^{-1} \{H^* - F\}, \quad X(0) = X^0. \end{aligned} \tag{8.77}$$

From (7.13), (8.36), and (8.75), the SMS of the output behavior

$$\begin{aligned} y_* &= F(Y, R) \\ &+ K_1^{-1} D_0 \{K_1^{-1} D_0 + G^*(t, X) K_0\}^{-1} \{H^*(t, X) - F\} \end{aligned} \tag{8.78}$$

results. Note that, on one hand, by (8.20), (8.35), and (8.78), the equation for $y_i(t)$ tends to the reference equation (7.23) as $k_i \to \infty$ if $d_{i,0} = 1$. On the other hand, the equation for $y_i(t)$ is the same as the reference equation (7.23) if $d_{i,0} = 0$.

The main results of the above can be expressed by the following theorems.

Theorem 8.8 *If in the closed-loop system (8.71)–(8.72) the FMS (8.73) is stable and $\mu \to 0$, then the SMS is given by (8.77) and the output behavior is described by (8.78).*

Theorem 8.9 *Let (8.20) and (8.35) hold and $d_{i0} \neq 0$. Then the equation for $y_i(t)$ in (8.78) tends to the reference equation (7.23) as $k_i \to \infty$. If $d_{i0} = 0$, then the equation for $y_i(t)$ in (8.78) is identical to the reference equation (7.23) and, accordingly, the integral action is incorporated in the control loop by virtue of (7.24). That is, robust zero steady-state error is provided.*

Remark 8.12 *Note that the SMS (8.77) tends to, or is the same as, the system (7.30). Therefore, Assumption 8.6 regarding internal stability is the essential point to obtain the desired input-output mapping, since the observable effect of the internal dynamics can be canceled in (8.78) only if those dynamics are smooth and bounded in a specified region of the state space.*

8.4.3 *Example*

Let us consider the system (7.73), where the realizability of the desired output behavior is satisfied in a specified region of the state space as shown above (see p. 170).

In accordance with the discussed design methodology, let us find the relative degrees of (7.73). We obtain $\alpha_1 = 2$ and $\alpha_2 = 1$. Let $q_1 = q_2 = 2$. Then a control law structure can be chosen such as

$$\mu_1^2\tilde{u}_1^{(2)} + d_{11}\mu_1\tilde{u}_1^{(1)} + d_{10}\tilde{u}_1 = k_1\left\{-y_1^{(2)} - \frac{a_{11}^d}{T_1}y_1^{(1)} + \frac{b_{11}^d}{T_1}r^{(1)} + \frac{1}{T_1^2}[r_1 - y_1]\right\}, \tag{8.79}$$

$$\mu_2^2\tilde{u}_2^{(2)} + d_{21}\mu_2\tilde{u}_2^{(1)} + d_{20}\tilde{u}_2 = k_2\left\{-y_2^{(1)} + \frac{1}{T_2}[r_2 - y_2]\right\}. \tag{8.80}$$

In accordance with (7.74), (8.36), and Remark 8.9, let us assume that

$$K_0 = \begin{bmatrix} -1/3 & 1/3 \\ 2/3 & 1/3 \end{bmatrix} \tag{8.81}$$

and consider the following values of the controller parameters:

$$T_1 = T_2 = 1 \text{ s}, \quad \mu_1 = \mu_2 = 0.3 \text{ s}, \quad d_{10} = d_{20} = 0,$$
$$a_{11}^d = 2, \quad b_{11}^d = 0, \quad d_{11} = 6, \quad d_{21} = 8, \quad k_1 = 30, \quad k_2 = 50.$$

Simulation results for the system (7.73) and (8.79)–(8.81) appear in Figs. 8.6–8.8, where $t \in [0, 15]$ s.

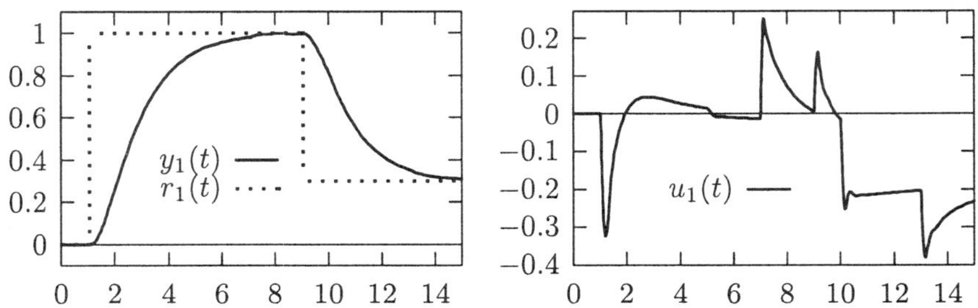

Fig. 8.6 Simulation results for $r_1(t), y_1(t), u_1(t)$ in the system (7.73) and (8.79)–(8.81).

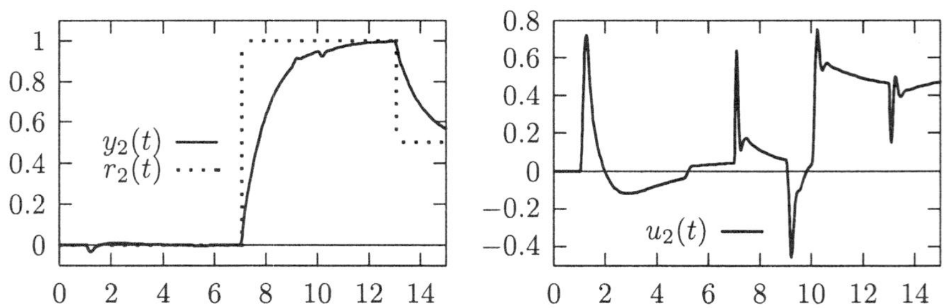

Fig. 8.7 Simulation results for $r_2(t), y_2(t), u_2(t)$ in the system (7.73) and (8.79)–(8.81).

8.5 Decentralized output feedback controller

The problem of decentralized controller design for large-scale interconnected linear and nonlinear systems is considered in a broad set of references, e.g., [Davison (1976); Siljak (1991); Tang *et al.* (2000); Zhong-Ping Jiang *et al.* (2001); Narendra and Oleng (2002)]. The decentralized output feedback control scheme corresponds to the system shown in Fig. 8.9. The particular feature of this scheme is that information exchange between the outputs of

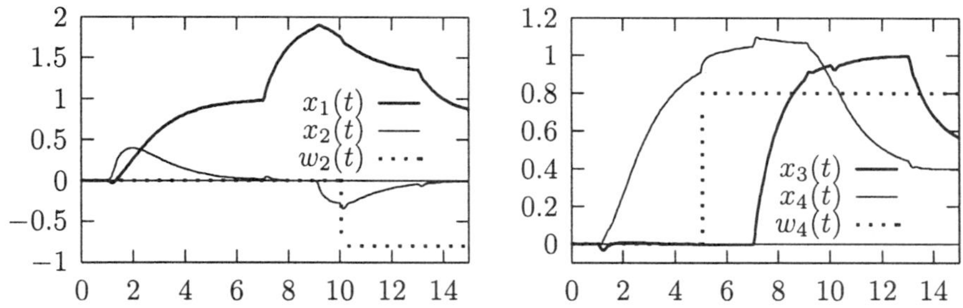

Fig. 8.8 Simulation results for $x_1(t), x_2(t), x_3(t), x_4(t), w_2(t), w_4(t)$ in the system (7.73) and (8.79)–(8.81).

the MIMO system is not used. Hence, the controller of the MIMO system is decomposed into a set of p independent controllers.

From the above consideration we can see that the centralized output feedback controller with the relative highest derivatives in feedback for an MIMO system is given by the system of equations (8.67) generating the auxiliary control vector $\tilde{u}$ and accompanied by the matching matrix K_0 as shown in Fig. 8.5. It is clear that with $K_0 = I_p$ this scheme yields that of the decentralized output feedback controller in Fig. 8.9. So, in order to enable use of the decentralized output feedback controller with the relative highest derivatives in feedback for the MIMO system given by (8.1)–(8.2), we must be able to provide the stability of the FMS (8.73) given that the condition $K_0 = I_p$ holds.

Note that any permutation of the rows or columns of the identity matrix $K_0 = I_p$ maintains the decentralized scheme of the controller; at the same time, permutations have an influence on the stability of the FMS.

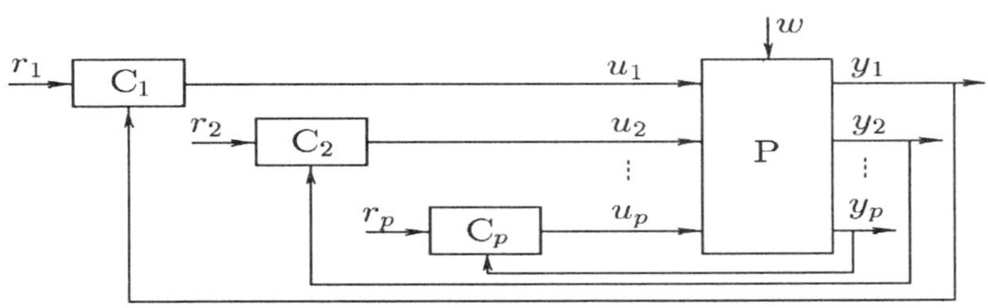

Fig. 8.9 Block diagram of the system with decentralized output feedback controller.

8.6 Notes

In order to design a feedback controller to stabilize a nonlinear system, a two-step approach is widely used: a state feedback controller is designed, and then a high-gain observer is constructed to estimate the state of the nonlinear system [Esfandiary and Khalil (1992); Khalil (1994); Teel and Praly (1995); Isidori (1995); Atassi and Khalil (1999)].

In contrast to the systems with state feedback controller [Isidori and Byrnes (1990); Nijmeijer and Schaft (1990); Esfandiary and Khalil (1992)], the discussed approach to control system design is based on the use of the relative highest output derivative (or derivative of the state of a nonlinear system) in the feedback loop [Vostrikov and Sarycheva (1982); Vostrikov *et al.* (1982); Vostrikov and Yurkevich (1991); Vostrikov and Yurkevich (1993a)].

The presented design procedure represents the development of results published in [Yurkevich (1994); Yurkevich (1995b)], and allows us to provide a desired output behavior for an MIMO control system despite the presence of unknown external disturbances and varying system parameters. The resulting dynamical output feedback controller with the relative highest derivatives of the output signal in feedback has a simple form consisting of p separate low-order linear dynamical filters accompanied by a matrix K_0. The proposed dynamical controller with the sufficiently small parameters μ_i induces the two-time-scale separation of the fast and slow modes in the closed-loop system, where, after damping of the stabilized fast transients, the behavior of the output vector $y(t)$ is desired and insensitive to variation of parameters of the plant model and external disturbances.

The main advantage of this method is that knowledge of the relative degrees and the matrix G^* suffice for controller design given that the previously presented realizability of the desired output behavior is satisfied. We need not know how the system parameters and external disturbances vary, or how they enter into the system dynamics.

The problem of flight controller design for an aircraft based on the presented approach was discussed in [Blachuta *et al.* (1995); Blachuta *et al.* (1999)], while the design of two-input two-output control system for a reactive ion etching (RIE) system was discussed in [Tudoroiu *et al.* (2003b)].

8.7 Exercises

8.1 Verify the invertibility and internal stability of the system given by

$$\dot{x}_1 = x_1 + 3x_2 + u_1 + u_2, \quad \dot{x}_2 = x_1 + x_2 + u_1 - 2u_2,$$
$$y_1 = x_1 + x_2, \quad y_2 = -x_1 + 2x_2.$$

Assume that the inequalities $|x_j(t)| \leq 2 \ \forall \ j$ and $|r(t)| \leq 1$ hold for all $t \in [0, \infty)$. Find the control law of the form (8.67) such that $\bar{\varepsilon}_{r1} = 0$, $\bar{\varepsilon}_{r2} = 0$, $t_{s1}^d \approx 1$ s, $\sigma_1^d \approx 0\%$, $t_{s2}^d \approx 3$ s, $\sigma_2^d \approx 0\%$. Compare simulation results for the step response of the closed-loop control system with the assignment.

8.2 Verify the invertibility and internal stability of the system given by

$$\dot{x}_1 = x_1^2 + x_1 x_2 + 0.5u_1 - \{1 + 0.2\sin(t)\}u_2,$$
$$\dot{x}_2 = x_1 + \sin(x_2) - u_1 - 2\{1 + 0.5\sin(2t)\}u_2,$$
$$y_1 = x_1 - x_2, \quad y_2 = x_1 + x_2.$$

Assume that the inequalities $|x_j(t)| \leq 2 \ \forall \ j$ and $|r(t)| \leq 1$ hold for all $t \in [0, \infty)$. Find the control law of the form (8.67) such that $\bar{\varepsilon}_{r1} = 0$, $\bar{\varepsilon}_{r2} = 0$, $t_{s1}^d \approx 3$ s, $\sigma_1^d \approx 0\%$, $t_{s2}^d \approx 3$ s, $\sigma_2^d \approx 0\%$. Compare simulation results for the step response of the closed-loop control system with the assignment.

8.3 The system is given by (7.141). Verify invertibility and internal stability. Assume that the inequalities $|x_j(t)| \leq 2 \ \forall \ j$ and $|r(t)| \leq 1$ hold for all $t \in [0, \infty)$. Find the control law of the form (8.67) such that $\bar{\varepsilon}_{r1} = 0$, $\bar{\varepsilon}_{r2} = 0$, $t_{s1}^d \approx 3$ s, $\sigma_1^d \approx 0\%$, $t_{s2}^d \approx 3$ s, $\sigma_2^d \approx 0\%$. Compare simulation results for the step response of the closed-loop control system with the assignment.

8.4 The system is given by (7.142). Verify invertibility and internal stability. Find the control law of the form (8.67) such that $\bar{\varepsilon}_{r1} = 0$, $\bar{\varepsilon}_{r2} = 0$, $t_{s1}^d \approx 2$ s, $\sigma_1^d \approx 0\%$, $t_{s2}^d \approx 1$ s, $\sigma_2^d \approx 0\%$. Compare simulation results for the step response of the closed-loop control system with the assignment.

8.5 The system is given by (7.143). Verify invertibility and internal stability. Assume that the inequalities $|x_j(t)| \leq 2 \ \forall \ j$ and $|r(t)| \leq 1$ hold for all $t \in [0, \infty)$. Find the control law of the form (8.67) such that $\bar{\varepsilon}_{r1} = 0$, $\bar{\varepsilon}_{r2} = 0$, $t_{s1}^d \approx 3$ s, $\sigma_1^d \approx 0\%$, $t_{s2}^d \approx 3$ s, $\sigma_2^d \approx 0\%$. Compare simulation results for the step response of the closed-loop control system with the assignment.

8.6 Verify the invertibility and internal stability of the system given by

$$\dot{x}_1 = x_1^2 + x_1 x_2 + \{1 + 0.5\sin(t)\}u_1 + 0.2u_2,$$
$$\dot{x}_2 = x_1 + \sin(x_2) - 0.3\sin(2t)u_1 - 2u_2,$$
$$y_1 = x_1 + x_2, \quad y_2 = x_1 + 2x_2.$$

Assume that the inequalities $|x_j(t)| \leq 2 \; \forall \; j$ and $|r(t)| \leq 1$ hold for all $t \in [0, \infty)$. Find the control law of the form (8.67) such that $\bar{\varepsilon}_{r1} = 0$, $\bar{\varepsilon}_{r2} = 0$, $t_{s1}^d \approx 3$ s, $\sigma_1^d \approx 0\%$, $t_{s2}^d \approx 3$ s, $\sigma_2^d \approx 0\%$. Compare simulation results for the step response of the closed-loop control system with the assignment.

8.7 Consider the system given by

$$\dot{x}_1 = x_1(1 - x_2^2) + g_{11}u_1 + g_{12}u_2,$$
$$\dot{x}_2 = x_2(1 - x_1^2) + g_{21}u_1 + g_{22}u_2,$$
$$y_1 = x_1, \quad y_2 = x_2.$$

Express by a formula the existence conditions for a decentralized output feedback controller of the form (8.67), where $q_1 = q_2 = 1$.

Chapter 9

Stabilization of internal dynamics

In the preceding chapter, the problem of output regulation was discussed for nonlinear time-varying systems with identical dimensions of the input and output vectors u, y. It was also assumed that the realizability of the desired behavior takes place, in particular, that the system is invertible and internal stability is fulfilled. This assumption is the essential point for controller design, and provides the range of applicability of the method.

This chapter is devoted to consideration of control system design where the dimension of the control vector u is as large as that of the output vector y. Redundant control variables enable internal dynamics stabilization, and this is the main subject matter. Note that the problem of internal dynamics stabilization for linear time-invariant (LTI) systems corresponds to the displacement of all zeroes of the transfer function in the left half of the complex plane.

9.1 Zero placement by redundant control

First, let us consider an LTI system given by

$$\dot{X} = AX + Bu, \quad X(0) = X^0, \tag{9.1}$$

$$y = CX, \tag{9.2}$$

where $y \in \mathbb{R}^p$, $u \in \mathbb{R}^m$, $m > p$, and

$$\operatorname{rank} CB = p. \tag{9.3}$$

That is, the right inverse of the system (9.1)–(9.2) exists.

The transformation (7.50) of the system (9.1)–(9.2) yields the normal form (7.52)–(7.53):

$$\dot{y} = A_{11}y + A_{12}z + B_1u, \tag{9.4}$$

$$\dot{z} = A_{21}y + A_{22}z + B_2u. \tag{9.5}$$

Let us introduce a new control vector consisting of two parts, $u_1 \in \mathbb{R}^p$ and $u_2 \in \mathbb{R}^{m-p}$, such that

$$u = \begin{bmatrix} P_1 & P_2 \end{bmatrix} \begin{bmatrix} u_1 \\ u_2 \end{bmatrix}. \tag{9.6}$$

Denote

$$\begin{bmatrix} B_1 \\ B_2 \end{bmatrix} \begin{bmatrix} P_1 & P_2 \end{bmatrix} = \begin{bmatrix} B_{11} & B_{12} \\ B_{21} & B_{22} \end{bmatrix}. \tag{9.7}$$

Then (9.4), (9.5), and (9.6) can be rewritten in the form

$$\dot{y} = A_{11}y + A_{12}z + B_{11}u_1 + B_{12}u_2, \tag{9.8}$$

$$\dot{z} = A_{21}y + A_{22}z + B_{21}u_1 + B_{22}u_2. \tag{9.9}$$

From (9.3) it follows that the matrix

$$P = \begin{bmatrix} P_1 & P_2 \end{bmatrix} \tag{9.10}$$

exists such that the conditions

$$u_1 \in \mathbb{R}^p \quad \text{and} \quad \det B_{11} \neq 0 \tag{9.11}$$

are satisfied.

The block diagram representation of the discussed control system is shown in Fig. 9.1(a), where the control vector $u_1(t)$ is used to provide the desired output behavior of (9.1)–(9.2) and the redundant control $u_2(t)$ can be used to stabilize the behavior of the internal variables $z(t)$.

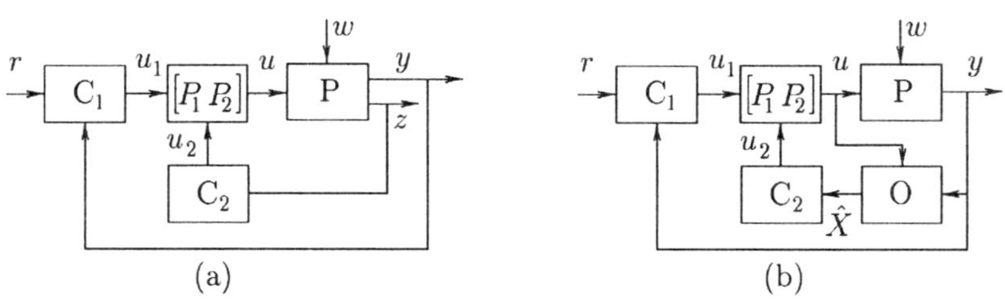

Fig. 9.1 Block diagram of the control system with the redundant control $u_2(t)$.

Let us consider the output regulation by the control vector $u_1(t)$ with the control law in the form (8.19):

$$D_q\mu^q u_1^{(q)} + D_{q-1}\mu^{q-1}u_1^{(q-1)} + \cdots + D_1\mu u_1^{(1)} \\ +D_0 u_1 = Ke^F, \quad U_1(0) = U_1^0, \tag{9.12}$$

where $q \geq 1$, $U_1 = \{u_1^T, [u_1^{(1)}]^T, \ldots, [u_1^{(q-1)}]^T\}^T$, and e^F is defined by (7.56). Then the closed-loop system is given by

$$\dot{y} = A_{11}y + A_{12}z + B_{11}u_1 + B_{12}u_2, \quad y(0) = y^0, \tag{9.13}$$

$$\dot{z} = A_{21}y + A_{22}z + B_{21}u_1 + B_{22}u_2, \quad z(0) = z^0, \tag{9.14}$$

$$D_q\mu^q u_1^{(q)} + D_{q-1}\mu^{q-1}u_1^{(q-1)} + \cdots \\ + D_1\mu u_1^{(1)} + D_0 u_1 = Ke^F, \quad U_1(0) = U_1^0, \tag{9.15}$$

where μ is the diagonal matrix of small parameters.

Further, let us introduce the new fast time scale $t_0 = t/\mu$ into the closed-loop system (9.13), (9.14), and (9.15) (see Remark 8.6 on p. 195), and find the limit as $\mu \to 0$. Then, by returning to the primary time scale $t = \mu t_0$, we obtain the FMS equation

$$D_q\mu^q u_1^{(q)} + D_{q-1}\mu^{q-1}u_1^{(q-1)} + \cdots + D_1\mu u_1^{(1)} + \{D_0 + KB_{11}\}u_1 \\ = K\{F(y,r) - A_{11}y - A_{12}z - B_{12}u_2\}, \quad U_1(0) = U_1^0 \tag{9.16}$$

as $\mu \to 0$, where y and z are the frozen variables.

For simplicity, assume the matrix K is given by $K = k_0 K_0$, where $k_0 > 0$. Assume also that the stability and sufficiently rapid decay of the FMS transients in (9.16) are provided by selection of the control law parameters D_j, K, μ. Let us consider the steady state (more precisely, quasi-steady state) of the FMS (9.16). In the limit as $\mu \to 0$ in (9.16), we obtain $u_1(t) = u_1^s(t)$ where

$$u_1^s = u_1^{LID} + k_0^{-1}\{k_0^{-1}D_0 + K_0 B_{11}\}^{-1} \\ \times D_0\{B_{11}\}^{-1}\{A_{11}y + A_{12}z + B_{12}u_2 - F(y,r)\} \tag{9.17}$$

and

$$u_1^{LID} = B_{11}^{-1}\{F(y,r) - A_{11}y - A_{12}z - B_{12}u_2\}. \tag{9.18}$$

From (9.17) we have

$$u_1^s(t) - u_1^{LID}(t) \to 0 \quad \text{as} \quad k_0 \to \infty$$

or $u_1^s = u_1^{LID}$ if the condition $D_0 = 0$ is satisfied.

In particular, let $D_0 = 0$. Then from the closed-loop system (9.13), (9.14), and (9.15), we get the SMS equations

$$\dot{y} = F(y, r), \tag{9.19}$$
$$\dot{z} = \bar{A}z + \bar{B}u_2 + \bar{f}_1(y, r), \tag{9.20}$$
$$u_1^{LID} = \bar{C}z + \bar{D}u_2 + \bar{f}_2(y, r), \tag{9.21}$$

where

$$\bar{A} = A_{22} - B_{21}B_{11}^{-1}A_{12}, \quad \bar{B} = B_{22} - B_{21}B_{11}^{-1}B_{12},$$

$$\bar{f}_1 = \{A_{21} - B_{21}B_{11}^{-1}A_{11}\}y + B_{21}B_{11}^{-1}F(y, r),$$

$$\bar{C} = -B_{11}^{-1}A_{12}, \quad \bar{D} = -B_{11}^{-1}B_{12}, \quad \bar{f}_2 = B_{11}^{-1}\{F(y, r) - A_{11}y\}.$$

We see that the desired stable output behavior assigned by (7.54) is maintained in the SMS system (9.19)–(9.21).

Let us consider (9.19)–(9.21) as the system where $u_2(t)$ is the control, $\bar{y} = u_1^{LID}$ is the output, and the vectors $\bar{f}_1(y, r)$ and $\bar{f}_2(y, r)$ play the role of external disturbances. Assume that the vector $z(t)$ is available for measurement and that $u_2 = u_2(z)$. Then from linear control system theory (see, e.g., [Brogan (1991)]) the following results are available.

Theorem 9.1 *Assume that the following conditions hold:*

(i) $z(t)$ is the vector available for measurement from the system described by (9.19), (9.20), and (9.21).
(ii) The pair $\{\bar{A}, \bar{B}\}$ is completely controllable.

Then there exists a matrix $\bar{K}$ such that the feedback control $u_2 = \bar{K}z$ allows the eigenvalues of $\{\bar{A} + \bar{B}\bar{K}\}$ to be arbitrarily assigned (or, in other words, allows the zeroes of $G_{yu_1}(s) = y(s)/u_1(s)$ to be arbitrarily assigned in the system (9.8)–(9.9)).

Theorem 9.2 *Assume that the following conditions hold:*

(i) The system is described by (9.19), (9.20), and (9.21), where $u_1^{LID}(t)$ is regarded as the output vector that is available for measurement.
(ii) The pair $\{\bar{C}, \bar{A}\}$ is completely observable.

Then an observer can be designed that will estimate the vector $z(t)$.

Theorem 9.3 *Suppose, for the system described by (9.19)–(9.21), the following conditions hold:*

(i) The pair $\{\bar{A}, \bar{B}\}$ is completely controllable.
(ii) The pair $\{\bar{C}, \bar{A}\}$ is completely observable.

Then the stability of the internal dynamics described by (9.20) can be provided.

Remark 9.1 *It is clear that the above approach for zero placement by redundant control variables can be extended to LTI systems (7.6)–(7.7), where right invertibility holds.*

Note that the above approach to internal dynamics stabilization via redundant control was considered for LTI systems for the sake of simplicity. It will be extended for nonlinear systems below.

9.2 Internal dynamics stabilization (particular case)

Consider the nonlinear time-varying system given by

$$\dot{X} = f(t, X) + G(t, X)u, \quad X(0) = X^0, \tag{9.22}$$

$$y = h(t, X), \tag{9.23}$$

where $X \in \mathbb{R}^n$, $y \in \mathbb{R}^p$, $u \in \mathbb{R}^m$, and $n \geq m > p$. First suppose the condition

$$\text{rank}\{\partial h(t, X)/\partial X\}G(t, X) = p, \quad \forall\ (t, X) \in \Omega_{t,X} \tag{9.24}$$

is maintained, where $\Omega_{t,X} = [0, \infty) \times \Omega_X$ and $\Omega_X \subset \mathbb{R}^n$.

It is clear that (9.24) is a particular case of the right invertibility condition given by (7.17), and corresponds to one in which the vector of the relative degree equals $\alpha = \{1, \ldots, 1\}$.

Let us consider the state space transformation (7.40) (see p. 163) where, in accordance with (9.24), we have

$$\begin{bmatrix} y \\ z \end{bmatrix} = Q(t, X) \tag{9.25}$$

and

$$Q(t, X) = \begin{bmatrix} Q_1(t, X) \\ Q_2(t, X) \end{bmatrix}, \quad Q_1(t, X) = h(t, X). \tag{9.26}$$

The vector function $Q_2(t, X)$ is chosen so that

$$\det\{\partial Q/\partial X\} \neq 0, \quad \forall\ (t, X)\ \in\ \Omega_{t,X}.$$

The transformation (9.26) with (9.6) of the system (9.22)–(9.23) gives the normal form representation

$$\dot{y} = f_1(t, y, z) + G_{11}(t, y, z)u_1 + G_{12}(t, y, z)u_2, \tag{9.27}$$

$$\dot{z} = f_2(t, y, z) + G_{21}(t, y, z)u_1 + G_{22}(t, y, z)u_2, \tag{9.28}$$

where the new control variable consists of two parts: $u_1 \in \mathbb{R}^p$ and $u_2 \in \mathbb{R}^{m-p}$. The transformation (9.6) is chosen so that

$$\det G_{11}(t, y, z) \neq 0, \quad \forall\ (t, y, z) \in \Omega_{t,y,z}.$$

The control vector $u_1(t)$ will be used later to provide the desired output behavior of the system (9.22)–(9.23). In addition, the redundant control variables $u_2(t)$ can be used to stabilize the behavior of the internal variables $z(t)$.

From (7.54) and (9.27) it follows that the solution of (7.55) is the NID control function given by

$$u_1^{NID}(t) = \{G_{11}(t, y, z)\}^{-1}\{F(y, r) - f_1(t, y, z) - G_{12}(t, y, z)u_2\}. \tag{9.29}$$

Substituting (9.29) into (9.27)–(9.28), we get

$$\dot{y} = F(y, r), \tag{9.30}$$

$$\dot{z} = f_2 + G_{21}G_{11}^{-1}\{F(y, r) - f_1\} + \{G_{22} - G_{21}G_{11}^{-1}\}u_2. \tag{9.31}$$

Here the desired stable output behavior assigned by (7.54) is satisfied. We see that the internal dynamics are represented by (9.31), where the redundant control variables $u_2(t)$ can be used for internal dynamics stabilization.

9.3 Internal dynamics stabilization (generalized case)

The above approach to internal dynamics stabilization by the redundant control variables can be extended for the nonlinear time-varying systems where the right invertibility condition is satisfied.

Consider the nonlinear time-varying system given by (9.22)–(9.23):

$$\dot{X} = f(t, X) + G(t, X)u, \quad X(0) = X^0,$$

$$y = h(t, X),$$

where $X \in \mathbb{R}^n$, $y \in \mathbb{R}^p$, $u \in \mathbb{R}^m$, and $n \geq m > p$. Assume that the right invertibility condition (7.17) holds.

The transformation (7.40) with (9.6) of the system (9.22)–(9.23) gives the following normal form representation:

$$y_* = H^*(t,Y,z) + G_1^*(t,Y,z)u_1 + G_2^*(t,Y,z)u_2, \tag{9.32}$$
$$\dot{z} = \tilde{f}(t,Y,z) + \tilde{G}_1(t,Y,z)u_1 + \tilde{G}_2(t,Y,z)u_2. \tag{9.33}$$

The system (9.32)–(9.33) is the counterpart of (7.44)–(7.45), where the control vector is divided into $u_1(t)$ and $u_2(t)$ such that $u_1 \in \mathbb{R}^p$ and

$$\det G_1^*(t,Y,z) \neq 0, \quad \forall\ (t,Y,z) \in \Omega_{t,Y,z}. \tag{9.34}$$

In order to provide the desired output behavior of (9.22)–(9.23), let us construct the control law of $u_1(t)$ in the form (8.68):

$$\begin{aligned} D_q\boldsymbol{\mu}^q\tilde{u}_1^{(q)} + D_{q-1}\boldsymbol{\mu}^{q-1}\tilde{u}_1^{(q-1)} + \cdots + D_1\boldsymbol{\mu}\tilde{u}_1^{(1)} + D_0\tilde{u}_1 \\ = K_1\boldsymbol{e}^F, \quad \tilde{U}_1(0) = \tilde{U}_1^0, \end{aligned} \tag{9.35}$$

where

$$u_1 = K_0\tilde{u}_1, \quad K_1 = \text{diag}\{k_1, k_2, \ldots, k_p\}.$$

Assume also that the vector $z(t)$ is available for measurement and $u_2 = u_2(z)$. Then the closed-loop system is given by

$$y_* = H^*(t,Y,z) + G_1^*(t,Y,z)K_0\tilde{u}_1 + G_2^*(t,Y,z)u_2(z), \tag{9.36}$$
$$\dot{z} = \tilde{f}(t,Y,z) + \tilde{G}_1(t,Y,z)K_0\tilde{u}_1 + \tilde{G}_2(t,Y,z)u_2(z), \tag{9.37}$$
$$D_q\boldsymbol{\mu}^q\tilde{u}_1^{(q)} + D_{q-1}\boldsymbol{\mu}^{q-1}\tilde{u}_1^{(q-1)} + \cdots + D_1\boldsymbol{\mu}\tilde{u}_1^{(1)} + D_0\tilde{u}_1 = K_1\boldsymbol{e}^F. \tag{9.38}$$

The system (9.36)–(9.38) can be rewritten in the form

$$y_* = H^*(t,Y,z) + G_1^*(t,Y,z)K_0\tilde{u}_1 + G_2^*(t,Y,z)u_2(z), \tag{9.39}$$
$$\dot{z} = \tilde{f}(t,Y,z) + \tilde{G}_1(t,Y,z)K_0\tilde{u}_1 + \tilde{G}_2(t,Y,z)u_2(z), \tag{9.40}$$
$$\begin{aligned} D_q\boldsymbol{\mu}^q\tilde{u}_1^{(q)} + D_{q-1}\boldsymbol{\mu}^{q-1}\tilde{u}_1^{(q-1)} + \cdots + D_1\boldsymbol{\mu}\tilde{u}_1^{(1)} + \Gamma_1(t,Y,z)\tilde{u}_1 \\ = K_1\{F - H^*(t,Y,z) - G_2^*(t,Y,z)u_2(z)\}, \end{aligned} \tag{9.41}$$

where

$$\Gamma_1(t,Y,z) = \{D_0 + K_1G_1^*(t,Y,z)K_0\}.$$

From (9.39)–(9.41), we obtain the FMS equation

$$\begin{aligned} D_q\boldsymbol{\mu}^q\tilde{u}_1^{(q)} + D_{q-1}\boldsymbol{\mu}^{q-1}\tilde{u}_1^{(q-1)} + \cdots + D_1\boldsymbol{\mu}\tilde{u}_1^{(1)} + \Gamma_1(t,Y,z)\tilde{u}_1 \\ = K_1\{F - H^*(t,Y,z) - G_2^*(t,Y,z)u_2(z)\}, \quad \tilde{U}_1(0) = \tilde{U}_1^0, \end{aligned} \tag{9.42}$$

where all elements of the matrix $\Gamma(t, Y, z)$ are frozen during the transients in the FMS (9.42).

Assumption 9.1 Stability and sufficiently rapid decay of the FMS transients in (9.42) are provided for all $(t, Y, z) \in \Omega_{t,Y,z}$ by selection of the control law parameters $D_j, K_1, K_0, \boldsymbol{\mu}$.

In accordance with Assumption 9.1, let us consider the steady state (more precisely, quasi-steady state) of the FMS (9.42). Letting $\boldsymbol{\mu} \to 0$ in (9.42) we obtain

$$\tilde{u}_1(t) = \tilde{u}_1^s(t),$$

where

$$\tilde{u}_1^s = \Gamma_1^{-1} K_1 \{F - H^*(t, Y, z) - G_2^*(t, Y, z) u_2(z)\} \tag{9.43}$$

and $\tilde{u}_1^s(t)$ is the control function of $\tilde{u}_1(t)$ in the closed-loop system which corresponds to the quasi-steady state of the FMS (9.42). Let us rewrite (9.43) in the form

$$\begin{aligned} \tilde{u}_1^s = \tilde{u}_1^{NID} &+ \{G_1^* K_0\}^{-1} K_1^{-1} D_0 \{K_1^{-1} D_0 + G_1^* K_0\}^{-1} \\ &\times \{H^*(t, Y, z) + G_2^*(t, Y, z) u_2(z) - F\}. \end{aligned} \tag{9.44}$$

From (7.25), (7.26), and (9.39) it follows that the solution of (7.27) is given by

$$\tilde{u}_1^{NID} = \{G_1^*(t, Y, z) K_0\}^{-1} \{F - H^*(t, Y, z) - G_2^*(t, Y, z) u_2(z)\}. \tag{9.45}$$

Remark 9.2 *From (9.44) it follows that* $\tilde{u}_1^s(t) - \tilde{u}_1^{NID}(t) \to 0$ *as* $K_1 \to \infty$ *and* $\tilde{u}_1^s(t) = \tilde{u}_1^{NID}(t)$ *if* $D_0 = 0$.

Let $D_0 = 0$. Then from (9.39)–(9.41) we get the SMS equations

$$y_* = F(Y, R), \tag{9.46}$$

$$\begin{aligned} \dot{z} = \tilde{f}(t, Y, z) &+ \tilde{G}_1(t, Y, z) \{G_1^*(t, Y, z)\}^{-1} \\ \times \{F - H^*(t, Y, z) &- G_2^*(t, Y, z) u_2(z)\} + \tilde{G}_2(t, Y, z) u_2(z) \end{aligned} \tag{9.47}$$

as $\boldsymbol{\mu} \to 0$. Note that in the system (9.46)–(9.47), the condition (7.27) holds; that is, the desired output behavior is provided.

It is easy to see that the redundant control variables u_2 can be used in order to reach the boundedness (stability) of the transients in the internal subsystem (9.47), and hence the boundedness of the control function $\tilde{u}_1^s(t)$

is obtained. Note that there are no special requirements for time-domain specifications of the transients in the internal subsystem (9.47).

Remark 9.3 *If the vector $z(t)$ is available for measurement, then any known technique can be used in order to design a static or dynamic control law for the control $u_2 = u_2(z)$. Otherwise, the control system with an observer (O) can be applied as shown in Fig. 9.1(b).*

9.4 Stabilization of degenerated mode and zero dynamics

Let us assume that the steady state of the output variable $y(t)$ is reached in the system (9.46)–(9.47), that is,

$$y(t) = r = \text{const} \quad \forall\, t \in [0, \infty).$$

Then the state vector belongs to the manifold (7.67) in the state space Y, z (or to the manifold (7.68) in the state space X).

From (9.46)–(9.47) and (7.70), the behavior on the manifold (7.67) (degenerated mode) is described by the following equation of the $(n - l)$th order:

$$\begin{aligned} \dot{z} = \tilde{f}(t, Y_r, z) - \tilde{G}_1(t, Y_r, z)\{G_1^*(t, Y_r, z)\}^{-1} \\ \times\{H^*(t, Y_r, z) + G_2^*(t, Y_r, z)u_2(z)\} + \tilde{G}_2(t, Y_r, z)u_2(z). \end{aligned} \tag{9.48}$$

The expression (9.45) yields

$$\tilde{u}_1^{NID} = -\{G_1^*(t, Y_r, z)K_0\}^{-1}\{H^*(t, Y_r, z) + G_2^*(t, Y_r, z)u_2(z)\}. \tag{9.49}$$

In particular, if $y(t) = 0 = \text{const}, \quad \forall\, t \in [0, \infty)$, then (9.48) yields the zero dynamics equation. The stabilizing control law u_2 can be constructed by taking into account Remark 9.3.

9.5 Methods of internal dynamics stabilization

Internal dynamics stabilization by redundant control

Our procedure, developed above for internal dynamics stabilization by redundant control, can be stated as the following sequence of steps:

Step 1. The system given by (9.22)–(9.23) is transformed to the normal form (9.32)–(9.33).

Step 2. The transformation (9.6) is constructed such that the requirement (9.34) is met.

Step 3. The SMS equations (9.46)–(9.47) are derived, and then the stabilizing control law $u_2(z)$ is constructed for the internal subsystem (9.47).

Step 4. In accordance with (7.40), the control law in the form $u_2(Q_2(t, X))$ is executed in order to obtain the stability of the internal dynamics in the system (9.22)–(9.23) (in view of Remark 9.3).

Selective exclusion of redundant control variables

There is non-uniqueness of the 2nd step of the above procedure, where the transformation (9.6) is constructed such that (9.34) holds. Moreover, the internal subsystem (9.47) depends on the matrix P. Therefore, P can be selected so that the corresponding internal subsystem is stable. Such an approach provides the method of selective exclusion of redundant control variables.

Step 1. Select a subset of ζ components of the vector u, where $\zeta = m-p$. Then suppose that $u_j \equiv 0$ for all $j = j_1, j_2, \ldots, j_\zeta$.

Step 2. Check the invertibility and internal stability of the system with the remaining control variables.

The enumeration of C_m^{m-p} possibilities is provided, where C_m^{m-p} is the number of combinations of m different components of $u(t)$, taken $m - p$ at a time, without repetition.

Insertion of supplementary conditions

Let us consider the insertion of supplementary conditions for an LTI system given by

$$\begin{aligned} \dot{X} &= AX + Bu, \quad X(0) = X^0, \\ y &= CX, \end{aligned}$$

where $X \in \mathbb{R}^n$, $u \in \mathbb{R}^m$, $y \in \mathbb{R}^p$, and $m > p$.

Differentiation of y_i gives

$$y_i^{(1)} = c_i AX + c_i Bu,$$

where $c_i B \in \mathbb{R}^m$. If $c_i B \neq 0$ (i.e., $\alpha_i = 1$), then let us insert the supplementary condition for control variables $u(t)$ such that

$$c_i Bu = 0.$$

This insertion increases the relative degree α_i. Then we can obtain

$$y_i^{(2)} = c_i A^2 X + c_i ABu.$$

If $c_i AB \neq 0$, then a second supplementary condition of the form $c_i ABu = 0$ can be inserted. Each new supplementary condition increases the relative degree α_i.

Let us assume that a set of supplementary conditions may be inserted such that $n = \alpha_1 + \alpha_2 + \cdots + \alpha_p$. Then the internal subsystem (9.47) disappears, and right invertibility of the system is sufficient for realizability of the desired output behavior.

Insertion of redundant control variables

Let us consider an LTI system (7.47)–(7.48):

$$\begin{aligned} \dot{X} &= AX + Bu, \quad X(0) = X^0, \\ y &= CX, \end{aligned}$$

where $X \in \mathbb{R}^n$, $u \in \mathbb{R}^p$, $y \in \mathbb{R}^p$. We assume that the condition (7.49) is satisfied, that is, the system (7.47)–(7.48) is invertible and the vector of the relative degree equals $\alpha = \{1, \ldots, 1\}$.

Let the characteristic polynomial (7.60) of the internal subsystem (7.59) have at least one root with nonnegative real part. So the condition of internal stability of the system (7.47)–(7.48) is not satisfied.

Let us insert redundant control variables $\hat{u}$ and consider the system given by

$$\dot{X} = AX + Bu + \tilde{B}\hat{u}, \quad X(0) = X^0, \tag{9.50}$$

$$y = CX, \tag{9.51}$$

where

$$\hat{u} = \hat{K}X. \tag{9.52}$$

We consider the procedure for selecting the matrices $\tilde{B}$ and $\hat{K}$ in order to obtain an allowable root placement for the internal subsystem of the system given by

$$\dot{X} = [A + \tilde{B}\hat{K}]X + Bu, \quad X(0) = X^0, \tag{9.53}$$

$$y = CX. \tag{9.54}$$

This procedure can be stated as the following sequence of steps:

Step 1. The system given by (7.47)–(7.48) is transformed by matrix Q defined by (7.50) to the normal form (7.52)–(7.53).

Step 2. From (7.52)–(7.53) the system (7.58)–(7.59) can be found.

Step 3. Insert redundant control variables $\hat{u}$ in the system (7.58)–(7.59). As a result we get

$$\begin{aligned}\dot{y} &= F(y,r),\\ \dot{z} &= \{A_{22} - B_2B_1^{-1}A_{12}\}z + \{A_{21} - B_2B_1^{-1}A_{11}\}y\\ &\quad + B_2B_1^{-1}F(y,r) + \hat{B}\hat{u}.\end{aligned} \tag{9.55}$$

Step 4. The matrices $\hat{B}$ and $\hat{K}$ are selected such that the characteristic polynomial

$$\det\{sI_{n-p} - \bar{A} - \bar{B}\bar{K}\}$$

has an allowable root placement, where $\bar{A} = A_{22} - B_2B_1^{-1}A_{12}$. As a result we get

$$\tilde{B} = Q^{-1}\begin{bmatrix} 0 \\ \hat{B} \end{bmatrix}, \qquad \hat{u} = \hat{K}Q_2X.$$

9.6 Example

Let us consider an LTI system given by

$$\begin{aligned}\dot{x}_1 &= x_1 + x_2 + u_1,\\ \dot{x}_2 &= x_1 + 0.5x_2 + 2x_3 + u_1 + u_2,\\ \dot{x}_3 &= x_1 + x_2 + x_3 + u_1 + u_2,\\ y &= x_1.\end{aligned} \tag{9.56}$$

Selective exclusion of redundant control variables. (i) The control variable u_1 can be excluded by taking $u_1 \overset{t}{\equiv} 0$. Let us consider the possibility of using the control variable u_2 to provide the desired behavior of the output $y = x_1$. It is easy to see that the relative degree of $y(t)$ with respect to the control variable u_2 is $\alpha = 2$. Then let us introduce the new state variables y_0, y_1, z in the following way:

$$y_0 = y = x_1, \quad y_1 = y_0^{(1)}, \quad z = x_3. \tag{9.57}$$

By transformation (9.57), from (9.56) the system

$$\begin{aligned}\dot{y}_0 &= y_1, \\ \dot{y}_1 &= F(y_1, y_0, R), \\ \dot{z} &= -z + F(y_1, y_0, R) - 0.5y_0 - 0.5y_1,\end{aligned}$$

follows, which is the counterpart of (7.62)–(7.63), where $R = \{r, r^{(1)}\}^T$.

It is easy to see that if the transients of $z(t)$ are stable, then the control law structure of the form

$$\mu^2 u_2^{(2)} + d_1 \mu u_2^{(1)} + d_0 u_2 = k_0\{F(y^{(1)}, y, r^{(1)}, r) - y^{(2)}\}$$

can be used in order to provide the desired behavior of the output variable $y(t)$ given that $u_1 \stackrel{t}{\equiv} 0$.

(ii) Exclude the control variable u_2 by taking $u_2 \stackrel{t}{\equiv} 0$ and consider the possibility of using the control variable u_1 in order to provide the desired behavior of $y(t)$, where the relative degree of $y(t)$ with respect to the control variable u_1 is $\alpha = 1$.

By the transformation

$$y = x_1, \quad z_1 = x_2, \quad z_2 = x_3,$$

from (9.56) the counterpart of (7.62)–(7.63) can be derived. It is

$$\begin{aligned}\dot{y} &= F(y, r), \\ \dot{z}_1 &= -0.5z_1 + 2z_2 + F(y, r), \\ \dot{z}_2 &= z_2 + F(y, r).\end{aligned}$$

Here $(s + 0.5)(s - 1)$ is the characteristic polynomial of the internal subsystem. The desired behavior of $y(t)$ cannot be provided by the control variable u_1, since the internal subsystem is unstable given that $u_2 \stackrel{t}{\equiv} 0$.

Internal dynamics stabilization by redundant control. By the transformation

$$y = x_1, \quad z_1 = x_2, \quad z_2 = x_3,$$

from (9.56) the system

$$\begin{aligned}\dot{y} &= y + z_1 + u_1, \\ \dot{z}_1 &= y + 0.5z_1 + 2z_2 + u_1 + u_2, \\ \dot{z}_2 &= y + z_1 + z_2 + u_1 + u_2,\end{aligned} \tag{9.58}$$

results. This is the counterpart of (9.8)–(9.9).

From (9.58), we can find that the relative degree of $y(t)$ with respect to the control variable u_1 is $\alpha = 1$, and that the condition (9.34) holds. Assume that u_1 is used to provide the desired behavior of $y(t)$ and that the control law is given by

$$\mu u_1^{(1)} + d_0 u_1 = k_0\{F(y,r) - y^{(1)}\}. \tag{9.59}$$

Then u_2 is considered as the redundant control variable and will be used for internal dynamics stabilization.

The counterpart of the system (9.19)–(9.21) for the discussed example has the following form:

$$\begin{aligned} \dot{y} &= F(y,r), \\ \dot{z}_1 &= -0.5z_1 + 2z_2 + F(y,r) + u_2, && (9.60) \\ \dot{z}_2 &= z_2 + F(y,r) + u_2, && (9.61) \\ u_1^{LID} &= -z_1 + F(y,r) - y, && (9.62) \end{aligned}$$

where

$$\bar{A} = \begin{bmatrix} -0.5 & 2 \\ 0 & 1 \end{bmatrix}, \quad \bar{B} = \begin{bmatrix} 1 \\ 1 \end{bmatrix}, \quad \bar{C} = \begin{bmatrix} -1 & 0 \end{bmatrix}. \tag{9.63}$$

We can check that the internal subsystem described by (9.60)–(9.61) is fully controllable by u_2, and fully observable if $\bar{y} = u_1^{LID}$ is regarded as the measured output.

First, assume that the vector $z(t)$ is available for measurement. Then the control law

$$u_2 = k_1 z_1 + k_2 z_2 \tag{9.64}$$

gives the stable internal subsystem (9.60)–(9.61) if k_1, k_2 are defined by the condition

$$\det\{sI - \bar{A} - \bar{B}\bar{K}\} = A^d_{Int}(s)$$

where $A^d_{Int}(s)$ is the desired stable characteristic polynomial of the internal subsystem. If, for instance, we have

$$A^d_{Int}(s) = s^2 + 2s + 2, \tag{9.65}$$

then $k_1 = -2.5$ and $k_2 = 0$.

Second, since $k_2 = 0$; hence, from (9.62) we obtain that the control law

$$u_2 = -k_1 u_1^{LID} \tag{9.66}$$

gives the same characteristic polynomial (9.65) of the internal subsystem.

Finally, replace u_1^{LID} by u_1 in (9.66), and from (9.59), as a result, we get the control law given by

$$\mu u_1^{(1)} + d_0 u_1 = k_0\{F(y, r) - y^{(1)}\}, \tag{9.67}$$

$$u_2 = -k_1 u_1, \tag{9.68}$$

which executes the desired output behavior and stabilizes the internal dynamics.

Simulation results for the system (9.56) controlled by the algorithm (9.67)–(9.68) are displayed in Fig. 9.2, where the initial conditions are zero and $k_0 = 10$, $k_1 = -2.5$, $k_2 = 0$, $d_0 = 0$, $\mu = 0.1$ s, and $t \in [0, 6]$ s. The desired output dynamics $F(y, r)$ are given by (4.6) with the parameter $T = 1$ s.

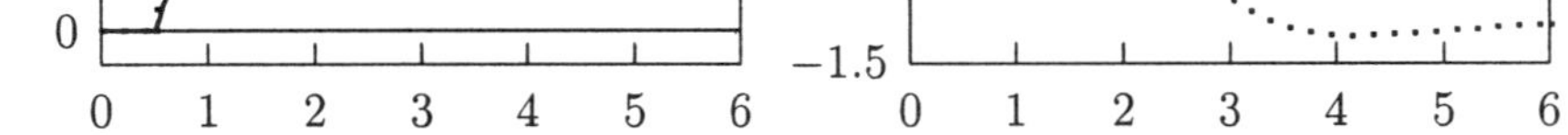

Fig. 9.2 Simulation results for the system (9.56), (9.67), and (9.68).

9.7 Notes

Our main purpose above was to explain the various procedures for internal dynamics stabilization that apply if the dimension of the control vector u is as large as that of the output vector y. It was shown that redundant control variables can provide for such stabilization and, consequently, that the range of application of the discussed design methodology can be extended.

9.8 Exercises

9.1 Stabilize the internal subsystem by selective exclusion of redundant control variables in the system given by

$$\begin{aligned} \dot{x}_1 &= x_1 + x_2 + u_1 + u_2, \\ \dot{x}_2 &= x_1 + 3x_2 - 2u_1, \\ y &= x_1. \end{aligned} \tag{9.69}$$

9.2 Consider the system given by (9.69). Stabilize the internal subsystem by insertion of supplementary conditions.

9.3 Stabilize the internal subsystem by selective exclusion of redundant control variables in the system given by

$$\begin{aligned} \dot{x}_1 &= x_1 + 2x_2 + u_1 + 2u_2, \\ \dot{x}_2 &= x_1 + 1.5x_2 + u_1 + u_2, \\ y &= x_1. \end{aligned} \tag{9.70}$$

9.4 Consider the system given by (9.70). Stabilize the internal subsystem by insertion of supplementary conditions.

9.5 Stabilize the internal subsystem by selective exclusion of redundant control variables in the system given by

$$\begin{aligned} \dot{x}_1 &= x_1 + x_2 + x_3 + u_1 + u_2, \\ \dot{x}_2 &= x_3, \\ \dot{x}_3 &= x_1 + x_2 + 2x_3 + u_1, \\ y &= x_1 + x_2. \end{aligned} \tag{9.71}$$

9.6 Consider the system given by (9.71). Stabilize the internal subsystem by insertion of supplementary conditions.

9.7 Stabilize the internal subsystem in the system given by

$$\dot{x}_1 = x_1^2 - x_2^3 + u_1 + u_2, \quad \dot{x}_2 = |x_1| + u_1 + 2u_2, \quad y = x_1.$$

9.8 Stabilize the internal subsystem in the system given by

$$\dot{x}_1 = x_1^3 - |x_2| + u_1 + u_2, \quad \dot{x}_2 = |x_1| + u_1 - u_2, \quad y = x_1.$$

Chapter 10

Digital controller design based on pseudo-continuous approach

Because the implementation of modern controllers is usually based on the use of a computer or digital signal processor, the next few chapters are devoted to the problem of digital controller design for continuous nonlinear time-varying systems. In particular, in this chapter, the design of digital controllers based on the so-called pseudo-continuous approach is presented, where the digital controller is the result of continuous-time controller discretization. The continuous-time controller with the relative highest output derivatives in feedback is used, and the effect of discretization at discrete time instants has been taken into account by inclusion of a zero-order hold (ZOH) transfer function. A distinguishing feature of the approach is that a pseudo-continuous-time model of the control loop with a pure time delay is used, where the pure delay is a result of the ZOH transfer function approximation. Finally, a control law with compensation of the pure time delay is presented, which allows us to increase the sampling period.

10.1 Continuous system preceded by zero-order hold

10.1.1 *Control problem*

This chapter is concerned with the control system shown in Fig. 10.1. Here the continuous system is preceded by a digital-analog converter (DAC), and the output is sampled by an analog-digital converter (ADC). Note that the output signals of the DAC and ADC are to be taken as sampled-data signals; that is, the effect of the sampling process at discrete time instants is taken into account, while the quantization of signals in amplitude is not discussed. Assume that the continuous-time system is the nonlinear time-

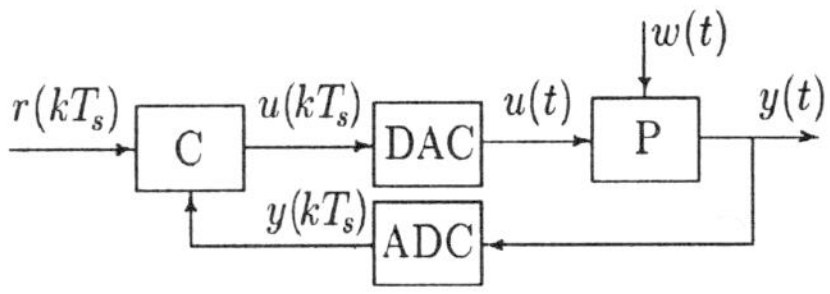

Fig. 10.1 Block diagram of digital control system.

varying system given by (8.1)–(8.2):

$$\dot{X} = f(t, X) + G(t, X)u, \quad X(0) = X^0,$$
$$y = h(t, X),$$

where $X \subseteq \mathbb{R}^n$, $u \subseteq \mathbb{R}^p$, $y \subseteq \mathbb{R}^p$, and $p \leq n$.

We seek a control system for which

$$\lim_{t \to \infty} e(t) = 0,$$

where $e(t)$ is the error of the reference input realization; $e(t) = r(t) - y(t)$; $r(t)$ is the reference input; $r = \{r_1, r_2, \ldots, r_p\}^T$.

Moreover, the controlled transients $e_i(t) \to 0$ should have a desired behavior, where $i = 1, \ldots, p$. These transients should not depend on the external disturbances and varying parameters of the system (8.1)–(8.2).

Assumption 10.1 The invertibility condition (7.20) [Porter (1970)] is satisfied for the system (8.1)–(8.2).

Assumption 10.2 The internal stability of the system (8.1)–(8.2) is maintained; that is, the condition (7.32) is satisfied in a specified region of the state space. In particular, if $y(t) = r(t) = 0$ for all $t \in [0, \infty)$, then the zero-dynamics in (8.1)–(8.2) are stable.

Assumption 10.3 Consider the DAC as a ZOH. Let a series connection of a ZOH and the continuous-time system (8.1)–(8.2) take place, where for a discrete input sequence $\{u_k\}_{k=0}^{\infty}$ we have

$$u(t) = u_k, \ \forall\, t \in [kT_s, (k+1)T_s).$$

10.1.2 *Pseudo-continuous-time model with pure delay*

The transfer function of the ZOH device has the form

$$G_{ZOH}(T_s, s) = k_{ZOH}\frac{1 - e^{-T_s s}}{s}, \quad k_{ZOH} = \frac{1}{T_s}, \tag{10.1}$$

where T_s is the sampling period.

Based on the representation of the transfer function $G_{ZOH}(T_s, s)$ in terms of its Maclaurin series, it follows that

$$\lim_{T_s s \to 0} \{G_{ZOH}(T_s, s) - e^{-T_s s/2}\} = 0. \tag{10.2}$$

In accordance with (10.2), a pure time delay may be used as a result of the ZOH transfer function approximation. The delay time depends on the sampling period. This approximation is widely used in various design techniques for linear control systems, e.g., [Åström and Wittenmark (1997)], and was applied to nonlinear digital control system design in [Blachuta *et al.* (1997); Yurkevich *et al.* (1997)].

Provided that a digital controller fitted with a ZOH device is used, we can assume, instead of (8.1)–(8.2), a nonlinear pseudo-continuous-time model

$$\dot{X}(t) = f(t, X(t)) + G(t, X(t))u(t-\tau), \quad X(0) = X^0, \tag{10.3}$$

$$y(t) = h(t, X(t)), \tag{10.4}$$

with a delay $\tau = T_s/2$ taken into account.

10.2 Digital controller design

10.2.1 *Insensitivity condition*

From (10.3)–(10.4) it follows that the expression (7.13) turns into

$$y_*(t) = H^*(t, X(t)) + G^*(t, X(t))u(t-\tau), \tag{10.5}$$

where

$$\det G^*(t, X(t)) \neq 0, \quad \forall\ (t, X)\ \in\ \Omega_{t,X}.$$

By (10.5) we see that $y_i^{(\alpha_i)}$ depends explicitly on the input $u(t-\tau)$. The control problem can be solved if $y_i^{(\alpha_i)}$ fulfills the reference model defined by the stable differential equation (7.23):

$$y_i^{(\alpha_i)} = F_i(Y_i, R_i),$$

where F_i represents the desired dynamics of $y_i(t)$. Here $Y_i = [y_i, y_i^{(1)}, \dots, y_i^{(\alpha_i-1)}]^T$, $R_i = [r_i, r_i^{(1)}, \dots, r_i^{(\rho_i-1)}]^T$, $\rho_i < \alpha_i$, and $r_i = y_i$ at equilibrium for $r_i = \text{const}$.

As a result, the reference model for the transients of $y(t)$ is given in the form of the vector differential equation (7.25):

$$y_* = F(Y, R),$$

where
$Y \in \mathbb{R}^l$; $l = \sum_{i=1}^{p} \alpha_i$; $l \leq n$; $Y = [y_1, y_1^{(1)}, \dots, y_1^{(\alpha_1 - 1)}, y_2, \dots, y_p^{(\alpha_p - 1)}]^T$;
$R = [r_1, r_1^{(1)}, \dots, r_1^{(\rho_1 - 1)}, r_2, \dots, r_p^{(\rho_p - 1)}]^T$.

Let us consider the error of the desired dynamics realization given by (7.26):

$$e^F = F(Y, R) - y_*.$$

Then (7.25), defining the desired behaviour of $y(t)$, is fulfilled if and only if

$$\mathbf{e}^F(t, u(t - \tau)) = 0. \tag{10.6}$$

So the control problem has been reformulated as the insensitivity condition (10.6), which should be maintained in the closed-loop system given that the external disturbances and varying parameters of the system (8.1)–(8.2) are unknown.

10.2.2 *Pseudo-continuous closed-loop system*

In order to satisfy (10.6), let us consider the control law in the form of the following system of differential equations (8.68) (see p. 208):

$$D_q \boldsymbol{\mu}^q \tilde{u}^{(q)} + D_{q-1} \boldsymbol{\mu}^{q-1} \tilde{u}^{(q-1)} + \dots + D_1 \boldsymbol{\mu} \tilde{u}^{(1)} + D_0 \tilde{u} = K_1 \mathbf{e}^F,$$

where

$$u = K_0 \tilde{u},$$
$$\boldsymbol{\mu} = \mathrm{diag}\{\mu_1, \dots, \mu_p\}, \quad K_1 = \mathrm{diag}\{k_1, \dots, k_p\},$$
$$D_j = \mathrm{diag}\{d_{1,j}, \dots, d_{p,j}\}, \quad \forall\, j = 0, 1, \dots, q.$$

From (10.5) and (8.68) it follows that the closed-loop input-output equations are given by

$$y_*(t) = H^*(t, X(t)) + G^*(t, X(t)) K_0 \tilde{u}(t - \tau), \tag{10.7}$$

$$D_q \boldsymbol{\mu}^q \tilde{u}^{(q)}(t) + D_{q-1} \boldsymbol{\mu}^{q-1} \tilde{u}^{(q-1)}(t) + \dots + D_1 \boldsymbol{\mu} \tilde{u}^{(1)}(t) + D_0 \tilde{u}(t) = K_1 \{F(Y(t), R(t)) - y_*(t)\}. \tag{10.8}$$

Similar to the expression (6.8) (see p. 118), in order to enable usage of the above technique for two-time-scale motion analysis, let us represent the time delay τ in the form

$$\tau = \tau_0 \mu,$$

where τ_0 is the normalized delay and μ is defined by (8.26). As a result, from (10.7)–(10.8) we obtain the FMS

$$\begin{aligned} &D_q \mu^q \tilde{u}^{(q)}(t) + D_{q-1}\mu^{q-1}\tilde{u}^{(q-1)}(t) + \cdots + D_1 \mu \tilde{u}^{(1)}(t) + D_0 \tilde{u}(t) \\ &+K_1 G^*(t, X(t)) K_0 \tilde{u}(t-\tau) = K_1\{F(Y(t), R(t)) - H^*(t, X(t))\} \end{aligned} \quad (10.9)$$

as $\mu \to 0$, where $X(t)$ is the frozen variable during the transients in the FMS (10.9).

Assumption 10.4 The stability and sufficiently rapid decay of the FMS transients in (10.9) are provided for all $(t, X) \in \Omega_{t,X}$ by selection of the control law parameters D_j, K_1, μ.

In accordance with Assumption 10.4, let us consider the steady state (quasi-steady state) of the FMS (10.9). Letting $\mu \to 0$ in (10.9) with (6.8) (i.e., $T_s \to 0$), we obtain $\tilde{u}(t) = \tilde{u}^s(t)$ where $\tilde{u}^s(t)$ is given by (8.75) and the corresponding SMS is given by (8.77).

So, after damping of the stabilized fast transients in the closed-loop system (10.7)–(10.8), the behavior of the output $y(t)$ corresponds to the reference model (7.25) and is insensitive to external disturbances and parameter variations in the system (10.3)–(10.4). The influence of the time delay or, in other words, the effect of discretization, can be neglected in the SMS. At the same time, the influence of the time delay should be taken into consideration in FMS stability.

10.2.3 *Influence of sampling period*

The delay τ caused by discretization alters the stability of the FMS (10.9), and degrades the transient performance indices of the control variable $\tilde{u}(t)$ in the closed-loop system. Hence, the control law parameters should be selected to maintain quality of the control transients $\tilde{u}(t)$ in the FMS (10.9) in the presence of quantized feedback. In particular, the requirements of Assumption 10.4 should be met.

In order to simplify the analysis of the FMS (10.9), let us make the following assumption.

Assumption 10.5 We have

$$K_0 \approx \{G^*\}^{-1}. \tag{10.10}$$

Due to Assumption 10.5 and conditions (8.20), (8.35), and (10.10), the FMS (10.9) decouples into the p equations

$$\begin{aligned} &\mu_i^{q_i}\tilde{u}_i^{(q_i)}(t)+\mu_i^{q_i-1}d_{i,q_i-1}\tilde{u}_i^{(q_i-1)}(t)+\cdots+\mu_i d_{i,1}\tilde{u}_i^{(1)}(t)+d_{i,0}\tilde{u}_i(t)\\ &+k_i\tilde{u}_i(t-\tau)=k_i\{F_i(Y_i(t),R_i(t))-h_i^*(t,X(t))\},\ i=1,\ldots,p. \end{aligned} \tag{10.11}$$

Here the FMS equation of the ith control variable is the direct counterpart of (6.11) (see p. 119). Therefore, the above technique for controller design in the presence of a pure time delay in control may be used in order to select the controller parameters.

By comparing (6.11) with (10.11), and similar to (6.12), we see that the corresponding transfer function of the ith open-loop FMS with time delay is given by

$$G_{i,FMS}^{o}(s)=\frac{k_i\exp(-\tau s)}{D_i(s)}, \tag{10.12}$$

where

$$D_i(s)=\mu_i^{q_i}s^{q_i}+\mu_i^{q_i-1}d_{i,q_i-1}s^{q_i-1}+\ldots+\mu_i d_{i,1}s+d_{i,0}.$$

Denote by $\omega_{i,c}$ the crossover frequency on the Nyquist plot of the ith FMS (10.11). In equation form we have

$$|G_{i,FMS}^{o}(j\omega_{i,c},\mu_i)|=1. \tag{10.13}$$

From (10.12) and (10.13) the relationship

$$|D_i(j\mu_i\omega_{i,c})|=k_i \tag{10.14}$$

follows, which may be used to obtain the value of $\omega_{i,c}$.

Denote by φ_i the value of the phase margin of the ith FMS (10.11). Then the following can be formulated based on the Nyquist stability criterion and (6.16).

Theorem 10.1 *The requirement*

$$\varphi_i \ge \varphi_i^d > 0 \tag{10.15}$$

holds if the condition (10.10) and the inequality $T_s \leq T_{i,s}$ are satisfied, where

$$T_{i,s} = 2\{\pi - \varphi_i^d - ArgD_i(j\mu_i\omega_{i,c})\}/\omega_{i,c}. \tag{10.16}$$

Remark 10.1 *The requirement $\varphi_i \geq \varphi_i^d > 0$, $\forall\ i = 1, \ldots, p$ holds if the condition*

$$T_s \leq \min_{i=1,\ldots,p} T_{i,s}(\varphi_i^d) \tag{10.17}$$

is satisfied.

In this way, the discussed approach allows us to find an upper bound for the sampling period T_s in accordance with the desired value of the phase margin of the FMS (10.11), and to analyze the deterioration of the quality of the control transients in the FMS due to sampling.

10.2.4 *Digital realization of continuous controller*

If (8.20) and (8.35) are satisfied, that is, μ, D_j are diagonal matrices, then the continuous control law (8.68) can be represented as the system of the decomposed linear differential equations (8.67):

$$\begin{aligned}
&\mu_i^{q_i}\tilde{u}_i^{(q_i)} + d_{i,q_i-1}\mu_i^{q_i-1}\tilde{u}_i^{(q_i-1)} + \cdots + d_{i,1}\mu_i\tilde{u}_i^{(1)} + d_{i,0}\tilde{u}_i \\
&\quad = k_i\{-y_i^{(\alpha_i)} - a_{i,\alpha_i-1}y_i^{(\alpha_i-1)} - \cdots - a_{i,1}y_i^{(1)} - a_{i,0}y_i \\
&\quad +b_{i,\rho_i}r^{(\rho_i)} + b_{i,\rho_i-1}r_i^{(\rho_i-1)} + \cdots + b_{i,1}r_i^{(1)} + b_{i,0}r_i\},
\end{aligned} \tag{10.18}$$

where

$$\mu_i > 0, \quad k_i > 0, \qquad q_i \geq \alpha_i \geq \rho_i, \quad i = 1, \ldots, p.$$

From (10.18) the continuous-time controller of the ith channel in the form

$$D_i(s)\tilde{u}_i(s) = -k_i[A_i(s)y_i(s) - B_i(s)r_i(s)] \tag{10.19}$$

results, where

$$D_i(s) = \mu_i^{q_i}s^{q_i} + d_{i,q_i-1}\mu_i^{q_i-1}s^{q_i-1} + \cdots + d_{i,1}\mu_i s + d_{i,0}, \tag{10.20}$$

$$A_i(p) = s^{\alpha_i} + a_{i,\alpha_i-1}s^{\alpha_i-1} + \cdots + a_{i,1}s + a_{i,0}, \tag{10.21}$$

$$B_i(s) = b_{i,\rho_i}s^{\rho_i} + b_{i,\rho_i-1}s^{\rho_i-1} + \cdots + b_{i,1}s + b_{i,0}. \tag{10.22}$$

The digital realization of the continuous-time controller (10.19) can be found via the $\mathcal{Z}$-transform. Alternatively, for simplicity, various approximations of the function $z = \exp(T_s s)$ can be used. In particular, the Tustin transformation [Tustin (1947)], which maintains the stability conditions of the continuous-time system, can be applied to (10.19) to obtain its digital approximation by a substitution

$$s = \frac{2}{T_s}\frac{z-1}{z+1}. \tag{10.23}$$

As a result of (10.23), from (10.19) we obtain a digital controller

$$\tilde{D}_i(z)\tilde{u}_i(z) = -k_i[\tilde{A}_i(z)y_i(z) - \tilde{B}_i(z)r_i(z)]. \tag{10.24}$$

This can also be presented as the difference equation

$$\tilde{u}_{i,k} = \sum_{j=1}^{q_i} \bar{d}_{ij}\tilde{u}_{i,k-j} + \sum_{j=0}^{q_i} \bar{a}_{ij}y_{i,k-j} + \sum_{j=0}^{q_i} \bar{b}_{ij}r_{i,k-j}, \tag{10.25}$$

where $\tilde{u}_i(t) = \tilde{u}_{i,k}$ for $kT_s \le t < (k+1)T_s$.

The digital approximation of (10.18) by (10.25) based on the Tustin transformation (10.23) allows us to obtain analytical expressions for the digital control law parameters. For instance, let us consider the particular case of (10.18) given by

$$\begin{aligned}&\mu_i^2\tilde{u}_i^{(2)} + d_{i,1}\mu_i\tilde{u}_i^{(1)} + d_{i,0}\tilde{u}_i \\ &= k_i\{-a_{i,2}y_i^{(2)} - a_{i,1}y_i^{(1)} - a_{i,0}y_i + b_{i,1}r_i^{(1)} + b_{i,0}r_i\},\end{aligned} \tag{10.26}$$

where $a_{i,2} = 1$.

From the requirement (8.12) (see p. 150) it follows that $b_{i,0} = a_{i,0}$. If the reference model of the desired output behavior is given by (2.10), then

$$b_{i,1} = \frac{b_{i,1}^d}{T_i}, \quad a_{i,1} = \frac{a_{i,1}^d}{T_i}, \quad b_{i,0} = a_{i,0} = \frac{1}{T_i^2}. \tag{10.27}$$

Let $a_{i,2} = 0$, $b_{i,1} = 0$, and $a_{i,1} = 1$. Then from (10.26) the control law

$$\mu_i^2\tilde{u}_i^{(2)} + d_{i,1}\mu_i\tilde{u}_i^{(1)} + d_{i,0}\tilde{u}_i = k_i\{-y_i^{(1)} - a_{i,0}y_i + b_{i,0}r_i\} \tag{10.28}$$

results. This corresponds to the first-order reference model given by (4.6) where $b_{i,0} = a_{i,0} = 1/T_i$.

By the transformation (10.23), from (10.26) the control law in the form of the difference equation

$$\begin{aligned}\tilde{u}_{i,k} = {} & \bar{d}_{i1}\tilde{u}_{i,k-1} + \bar{d}_{i2}\tilde{u}_{i,k-2} + \bar{a}_{i0}y_{i,k} + \bar{a}_{i1}y_{i,k-1} + \bar{a}_{i2}y_{i,k-2} \\ & + \bar{b}_{i0}r_{i,k} + \bar{b}_{i1}r_{i,k-1} + \bar{b}_{i2}r_{i,k-2}\end{aligned} \tag{10.29}$$

results, where

$$\begin{aligned}\bar{d}_{i0} &= 4\mu_i^2 + 2\mu_i d_{i,1}T_s + d_{i,0}T_s^2, \\ \bar{d}_{i1} &= \{8\mu_i^2 - 2d_{i,0}T_s^2\}/\bar{d}_{i0}, \\ \bar{d}_{i2} &= -\{4\mu_i^2 - 2\mu_i d_{i,1}T_s + d_{i,0}T_s^2\}/\bar{d}_{i0}, \\ \bar{a}_{i0} &= -k_i\{4a_{i,2} + 2a_{i,1}T_s + a_{i,0}T_s^2\}/\bar{d}_{i0}, \\ \bar{a}_{i1} &= 2k_i\{4a_{i,2} - a_{i,0}T_s^2\}/\bar{d}_{i0}, \\ \bar{a}_{i2} &= -k_i\{4a_{i,2} - 2a_{i,1}T_s + a_{i,0}T_s^2\}/\bar{d}_{i0}, \\ \bar{b}_{i0} &= k_i\{2b_{i,1}T_s + b_{i,0}T_s^2\}/\bar{d}_{i0}, \\ \bar{b}_{i1} &= 2k_i b_{i,0}T_s^2/\bar{d}_{i0}, \\ \bar{b}_{i2} &= -k_i\{2b_{i,1}T_s - b_{i,0}T_s^2\}/\bar{d}_{i0}.\end{aligned} \tag{10.30}$$

Then, from (10.29), the block diagram can be obtained as shown in Fig. 10.2.

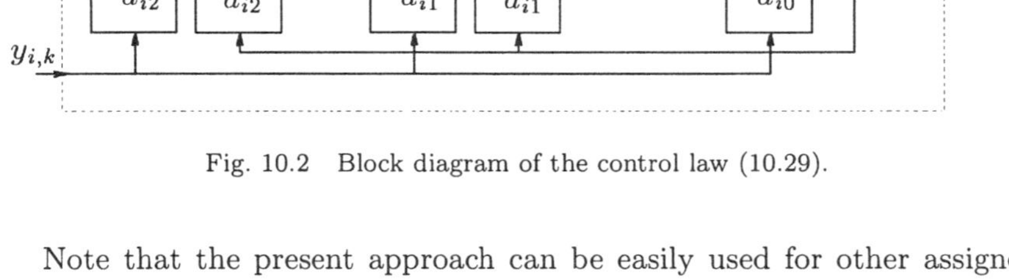

Fig. 10.2 Block diagram of the control law (10.29).

Note that the present approach can be easily used for other assigned degrees of the polynomials $D_i(s)$, $A_i(s)$, $B_i(s)$ in (10.19). The advantage of the method is that we have the analytical expressions for the digital control law parameters, which depend explicitly on the desired specifications of the output behavior through the parameters of (10.19).

10.2.5 *Example*

We consider the nonlinear system given by (5.80) (see p. 109):

$$x^{(2)} = [1 + \sin(x^{(1)})]x + [1 - 0.5\sin(x)]u + 0.5w(t)$$

with continuous-time controller in the form

$$\mu^2 u^{(2)} + d_1 \mu u^{(1)} + d_0 u = k_0\{-a_2 y^{(2)} - a_1 y^{(1)} - a_0 y + b_1 r^{(1)} + b_0 r\}, \quad (10.31)$$

where $\mu = 0.2$ s, $d_1 = 4$, $d_0 = 0$, $k_0 = 10$, $a_2 = 1$, $a_1 = 1.4$, $a_0 = b_0 = 1$, $b_1 = 0$.

For the given parameters, the appropriate FMS (10.11) has phase margin $\varphi \geq \varphi^d = 0.175$ rad if the sampling period is $T_s = 0.1$ s. By transforming (10.23) of (10.31) the digital controller (10.29) was obtained, where its parameters are given by (10.30).

Simulation results for the system (5.80) controlled by the algorithm (10.29) to a step reference input $r(t)$ are displayed in Fig. 10.3, where the initial conditions are zero and $b_1 = 0$. Results for a ramp reference input $r(t)$ are displayed in Fig. 10.4, where $a_1 = b_1 = 1.4$ and the initial conditions are zero.

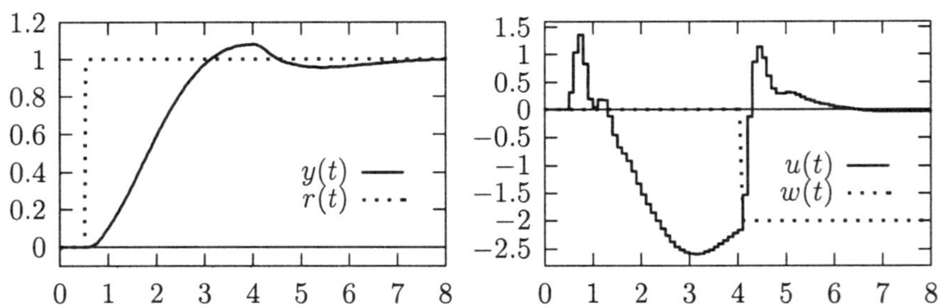

Fig. 10.3 Simulation results for the system (5.80) and (10.31), where $\mu = 0.2$ s, $d_1 = 4$, $d_0 = 0$, $k_0 = 10$, $a_2 = 1$, $a_1 = 1.4$, $a_0 = b_0 = 1$, $b_1 = 0$, $T_s = 0.1$ s, $t \in [0, 8]$ s.

10.3 Digital controller design with compensation of delay

10.3.1 *Control law structure*

There is a broad set of references where the idea of the time delay compensation scheme of the Smith predictor [Smith (1957)] is used for continuous-time as well as discrete-time systems, e.g., [Landau (1994);

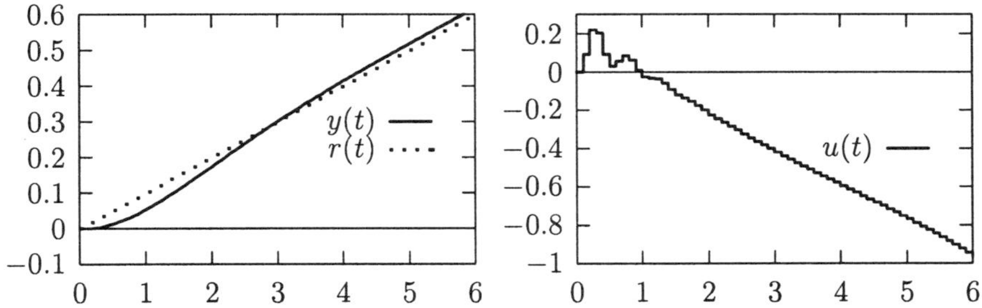

Fig. 10.4 Simulation results for the system (5.80) and (10.31), where $\mu = 0.2$ s, $d_1 = 4$, $d_0 = 0$, $k_0 = 10$, $a_2 = 1$, $a_1 = b_1 = 1.4$, $a_0 = b_0 = 1$, $T_s = 0.1$ s, $t \in [0, 6]$ s.

Palmor (1996)]. In context of control system design with the highest derivative in feedback, it was shown above that the modified control law in the form (6.19) allows us to reduce the effect that a pure time delay in the control loop has on FMS stability. The nature of the discussed below modification of the control law structure is related to the time delay compensation scheme of the Smith predictor.

Since the time delay in the pseudo-continuous-time model (10.3)–(10.4) is caused by sampling, and the sampling period is known, the modified control law (6.19) can easily be applied for the purpose of digital controller design [Yurkevich *et al.* (1998)]. Thus, instead of (8.68) as was discussed before, let us consider a modified control law of the form

$$\begin{aligned} &D_q\boldsymbol{\mu}^q\tilde{u}^{(q)}(t) + \boldsymbol{\mu}^{q-1}D_{q-1}\tilde{u}^{(q-1)}(t) + \cdots + \boldsymbol{\mu}D_1\tilde{u}^{(1)}(t) \\ &+D_0\tilde{u}(t) + \boldsymbol{\gamma}\tilde{u}(t) - \boldsymbol{\gamma}\tilde{u}(t-\tilde{\tau}) = K_1e^F,\ \tilde{U}(0) = \tilde{U}^0, \end{aligned} \tag{10.32}$$

where the auxiliary control vector $\tilde{u}(t)$ and actual control vector $u(t)$ are related by

$$u(t) = K_0\tilde{u}(t)$$

and $\tilde{U} = \{\tilde{u}^T, \tilde{u}^{(1)\ T}, \ldots, \tilde{u}^{(q-1)\ T}\}^T$, $\tilde{\tau} = T_s/2$, $\boldsymbol{\gamma} = K_1$.

Since K_1, D_j are diagonal matrices, the control law (10.32) can be represented as a decomposed system of linear differential equations:

$$\begin{aligned} &\mu_i^{q_i}\tilde{u}_i^{(q_i)}(t) + \mu_i^{q_i-1}d_{i,q_i-1}\tilde{u}_i^{(q_i-1)}(t) + \cdots + \mu d_{i,1}\tilde{u}_i^{(1)}(t) \\ &+d_{i,0}\tilde{u}_i(t) + \gamma_i\tilde{u}_i(t) - \gamma_i\tilde{u}_i(t-\tilde{\tau}) = k_ie_i^F,\ \tilde{U}_i(0) = \tilde{U}_i^0, \end{aligned} \tag{10.33}$$

where

$$i = 1, \ldots, p;\ \mu_i > 0;\ k_i > 0;\ \tilde{U}_i = \{\tilde{u}_i, \tilde{u}_i^{(1)}, \ldots, \tilde{u}_i^{(q_i-1)}\}^T;\ q_i \geq \alpha_i.$$

10.3.2 *Closed-loop system analysis*

Assume that $\mu \to 0$ and $\tau \to 0$, $\tilde{\tau} \to 0$; then from (10.3)–(10.5) and (10.32), the FMS in the form

$$\begin{aligned} & D_q \mu^q \tilde{u}^{(q)}(t) + D_{q-1}\mu^{q-1}\tilde{u}^{(q-1)}(t) + \cdots + D_1 \mu \tilde{u}^{(1)} \\ & + D_0 \tilde{u}(t) + \gamma \tilde{u}(t) - \gamma \tilde{u}(t-\tilde{\tau}) + K_1 G^*(t,X) K_0 \tilde{u}(t-\tau) \\ & \quad = K_1\{F - H^*(t,X)\}, \quad \tilde{U}(0) = \tilde{U}^0 \end{aligned} \tag{10.34}$$

results. By Remark 8.7, all elements of the matrix $G(t, X)$ are frozen variables during the transients in the FMS (10.34), i.e., $G(t, X) = \text{const}$.

Let $\tilde{\tau} = \tau$, $\gamma = K_1$, and $K_0 = \{G^*\}^{-1}$. Then, from (10.34), the FMS equation

$$\begin{aligned} & D_q \mu^q \tilde{u}^{(q)}(t) + D_{q-1}\mu^{q-1}\tilde{u}^{(q-1)}(t) + \cdots + D_1 \mu \tilde{u}^{(1)} \\ & + \{D_0 + K_1\}\tilde{u}(t) = K_1\{F - H^*(t,X)\}, \ \tilde{U}(0) = \tilde{U}^0 \end{aligned} \tag{10.35}$$

results. This does not depend on the delay τ.

Assume that the stability of the FMS (10.35) is provided by selection of the control law parameters, similar to what was discussed above, and let $\mu \to 0$. After damping of the stabilized fast transients in (10.35), we have the steady state (more precisely, quasi-steady state) of the FMS (10.35):

$$\tilde{u}(t) = \tilde{u}^s(t),$$

where

$$\tilde{u}^s(t) = \{D_0 + K_1\}^{-1} K_1 \{F - H^*(t,X)\}. \tag{10.36}$$

Substitution of (10.36) into (10.3), (10.4), and (10.5) yields the SMS equation

$$\dot{X} = f(\cdot) + G\{G^*\}^{-1}\tilde{u}^s(t-\tau), \quad X(0) = X^0. \tag{10.37}$$

In particular, if $D_0 = 0$ then from (10.5) and (10.37) the expression

$$\begin{aligned} \lim_{\mu \to 0} e^F(\mu) &= F(Y(t), R(t)) - H^*(t, X(t)) \\ & \quad - \{F(Y(t-\tau), R(t-\tau)) - H^*(t-\tau, X(t-\tau))\} \end{aligned} \tag{10.38}$$

follows.

This is the direct counterpart of (6.23) (see p. 123). So, the advantage of the modified control law (10.32) is that the value of the phase margin in the FMS (and, accordingly, the sampling period T_s) can be increased due to compensation of the delay. The disadvantage is the additional error (10.38) of the desired dynamics realization.

10.3.3 *Digital realization of continuous controller*

Similar to (10.19), from (10.33) the continuous-time controller of the ith channel in the form

$$\begin{aligned} D_i(s)\tilde{u}_i(s) + \gamma_i\tilde{u}_i(s) - \gamma_i \exp(-T_s s/2)\tilde{u}_i(s) \\ = -k_i[A_i(s)y_i(s) - B_i(s)r_i(s)] \end{aligned} \qquad (10.39)$$

results. Various kinds of digital approximation of (10.39) can be obtained given that

$$\gamma_i = k_i. \qquad (10.40)$$

For instance, based on the approximation

$$\exp(-T_s s/2) \Longrightarrow \frac{1 - e^{-T_s s}}{T_s s} \Longrightarrow \frac{1}{T_s} \frac{1 - \frac{1}{z}}{\frac{2}{T_s}\frac{z-1}{z+1}} = \frac{z+1}{2z}$$

along with the Tustin transformation (10.23), from (10.39) and (10.40) the digital controller

$$\tilde{D}_i(z)\tilde{u}_i(z) + k_i\frac{z-1}{2z}\tilde{u}_i(z) = -k_i[\tilde{A}_i(z)y_i(z) - \tilde{B}_i(z)r_i(z)] \qquad (10.41)$$

results. This yields a control law in the form of the difference equation

$$\tilde{u}_{i,k} = \sum_{j=1}^{q_i+1} \hat{d}_{ij}\tilde{u}_{i,k-j} + \sum_{j=0}^{q_i} \hat{a}_{ij}y_{i,k-j} + \sum_{j=0}^{q_i} \hat{b}_{ij}r_{i,k-j}, \qquad (10.42)$$

where

$$\tilde{u}_i(t) = \tilde{u}_{i,k}, \quad kT_s \le t < (k+1)T_s.$$

For instance, let us consider the particular case of (10.33) given by

$$\begin{aligned} \mu_i^2\tilde{u}_i^{(2)}(t) + d_{i,1}\mu_i\tilde{u}_i^{(1)}(t) + d_{i,0}\tilde{u}_i(t) + \gamma_i\tilde{u}_i(t) - \gamma_i\tilde{u}_i(t-\tilde{\tau}) \\ = k_i\{-a_{i,2}y_i^{(2)}(t) - a_{i,1}y_i^{(1)}(t) - a_{i,0}y_i(t) + b_{i,1}r_i^{(1)}(t) + b_{i,0}r_i(t)\}, \end{aligned} \qquad (10.43)$$

where $\tilde{\tau} = T_s/2$. Then the above digital approximation of (10.43) in the form (10.42) is given by

$$\begin{aligned}\tilde{u}_{i,k} &= \hat{d}_{i1}\tilde{u}_{i,k-1} + \hat{d}_{i2}\tilde{u}_{i,k-2} + \hat{d}_{i3}\tilde{u}_{i,k-3} + \hat{a}_{i0}y_{i,k} + \hat{a}_{i1}y_{i,k-1} \\ &\quad + \hat{a}_{i2}y_{i,k-2} + \hat{b}_{i0}r_{i,k} + \hat{b}_{i1}r_{i,k-1} + \hat{b}_{i2}r_{i,k-2},\end{aligned} \tag{10.44}$$

where

$$\begin{aligned}\hat{d}_{i0} &= 8\mu_i^2 + 4\mu_i d_{i,1}T_s + (2d_{i,0} + \gamma_i)T_s^2, \\ \hat{d}_{i1} &= \{16\mu_i^2 - (4d_{i,0} + \gamma_i)T_s^2\}/\hat{d}_{i0}, \\ \hat{d}_{i2} &= -\{8\mu_i^2 - 4\mu_i d_{i,1}T_s + (2d_{i,0} - \gamma_i)T_s^2\}/\hat{d}_{i0}, \\ \hat{d}_{i3} &= \gamma_i T_s^2/\hat{d}_{i0}, \\ \hat{a}_{i0} &= -2k_i\{4a_{i,2} + 2a_{i,1}T_s + a_{i,0}T_s^2\}/\hat{d}_{i0}, \\ \hat{a}_{i1} &= 4k_i\{4a_{i,2} - a_{i,0}T_s^2\}/\hat{d}_{i0}, \\ \hat{a}_{i2} &= -2k_i\{4a_{i,2} - 2a_{i,1}T_s + a_{i,0}T_s^2\}/\hat{d}_{i0}, \\ \hat{b}_{i0} &= 2k_i\{2b_{i,1}T_s + b_{i,0}T_s^2\}/\hat{d}_{i0}, \\ \hat{b}_{i1} &= 4k_i\, b_{i,0}T_s^2/\hat{d}_{i0}, \\ \hat{b}_{i2} &= -2k_i\{2b_{i,1}T_s - b_{i,0}T_s^2\}/\hat{d}_{i0}.\end{aligned} \tag{10.45}$$

10.3.4 *Example*

Let us apply these results to digital controller design for the manipulator model given by (8.46) (see p. 200). By analogy with (10.3), we consider the following pseudo-continuous-time model:

$$\begin{aligned}& m_3L^{(2)} - m_3\, g\, \sin(\theta) = M_1(t-\tau), \\ & \hat{q}\varphi^{(2)} + 0.5\, m_3\, L^{(1)}\, \varphi^{(1)} = M_2(t-\tau), \\ & \hat{q}^2\, \cos^2(\theta)\, \theta^{(2)} - \hat{q}^2\, \sin(\theta)\, \theta^{-2} + \\ & \quad + \hat{q}\, m_3\, \cos(\theta)\, \theta^{(1)}\, L^{(1)} - g\, \hat{q}\, \cos(\theta) = M_3(t-\tau),\end{aligned} \tag{10.46}$$

where $\{y_1, y_2, y_3\} = \{L, \varphi, \theta\}$ is the output vector, $\{u_1, u_2, u_3\} = \{M_1, M_2, M_3\}$ is the control vector (torques), $\hat{q} = l_1m_2 + (l_2 + L + l_3)m_3$, and $\tau = T_s/2$. The values of the manipulator parameters are represented on p. 200.

Assume that the reference model of the desired dynamics of $y(t)$ is assigned by

$$y_i^{(2)} = -a_{i,1}y_i^{(1)} - a_{i,0}y_i + b_{i,1}r_i^{(1)} + b_{i,0}r_i, \tag{10.47}$$

where the parameters of (10.47) are given by (10.27).

Consider the control law with compensation of delay given by (10.44) and (10.45), where

$$K_0 = \mathrm{diag}\{m_3,\ \hat{q},\ \hat{q}^2\cos^2(\theta)\},\quad k_1 = 27,\quad k_2 = 40,\quad k_3 = 45,$$

$$\mu_1 = \mu_2 = 0.5\ \mathrm{s},\ \mu_3 = 0.4\ \mathrm{s},\ d_{11} = 8,\ d_{21} = 10,\ d_{31} = 15,$$

$$d_{i0} = 0,\ T_i = 1\ \mathrm{s}, a_{i2} = 1,\ a_{i1}^d = 2,\ b_{i1}^d = 0,\ \forall\ i = 1,2,3.$$

Simulation results for the system (8.46) controlled by the algorithm (10.44) with a step reference input $r(t)$ are displayed in Figs. 10.5–10.7, where $t \in [0, 15]$ s. Here the initial conditions are zero, and the control law parameters are defined by (10.45) given that $T_s = 0.27$ s and $\gamma_i = 0,\ \forall\ i = 1, 2, 3$ (without compensation of the delay).

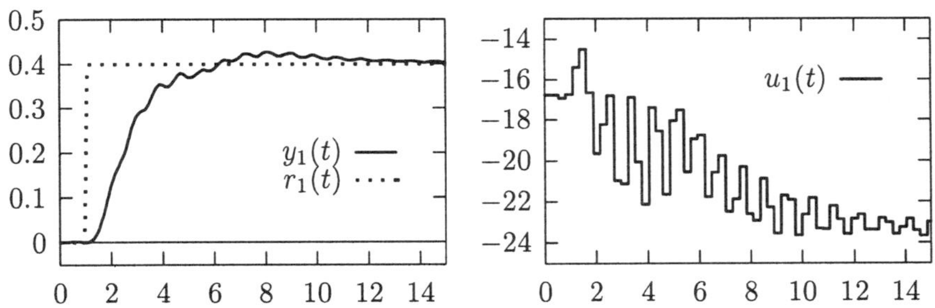

Fig. 10.5 Simulation results for $r_1(t), y_1(t), u_1(t)$ in the system (8.46) and (10.44) given that $\gamma_i = 0,\ \forall\ i = 1, 2, 3$ (without delay compensation).

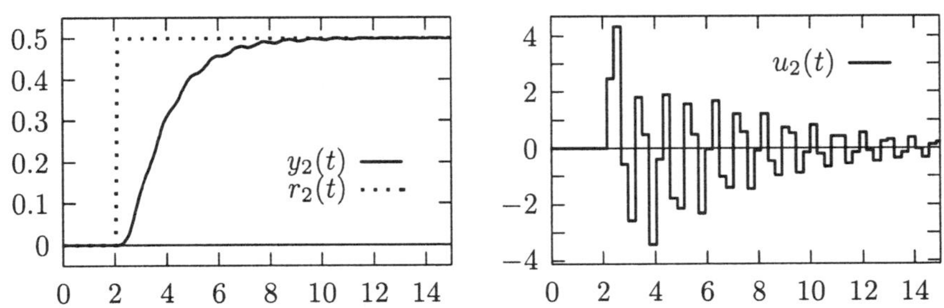

Fig. 10.6 Simulation results for $r_2(t), y_2(t), u_2(t)$ in the system (8.46) and (10.44) given that $\gamma_i = 0,\ \forall\ i = 1, 2, 3$ (without delay compensation).

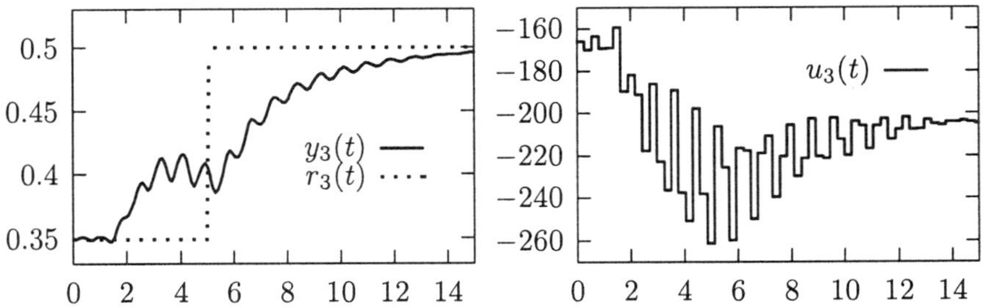

Fig. 10.7 Simulation results for $r_3(t), y_3(t), u_3(t)$ in the system (8.46) and (10.44) given that $\gamma_i = 0,\ \forall\, i = 1, 2, 3$ (without delay compensation).

Simulation results for the system (8.46) controlled by the algorithm (10.44) with a step reference input $r(t)$ are displayed in Figs. 10.8–10.10. Here the initial conditions are zero, and the control law parameters are defined by (10.45) given that $T_s = 0.27$ s and $\gamma_i = k_i,\ \forall\, i = 1, 2, 3$ (in the presence of delay compensation).

From these it is clear that the reduction in swings of the control variables reflects an increase in phase margin in the FMS due to time delay compensation effects.

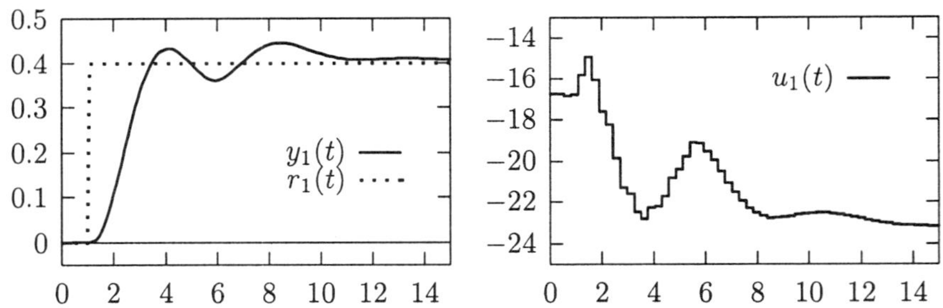

Fig. 10.8 Simulation results for $r_1(t), y_1(t), u_1(t)$ in the system (8.46) and (10.44) given that $\gamma_i = k_i,\ \forall\, i = 1, 2, 3$ (with delay compensation).

10.4 Notes

In this chapter the previous design methodology for a continuous-time controller with the highest output derivative in feedback was modified so that the effect of discretizing the control signal has been taken into consideration by inclusion of a ZOH transfer function. This, in turn, was approximated

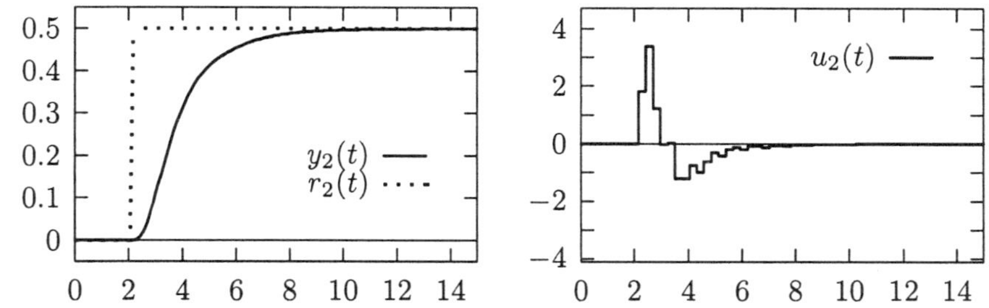

Fig. 10.9 Simulation results for $r_2(t), y_2(t), u_2(t)$ in the system (8.46) and (10.44) given that $\gamma_i = k_i,\ \forall\ i = 1, 2, 3$ (with delay compensation).

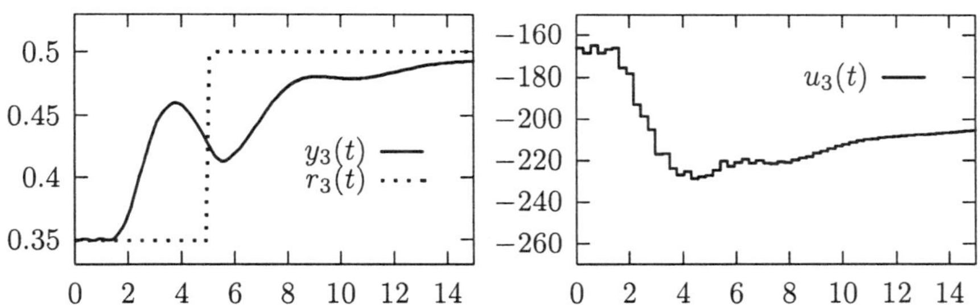

Fig. 10.10 Simulation results for $r_3(t), y_3(t), u_3(t)$ in the system (8.46) and (10.44) given that $\gamma_i = k_i,\ \forall\ i = 1, 2, 3$ (with delay compensation).

by a pure time delay. The procedure for digital controller design based on the pseudo-continuous approach consists of the following steps.

- Instead of the system (8.1)–(8.2), the nonlinear pseudo-continuous-time model (10.3)–(10.4) is considered.
- The continuous-time controller (8.68) with the highest derivative of the output signal in the feedback loop is designed in order to attain desired specifications on the output behavior for the system with delay (specifications such as accuracy, overshoot and settling time of the output response, stability of the fast transients, degree of time-scale separation between the fast and slow modes, and the phase margin of the FMS (10.12)).
- The sampling period T_s is determined in accordance with the required phase margin of the FMS (10.12)).
- The digital realization of the continuous-time controller (10.19) is found via the $\mathcal{Z}$-transform or, for instance, the Tustin transformation.

This design methodology may be used for a broad class of continuous nonlinear time-varying systems on the assumption of incomplete information about varying parameters of the plant model and unknown external disturbances. The advantage of the method is that analytical expressions for the digital control law parameters can be found, and depend explicitly on the specifications of the desired output behavior. The pseudo-continuous approach and its applications to the problem of aircraft flight controller design were discussed in [Yurkevich *et al.* (1997); Blachuta *et al.* (1997); Yurkevich *et al.* (1998)].

10.5 Exercises

10.1 The system is given by

$$\dot{x} = x^2 + 4u.$$

Find the parameters of the control law (10.25) to meet the following specifications: $\overline{\varepsilon}_r = 0$, $t_s^d \approx 3$ s, $\sigma^d \approx 0\%$, $q = 1$. Determine the sampling period T_s such that the phase margin of the FMS (10.12) will meet the requirement $\varphi(\tau) \geq 0.35$ rad. Compare simulation results of the step output response of the closed-loop control system with the assignment.

10.2 The system is given by

$$\dot{x} = x^2 + \{3 + \sin(t)\}u.$$

Find the parameters of the control law (10.25) to meet the following specifications: $\overline{\varepsilon}_r = 0$, $t_s^d \approx 3$ s, $\sigma^d \approx 0\%$, $q = 1$. Determine the sampling period T_s such that the phase margin of the FMS (10.12) will meet the requirement $\varphi(\tau) \geq 0.2$ rad. Compare simulation results of the step output response of the closed-loop control system with the assignment.

10.3 The system is given by (5.84). Find the parameters of the control law (10.25) to meet the following specifications: $\overline{\varepsilon}_r = 0$, $t_s^d \approx 1$ s, $\sigma^d \approx 20\%$, $q = 2$. Determine the sampling period T_s such that the phase margin of the FMS (10.12) will meet the requirement $\varphi(\tau) \geq 0.25$ rad. Compare simulation results of the step output response of the closed-loop control system with the assignment.

10.4 Consider the system given by

$$\begin{aligned}\dot{x}_1 &= x_2 + u + w,\\ \dot{x}_2 &= x_1 + x_2 + 2u + w,\\ y &= x_1.\end{aligned}$$

Find the parameters of the control law (10.25) to meet the following specifications: $\bar{\varepsilon}_r = 0$, $t_s^d \approx 3$ s, $\sigma^d \approx 0\%$, $q = 1$. Determine the sampling period T_s such that the phase margin of the FMS (10.12) will meet the requirement $\varphi(\tau) \geq 0.3$ rad. Compare simulation results of the step output response of the closed-loop control system with the assignment.

10.5 Consider the system given by

$$\begin{aligned}\dot{x}_1 &= x_2 + x_3 + u,\\ \dot{x}_2 &= x_1 - x_2 - x_3 - u,\\ \dot{x}_3 &= x_3 + 2u + w,\\ y &= x_1 + x_2.\end{aligned}$$

Find the parameters of the control law (10.25) to meet the following specifications: $\bar{\varepsilon}_r = 0$, $t_s^d \approx 6$ s, $\sigma^d \approx 10\%$. Determine the sampling period T_s such that the phase margin of the FMS (10.12) will meet the requirement $\varphi(\tau) \geq 0.25$ rad. Compare simulation results of the step output response of the closed-loop control system with the assignment.

10.6 Design the controller (10.25) for the system given by (7.139) that will give the following specifications: $\bar{\varepsilon}_r = 0$, $t_s^d \approx 3$ s, $\sigma^d \approx 0\%$. Determine the sampling period T_s such that the phase margin of the FMS (10.12) will meet the requirement $\varphi(\tau) \geq 0.25$ rad. Compare simulation results of the step output response of the closed-loop control system with the assignment.

10.7 Design the controller (10.25) for the system given by (7.140) that will give the following specifications: $\bar{\varepsilon}_r = 0$, $t_s^d \approx 5$ s, $\sigma^d \approx 0\%$. Determine the sampling period T_s such that the phase margin of the FMS (10.12) will meet the requirement $\varphi(\tau) \geq 0.2$ rad. Compare simulation results of the step output response of the closed-loop control system with the assignment.

10.8 Design the controller (10.25) for the system given by (7.141) that will give the following specifications: $\bar{\varepsilon}_{r1} = 0$, $\bar{\varepsilon}_{r2} = 0$, $t_{s1}^d \approx 1$ s, $\sigma_1^d \approx 0\%$, $t_{s2}^d \approx 3$ s, $\sigma_2^d \approx 0\%$. Determine the sampling period T_s such that the phase margin $\varphi_i(\tau)$ of the ith FMS (10.12) will meet the requirement $\varphi(\tau) \geq 0.3$ rad for $i = 1, 2$. Compare simulation results of the step output response of the closed-loop control system with the assignment.

10.9 Design the controller (10.25) for the system given by (7.142) that will give the following specifications: $\bar{\varepsilon}_{r1} = 0$, $\bar{\varepsilon}_{r2} = 0$, $t_{s1}^d \approx 3$ s, $\sigma_1^d \approx 0\%$, $t_{s2}^d \approx 3$ s, $\sigma_2^d \approx 0\%$. Determine the sampling period T_s such that the phase margin $\varphi_i(\tau)$ of the ith FMS (10.12) will meet the requirement $\varphi(\tau) \geq 0.2$ rad for $i = 1, 2$. Compare simulation results of the step output response of the closed-loop control system with the assignment.

10.10 Design the controller (10.25) for the system given by (7.143) that will give the following specifications: $\bar{\varepsilon}_{r1} = 0$, $\bar{\varepsilon}_{r2} = 0$, $t_{s1}^d \approx 3$ s, $\sigma_1^d \approx 0\%$, $t_{s2}^d \approx 2$ s, $\sigma_2^d \approx 0\%$. Determine the sampling period T_s such that the phase margin $\varphi_i(\tau)$ of the ith FMS (10.12) will meet the requirement $\varphi(\tau) \geq 0.2$ rad for $i = 1, 2$. Compare simulation results of the step output response of the closed-loop control system with the assignment.

10.11 Design the controller (10.25) for the system given by (7.73) that will give the following specifications: $\bar{\varepsilon}_{r1} = 0$, $\bar{\varepsilon}_{r2} = 0$, $t_{s1}^d \approx 3$ s, $\sigma_1^d \approx 10\%$, $t_{s2}^d \approx 6$ s, $\sigma_2^d \approx 0\%$. Determine the sampling period T_s such that the phase margin $\varphi_i(\tau)$ of the ith FMS (10.12) will meet the requirement $\varphi(\tau) \geq 0.2$ rad for $i = 1, 2$. Compare simulation results of the step output response of the closed-loop control system with the assignment.

10.12 Consider the system given by

$$\begin{aligned}
\dot{x}_1 &= \sin(x_1) + x_2^2 + x_1 x_2 + u_1 + u_2,\\
\dot{x}_2 &= x_1 + \sin(x_2) - u_1 + u_2,\\
y_1 &= 2x_1 + x_2,\\
y_2 &= -x_1 + 0.5x_2.
\end{aligned}$$

Design the controller (10.25) that will give the following specifications: $\bar{\varepsilon}_{r1} = 0$, $\bar{\varepsilon}_{r2} = 0$, $t_{s1}^d \approx 6$ s, $\sigma_1^d \approx 0\%$, $t_{s2}^d \approx 3$ s, $\sigma_2^d \approx 0\%$. Determine the sampling period T_s such that the phase margin $\varphi_i(\tau)$ of the ith FMS (10.12) will meet the requirement $\varphi(\tau) \geq 0.25$ rad for $i = 1, 2$. Compare simulation results of the step output response of the closed-loop control system with the assignment.

Chapter 11

Design of discrete-time control systems

This chapter is devoted to discrete-time control system design. The problem of forming desired output transients for a discrete-time system described by a difference equation is discussed. The insensitivity condition for the output transients with respect to varying parameters of the system and external disturbances is introduced, and a discrete-time control law is constructed. Desired output behavior with prescribed dynamics is achieved by inducing two-time-scale motions in the closed loop system, despite uncertainty in the system description. The singular perturbation method is used to analyze fast and slow motions in the discrete-time closed-loop control system. The approach may be considered as the discrete-time counterpart of the above design methodology for continuous-time control systems with the highest derivative in feedback. The chapter opens with explanations in simplified form, while various peculiarities associated with the sampling process will be discussed in later sections.

11.1 SISO two-time-scale discrete-time control systems

11.1.1 *Discrete-time systems*

Let us consider a discrete-time control system given by a difference equation of the form

$$y_k = \sum_{j=1}^{n} a_j y_{k-j} + \sum_{j=1}^{n} b_j u_{k-j} + \sum_{j=1}^{n} \hat{b}_j w_{k-j}, \tag{11.1}$$

where

k is the discrete time variable, $k = 0, 1, \ldots;$

y_k is the output, available for measurement;

u_k is the control;
w_k is the external disturbance, unavailable for measurement.

Assumption 11.1 For the unforced system (11.1) we have

$$y_k - y_{k-j} \approx 0, \quad \forall\, j = 1, \cdots, n. \tag{11.2}$$

In other words, only small variations of y_k occur during the settling time of the nth-order discrete-time system with deadbeat response.

Note that the requirement (11.2) can be easily provided for a sampled-data system preceded by a ZOH by decreasing the sampling period T_s. The condition (11.2) reflects the main qualitative performance of a sampled-data system preceded by a ZOH. From (11.2), the possibility of discrete-time control system design with two-time-scale motions arises as shown later.

Assumption 11.2 The roots of the polynomial

$$b_1 z^{n-1} + b_2 z^{n-2} + \cdots + b_{n-1} z + b_n$$

lie outside some neighborhood of 1, i.e.,

$$\sum_{j=1}^{n} b_j \neq 0. \tag{11.3}$$

11.1.2 *Control problem and insensitivity condition*

Our objective is to design a control system having

$$\lim_{k \to \infty} e_k = 0. \tag{11.4}$$

Here $e_k = r_k - y_k$ is the error of the reference input realization, r_k being the reference input. Moreover, the control transients $e_k \to 0$ should have desired performance indices such as overshoot, settling time, and system type. These transients of y_k should not depend on the external disturbances and varying parameters of the system (11.1).

Let us construct the continuous-time reference model (2.7) for the desired behavior of the output $y(t)$:

$$y = G^d_{yr}(s) r,$$

where the parameters of the nth-order stable continuous-time transfer function $G^d_{yr}(s)$ are selected based on the required output transient performance indices and such that

$$G_{yr}^d(s)\big|_{s=0} = 1.$$

By a $\mathcal{Z}$-transform of $G_{yr}^d(s)$ preceded by a ZOH, the desired pulse transfer function[1]

$$H_{yr}^d(z) = \frac{z-1}{z}\mathcal{Z}\left\{\mathcal{L}^{-1}\left[\frac{G_{yr}^d(s)}{s}\right]\bigg|_{t=kT_s}\right\} = \frac{B^d(z)}{A^d(z)} \tag{11.5}$$

can be found, where

$$H_{yr}^d(z)\big|_{z=1} = 1. \tag{11.6}$$

Hence, from (11.5), the desired stable difference equation

$$y_k = \sum_{j=1}^{n} a_j^d y_{k-j} + \sum_{j=1}^{n} b_j^d r_{k-j} \tag{11.7}$$

results, where

$$1 - \sum_{j=1}^{n} a_j^d = \sum_{j=1}^{n} b_j^d, \qquad \sum_{j=1}^{n} b_j^d \neq 0, \tag{11.8}$$

and the parameters of (11.7) correspond to the assigned output transient performance indices.

Let us rewrite, for short, the desired difference equation (11.7) as

$$y_k = F(Y_k, R_k), \tag{11.9}$$

where

$$Y_k = \{y_{k-n}, y_{k-n+1}, \ldots, y_{n-1}\}^T, \quad R_k = \{r_{k-n}, r_{k-n+1}, \ldots, r_{n-1}\}^T.$$

We have $r_k = y_k$ at the equilibrium of (11.9) for $r_k = \text{const}$, $\forall\, k$.

By definition, put $F_k = F(Y_k, R_k)$ and denote

$$e_k^F = F_k - y_k, \tag{11.10}$$

where e_k^F is the realization error of the desired dynamics assigned by (11.9). Accordingly, if for all $k = 0, 1, \ldots$ the condition

$$e_k^F = 0 \tag{11.11}$$

[1]Some additional details concerning the $\mathcal{Z}$-transform and calculation of pulse transfer functions can be found in [Jury (1958); Lindorff (1965); Chen (1993); Ogata (1994)].

holds, then the desired behavior of y_k with the prescribed dynamics of (11.9) is fulfilled. The expression (11.11) is the insensitivity condition for the output transient performance with respect to the external disturbances and varying parameters of the plant model (11.1). In other words, the control design problem (11.4) has been reformulated as the requirement (11.11).

Remark 11.1 *The insensitivity condition (11.11) is the discrete-time counterpart of (4.33) for the continuous-time system (4.27).*

11.1.3 *Discrete-time control law*

In order to fulfill (11.11), let us construct the control law as the difference equation

$$u_k = \sum_{j=1}^{n} d_j u_{k-j} + \lambda_0 e_k^F, \tag{11.12}$$

where

$$\sum_{j=1}^{n} d_j = 1 \quad \text{and} \quad \lambda_0 \neq 0. \tag{11.13}$$

From (11.13) it follows that the equilibrium of (11.12) corresponds to the insensitivity condition (11.11).

In accordance with (11.7) and (11.10), the control law (11.12) can be rewritten as the difference equation

$$u_k = \sum_{j=1}^{n} d_j u_{k-j} + \lambda_0 \left\{ -y_k + \sum_{j=1}^{n} a_j^d y_{k-j} + \sum_{j=1}^{n} b_j^d r_{k-j} \right\}. \tag{11.14}$$

Remark 11.2 *The control law (11.12) (as well as (11.14)) is the discrete-time counterpart of the continuous-time control law (4.38) (or (4.41)).*

Remark 11.3 *For a cost function $V(u) = \{e_k^F(u)\}^2$, equation (11.12) is related to the multistage optimization algorithm introduced in [Tsypkin (1971)].*

In particular, if $n = 2$, then (11.14) assume the form

$$\begin{aligned} u_k &= d_1 u_{k-1} + d_2 u_{k-2} \\ &\quad + \lambda_0 \left\{ -y_k + a_1^d y_{k-1} + a_2^d y_{k-2} + b_1^d r_{k-1} + b_2^d r_{k-2} \right\}. \end{aligned} \tag{11.15}$$

Let us rewrite the control law (11.15) in state-space form, e.g.,

$$\begin{aligned}\bar{u}_{1,k} &= \bar{u}_{2,k-1} + d_1\bar{u}_{1,k-1} + \lambda_0[a_1^d - d_1]y_{k-1} + \lambda_0 b_1^d r_{k-1},\\ \bar{u}_{2,k} &= d_2\bar{u}_{1,k-1} + \lambda_0[a_2^d - d_2]y_{k-1} + \lambda_0 b_2^d r_{k-1},\\ u_k &= \bar{u}_{1,k} - \lambda_0 y_k.\end{aligned} \tag{11.16}$$

Then, from (11.16), we get the block diagram as shown in Fig. 11.1.

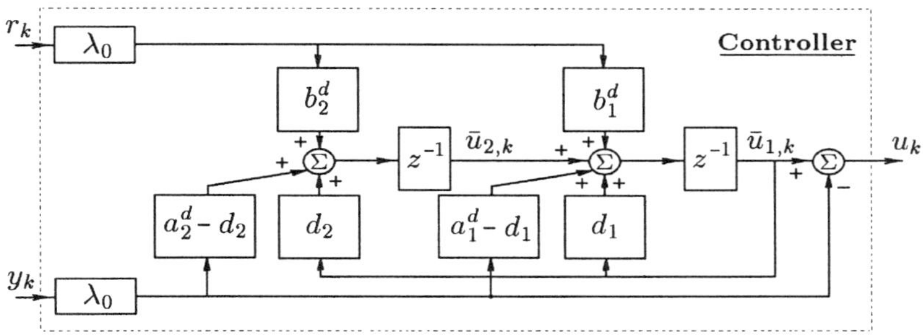

Fig. 11.1 Block diagram of the control law (11.15) represented in the form (11.16).

11.1.4 *Two-time-scale motion analysis*

The closed-loop system equations have the following form:

$$y_k = \sum_{j=1}^{n} a_j y_{k-j} + \sum_{j=1}^{n} b_j u_{k-j} + \sum_{j=1}^{n} \hat{b}_j w_{k-j}, \tag{11.17}$$

$$u_k = \sum_{j=1}^{n} d_j u_{k-j} + \lambda_0 \left\{ -y_k + \sum_{j=1}^{n} a_j^d y_{k-j} + \sum_{j=1}^{n} b_j^d r_{k-j} \right\}. \tag{11.18}$$

Substitution of (11.17) into (11.18) yields

$$y_k = \sum_{j=1}^{n} a_j y_{k-j} + \sum_{j=1}^{n} b_j u_{k-j} + \sum_{j=1}^{n} \hat{b}_j w_{k-j}, \tag{11.19}$$

$$u_k = \sum_{j=1}^{n} [d_j - \lambda_0 b_j] u_{k-j} + \lambda_0 \sum_{j=1}^{n} \{[a_j^d - a_j] y_{k-j} + b_j^d r_{k-j} - \hat{b}_j w_{k-j}\}. \tag{11.20}$$

First, note that the rate of the transients of u_k in (11.19)–(11.20) depends on the controller parameters $\lambda_0, d_1, \dots, d_n$. At the same time, in accordance with (11.2), we have a slow rate of the transients of y_k. Therefore, by choosing the controller parameters it is possible to induce two-time scale transients in the closed-loop system (11.19)–(11.20), where the rate of the transients of y_k is much smaller than that of u_k. Then, as an asymptotic limit, from the closed-loop system equations (11.19)–(11.20) it follows that the FMS is governed by

$$u_k = \sum_{j=1}^{n} [d_j - \lambda_0 b_j] u_{k-j} + \lambda_0 \sum_{j=1}^{n} \{[a_j^d - a_j] y_{k-j} + b_j^d r_{k-j} - \hat{b}_j w_{k-j}\}, \quad (11.21)$$

where $y_k - y_{k-j} \approx 0$, $\forall\, j = 1, 2, \ldots, n$, i.e., $y_k = \text{const}$ during the transients in the system (11.21).

Second, assume that the FMS (11.21) is stable[2] and consider its steady state (or more exactly quasi-steady state), i.e.,

$$u_k - u_{k-j} = 0, \quad \forall\, j = 1, \ldots, n. \quad (11.22)$$

Then, from (11.13), (11.21), and (11.22) we get

$$u_k = u_k^{LID},$$

where

$$u_k^{LID} = \left[\sum_{j=1}^{n} b_j\right]^{-1} \sum_{j=1}^{n} \{[a_j^d - a_j] y_{k-j} + b_j^d r_{k-j} - \hat{b}_j w_{k-j}\}. \quad (11.23)$$

Here, by analogy with the solution of the nonlinear inverse dynamics given by (3.13), the discrete-time function u_k^{LID} is called the solution of the linear inverse dynamics since the system at hand is linear.

Remark 11.4 *Note that the discrete-time control function u_k^{LID} given by (11.23) corresponds to the insensitivity condition (11.11), that is, u_k^{LID} is the discrete-time counterpart of the nonlinear inverse dynamics solution (3.13).*

Remark 11.5 *From (11.23) we know that if (11.3) holds, then u_k^{LID} exists. So (11.3) is the discrete-time counterpart of (3.12).*

Substitution of (11.22) into (11.19)–(11.20) yields the SMS of (11.19)–(11.20), which is the same as the desired difference equation (11.7).

[2]This means that the unique equilibrium point of (11.21) is asymptotically stable.

To ensure stability and fastest transient processes of u_k, let us choose the controller parameters λ_0 and d_j such that

$$\lambda_0 = \left[\sum_{j=1}^{n} b_j\right]^{-1} \quad \text{and} \quad d_j = \lambda_0 b_j, \quad \forall\, j = 1, \ldots, n. \tag{11.24}$$

That is, all roots of the characteristic polynomial of the FMS (11.21) are placed at the origin. Hence, the deadbeat response of the FMS (11.21) is provided. This, along with Assumption 11.1, justifies two-time-scale separation between the fast and slow motions.

The qualitative root distribution of the characteristic polynomial in the discrete-time closed-loop system with two-time-scale motions is shown in Fig. 11.2.

Fig. 11.2 Roots of the characteristic polynomial in the discrete-time closed-loop system with two-time-scale motions.

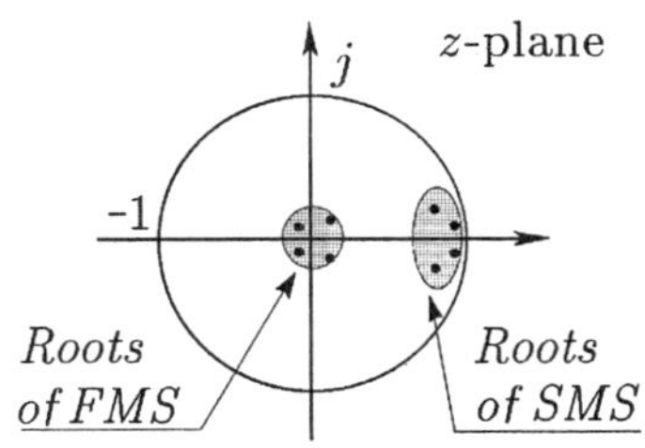

If the degree of time-scale separation between fast and slow motions in the closed-loop system (11.19)–(11.20) is sufficiently large and the FMS transients are stable, then after the fast transients have vanished the behavior of y_k tends to the solution of the reference model (11.9). Accordingly, the controlled output transient process meets the desired performance specifications.

The advantage of the method is that knowledge of the parameters $b_1, \ldots, b_n$ suffices for controller design; knowledge of external disturbances and other parameters of the system is not needed. In particular, the term due to $y_{k-1}, \ldots, y_{k-n}$ in the right member of (11.1) may be entered as some nonlinear function given that (11.2) is satisfied, and just such a numerical example will be presented later in this chapter.

Note that variations of the parameters $b_1, \ldots, b_n$ are possible within the domain where the FMS (11.21) is stable and the fast and slow motion separation is maintained. Some features related to this question are discussed in the next section.

11.1.5 Robustness of closed-loop system properties

In section we consider the influence of parameter variations of a discrete-time system and controller parameters on the properties of the closed-loop system; in other words, we consider the robustness of the closed-loop system properties against parameter variations.

Let us consider a discrete-time system given by (11.1):

$$y_k = \sum_{j=1}^{n} a_j y_{k-j} + \sum_{j=1}^{n} b_j u_{k-j} + \sum_{j=1}^{n} \hat{b}_j w_{k-j}.$$

Assume that the parameters a_j, b_j, $\hat{b}_j$ depend on the operating point and may vary in some bounded domain.

Denote by b_j^0 the known nominal value of the parameter b_j, and let

$$b_j = b_j^0 + \Delta b_j, \tag{11.25}$$

where Δb_j is the difference between b_j^0 and the unknown actual value of b_j.

Let

$$\Delta\lambda_0, \Delta d_1, \dots, \Delta d_n \tag{11.26}$$

denote the errors of the implementation of the calculated controller parameters in practice. Then assume that, instead of (11.24), the parameters of the controller (11.12) are chosen based on the known nominal values b_j^0 and used in practice with errors (11.26) such that

$$\lambda_0 = \lambda_0^0 + \Delta\lambda_0, \tag{11.27}$$

$$d_j = d_j^0 + \Delta d_j\,, \quad \forall\, j = 1, \dots, n, \tag{11.28}$$

where

$$\lambda_0^0 = \left[\sum_{j=1}^{n} b_j^0 \right]^{-1} \quad \text{and} \quad d_j^0 = \lambda_0^0 b_j^0. \tag{11.29}$$

Note that if $b_j = b_j^0$, $\forall\, j = 1, \dots, n$, then the parameters $\lambda_0 = \lambda_0^0$, $d_1 = d_1^0, \dots, d_n = d_n^0$ correspond to the deadbeat response of the FMS.

Let us consider the effect of $\Delta\lambda_0, \Delta d_1, \dots, \Delta d_n$ and $\Delta b_1, \dots, \Delta b_n$ on the behavior of the closed-loop system given by

$$y_k = \sum_{j=1}^{n} a_j y_{k-j} + \sum_{j=1}^{n} b_j u_{k-j} + \sum_{j=1}^{n} \hat{b}_j w_{k-j}, \tag{11.30}$$

$$u_k = \sum_{j=1}^{n} [d_j^0 + \Delta d_j] u_{k-j} + [\lambda_0^0 + \Delta\lambda_0] \left[-y_k + \sum_{j=1}^{n} a_j^d y_{k-j} + \sum_{j=1}^{n} b_j^d r_{k-j} \right]. \tag{11.31}$$

Substitution of (11.30) into (11.31) gives

$$y_k = \sum_{j=1}^{n} a_j y_{k-j} + \sum_{j=1}^{n} b_j u_{k-j} + \sum_{j=1}^{n} \hat{b}_j w_{k-j}, \tag{11.32}$$

$$u_k = \sum_{j=1}^{n} \tilde{\beta}_j u_{k-j} + \{\lambda_0^0 + \Delta\lambda_0\} \sum_{j=1}^{n} \{[a_j^d - a_j] y_{k-j} + b_j^d r_{k-j} - \hat{b}_j w_{k-j}\}, \tag{11.33}$$

where

$$\tilde{\beta}_j = \Delta d_j - \Delta\lambda_0 \{b_j^0 + \Delta b_j\} - \Delta b_j \, \lambda_0^0. \tag{11.34}$$

Similar to (11.21), from (11.32)–(11.33) we get the FMS

$$u_k = \sum_{j=1}^{n} \tilde{\beta}_j u_{k-j} + \{\lambda_0^0 + \Delta\lambda_0\} \sum_{j=1}^{n} \{[a_j^d - a_j] y_{k-j} + b_j^d r_{k-j} - \hat{b}_j w_{k-j}\}, \tag{11.35}$$

where $y_k - y_{k-j} \approx 0$, $\forall\, j = 1, \ldots, n$, i.e., $y_k = \text{const}$ during the transients in the system (11.35) because of the property (11.2).

In accordance with (11.35), the characteristic polynomial of the FMS (11.35) is

$$A_{FMS}(z) = z^n - \tilde{\beta}_1 z^{n-1} - \cdots - \tilde{\beta}_{n-1} z - \tilde{\beta}_n$$

with its parameters given by (11.34). Then the root-locus method (see, for instance, [Chen (1993); Ogata (1994)]) can be applied to investigate the effect of varying parameters on FMS stability.

Remark 11.6 *The range of allowable variations of* $\Delta\lambda_0, \Delta d_1, \ldots, \Delta d_n$ *and* $\Delta b_1, \ldots, \Delta b_n$ *is restricted by the requirement of FMS stability and the required degree of fast and slow motion rate separation in the closed-loop system.*

In order to find the SMS of the closed-loop system (11.32)–(11.33), let us consider the steady state of the FMS (11.35); that is, (11.22) is satisfied.

Then, from (11.35) and (11.27)–(11.28), we get the discrete-time control function u_k^s given by

$$u_k^s = \left[1 - \sum_{j=1}^{n} \tilde{\beta}_j\right]^{-1} \{\lambda_0^0 + \Delta\lambda_0\}$$
$$\times \sum_{j=1}^{n} \{[a_j^d - a_j]y_{k-j} + b_j^d r_{k-j} - \hat{b}_j w_{k-j}\}. \tag{11.36}$$

By substituting (11.22) into (11.32)–(11.33) (or (11.22) and (11.36) into (11.32)), we obtain the SMS

$$y_k = \sum_{j=1}^{n} a_j^d y_{k-j} + \sum_{j=1}^{n} b_j^d r_{k-j}$$
$$+\Delta d_s \left\{1 - \Delta d_s + \Delta\lambda_0 \sum_{j=1}^{n}(b_j^0 + \Delta b_j) + \lambda_0^0 \sum_{j=1}^{n} \Delta b_j\right\}^{-1} \tag{11.37}$$
$$\times \sum_{j=1}^{n} \{[a_j^d - a_j]y_{k-j} + b_j^d r_{k-j} - \hat{b}_j w_{k-j}\},$$

where

$$\Delta d_s = \sum_{j=1}^{n} \Delta d_j. \tag{11.38}$$

Proof of (11.37) see in Appendix A.4.

Remark 11.7 *This is the same as the desired difference equation (11.7) if $\Delta d_s = 0$; that is, the robust zero steady-state error of the reference input realization is maintained and the deviations Δb_j do not alter the SMS (11.37).*

11.1.6 *Control accuracy*

Steady-state error

In this section the steady-state error and velocity error are considered for the discrete-time closed-loop system given by (11.32)–(11.33).

Assume that the conditions

$$r_k = r = \text{const},\ y_k = y^s = \text{const},\ w_k = w^s = \text{const},\ \forall\ k = 0, 1, \ldots \tag{11.39}$$

are satisfied. Denote by

$$e^s = r - y^s$$

the steady-state error (or static position error) in the closed-loop system.

Since the system at hand is linear, e^s can be divided into two parts:

$$e^s = e^s_r + e^s_w,$$

where e^s_r and e^s_w are the steady-state errors due to the reference input r and the disturbance signal $w_k = w^s$, respectively.

From (11.37) we can obtain

$$e^s_r = -\Delta d_s \left[1 - \sum_{j=1}^{n} a_j\right] \phi_1^{-1} r, \tag{11.40}$$

$$e^s_w = \Delta d_s \left[\sum_{j=1}^{n} \hat{b}_j\right] \phi_1^{-1} w^s, \tag{11.41}$$

where

$$\phi_1 = \left[1 - \sum_{j=1}^{n} a^d_j\right] \left[1 - \sum_{j=1}^{n} \tilde{\beta}_j\right] - \Delta d_s \sum_{j=1}^{n} [a^d_j - a_j]. \tag{11.42}$$

Proof of (11.40)–(11.41) see in Appendix A.5. From (11.40)–(11.41) it follows that if $\Delta d_s = 0$ and the conditions (11.39) are satisfied, then $e^s_r = e^s_w = 0$. Therefore, the robust zero steady-state error is maintained in the closed-loop system.

Velocity error

Assume that

$$\Delta d_s = 0 \tag{11.43}$$

and consider the ramp disturbance

$$w_k = w^v T_s k, \quad k = 0, 1, \ldots \tag{11.44}$$

given that

$$r_k = r = \text{const}, \quad y_k = y^v = \text{const}, \quad k = 0, 1, \ldots. \tag{11.45}$$

Let

$$e_w^v = r - y^v, \tag{11.46}$$

where e_w^v is the velocity error due to the ramp disturbance signal (11.44). Then from (11.32)–(11.33), by taking into account (11.44)–(11.45), we obtain

$$e_w^v = -T_s \left[\sum_{j=1}^{n} \hat{b}_j \right] \left[1 - \sum_{j=1}^{n} a_j^d \right]^{-1} \phi_2 w^v, \tag{11.47}$$

where

$$\phi_2 = \left[\sum_{j=1}^{n} j\tilde{\beta}_j \right] \left[1 - \sum_{j=1}^{n} \tilde{\beta}_j \right]^{-1} + \left[\sum_{j=1}^{n} j b_j \right] \left[\sum_{j=1}^{n} b_j \right]^{-1}.$$

Proof of (11.47) see in Appendix A.6.

Similar to the above, assume that $\Delta d_s = 0$ and consider the ramp reference input

$$r_k = r^v T_s k, \quad k = 0, 1, \ldots \tag{11.48}$$

given that

$$w_k = w^s = \text{const}, \quad k = 0, 1, \ldots. \tag{11.49}$$

Denote by e_r^v the velocity error associated with (11.48):

$$e_r^v = \lim_{k \to \infty} \{ r_k - y_k \}, \tag{11.50}$$

where $e_r^v = \text{const}$. From the linear discrete-time control system (11.1) and (11.12), by taking into account (11.48)–(11.49), we obtain

$$e_r^v = T_s \left[1 - \sum_{j=1}^{n} a_j^d \right]^{-1} \left\{ \left[\sum_{j=1}^{n} j[a_j^d + b_j^d] \right] + \left[1 - \sum_{j=1}^{n} a_j \right] \phi_2 \right\} r^v. \tag{11.51}$$

Proof of (11.51) see in Appendix A.7. Note that relationships (11.47)–(11.51) may be used to find the sampling period T_s in accordance with the required velocity error.

11.1.7 *Example*

Let us consider an unstable nonlinear discrete-time system given by

$$y_k = 0.03y_{k-1}y_{k-2} + 1.95y_{k-1} - 0.951y_{k-2} + 0.2u_{k-1} + 0.3u_{k-2} \quad (11.52)$$

with the reference model (11.7) for $n = 2$ in the form

$$y_k = a_1^d y_{k-1} + a_2^d y_{k-2} + b_1^d r_{k-1} + b_2^d r_{k-2}, \quad (11.53)$$

where $a_1^d = 1.7$, $a_2^d = -0.72$, $b_1^d = 0$, $b_2^d = 0.02$. Then the pulse transfer function

$$H^d(z) = \frac{0.02}{z^2 - 1.7z + 0.72} = \frac{0.02}{(z-0.8)(z-0.9)}$$

corresponds to (11.53), where (11.6) is satisfied and the system (11.53) is of type 1.

The control law (11.14) with (11.24) gives

$$\begin{aligned} u_k &= 0.4u_{k-1} + 0.6u_{k-2} \\ &\quad + 2\{-y_k + a_1^d y_{k-1} + a_2^d y_{k-2} + b_1^d r_{k-1} + b_2^d r_{k-2}\}, \end{aligned} \quad (11.54)$$

where $d_1 = 0.4$, $d_2 = 0.6$, $\lambda_0 = 2$, which correspond to the deadbeat response of the FMS.

Simulation results for the output response of the system (11.52) controlled by the algorithm (11.54) for a step reference input r_k are displayed in Fig. 11.3 for the time interval $t \in [0, 6]$ s and with the sampling period $T_s = 0.1$ s. Results for the output response of (11.52) controlled by the algorithm (11.54) for a ramp reference input r_k are displayed in Fig. 11.4, where $r_k = 0.1T_s k$.

By substituting $a_1^d = 1.7$, $a_2^d = -0.72$, $b_1^d = 0.3$, and $b_2^d = -0.28$ into (11.53), we obtain the reference model, a system of type 2. Simulation results for the output response of the system (11.52) controlled by the algorithm (11.54) with the new parameters of (11.53) for a ramp reference input r_k are displayed in Fig. 11.5 where $r_k = 0.1T_s k$.

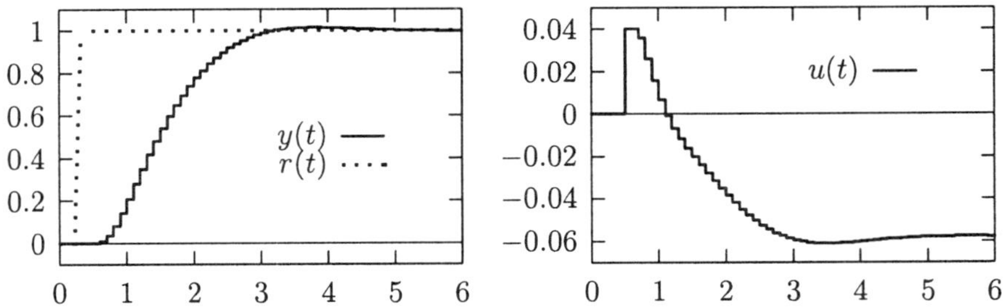

Fig. 11.3 Output response of the system (11.52) and (11.54) for a step reference input $r(t)$, where $a_1^d = 1.7$, $a_2^d = -0.72$, $b_1^d = 0$, $b_2^d = 0.02$.

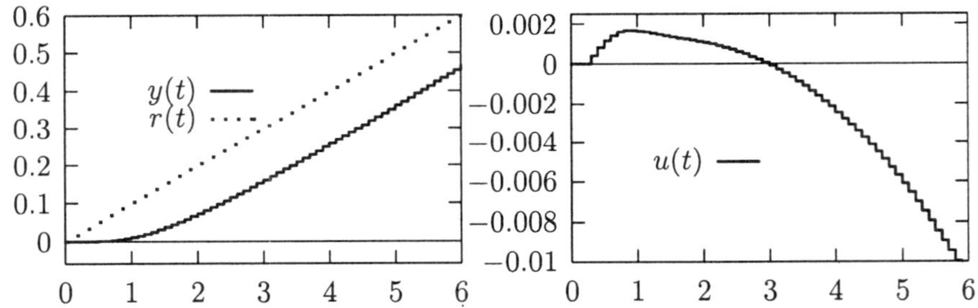

Fig. 11.4 Output response of the system (11.52) and (11.54) for a ramp reference input $r(t)$, where $a_1^d = 1.7$, $a_2^d = -0.72$, $b_1^d = 0$, $b_2^d = 0.02$.

11.2 SISO discrete-time control systems with small parameter

11.2.1 *System with small parameter*

Reasonableness of the two-time-scale motion separation is the weak point of the above design methodology, since the procedure for fast and slow motion separation in the closed-loop system is provided without the presence of some parameter that can play a role similar to that played by μ in the continuous-time case. We now introduce such a parameter artificially [Yurkevich (1997)]. Justification will be provided later, in the next chapter, for sampled-data control systems.

Let us consider a nonlinear time-varying discrete-time system

$$y_k = f(k, Y_k, W_k) + \sum_{j=1}^{n} b_j u_{k-j}, \quad Y_0 = Y^0, \tag{11.55}$$

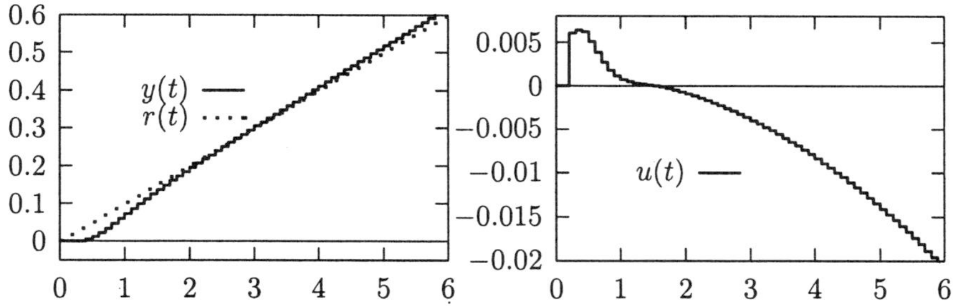

Fig. 11.5 Output response of the system (11.52) and (11.54) for a ramp reference input $r(t)$, where $a_1^d = 1.7$, $a_2^d = -0.72$, $b_1^d = 0.3$, $b_2^d = -0.28$.

where $k = 0, 1, \dots$ is discrete time, y_k is the output (available for measurement), u_k is the control, and w_k is the external disturbance (unavailable for measurement). Here $f(\cdot)$ is a function continuous in y_{k-j}, w_{k-j} for all $j = 1, 2, \dots, n$, and

$$Y_k = \{y_{k-n}, y_{k-n+1}, \dots, y_{k-1}\}^T, \quad W_k = \{w_{k-n}, w_{k-n+1}, \dots, w_{k-1}\}^T.$$

Assumption 11.3 The system (11.55) may be rewritten in the form of the following difference equation:

$$y_k = \sum_{j=1}^{n} a_{n,j} y_{k-j} + \mu \left\{ \tilde{f}(k, Y_k, W_k) + \sum_{j=1}^{n} \tilde{b}_j u_{k-j} \right\}, \; Y_0 = Y^0(\mu), \quad (11.56)$$

where μ is a small parameter, $Y_\mu^0 = Y^0(\mu)$ is the initial state, and

$$a_{n,j} = (-1)^{j+1} \frac{n!}{(n-j)!\, j!}, \quad \sum_{j=1}^{n} a_{n,j} = 1, \quad b_j = \mu \tilde{b}_j,$$
$$\lim_{\mu \to 0} Y^0(\mu) = Y_0^0, \; Y_0^0 = \{y_0^0, y_0^0, \dots, y_0^0\}^T. \quad (11.57)$$

Note that the parameters $a_{n,1}, \dots, a_{n,n}$ are the coefficients of the polynomial

$$z^n - a_{n,1} z^{n-1} - \dots - a_{n,n-1} z - a_{n,n} = (z-1)^n. \quad (11.58)$$

The difference equation (11.56) can be rewritten in the form of the discrete-time state equation

$$Y_{k+1} = A_n Y_k + \mu B_n \{\tilde{f}(k, Y_k, W_k) + \tilde{b}^T \tilde{U}_k\}, \quad Y_0 = Y^0(\mu), \quad (11.59)$$

where $A_n \in \mathbb{R}^{n\times n}$, $B_n \in \mathbb{R}^{n\times 1}$, and

$$A_n = \begin{bmatrix} 0 & 1 & 0 & \cdots & 0 \\ 0 & 0 & 1 & \cdots & 0 \\ \vdots & \vdots & \vdots & & \vdots \\ 0 & 0 & 0 & \cdots & 1 \\ a_{n,n} & a_{n,n-1} & a_{n,n-2} & \cdots & a_{n,1} \end{bmatrix}, \ B_n = \begin{bmatrix} 0 \\ 0 \\ \vdots \\ 0 \\ 1 \end{bmatrix}, \ \tilde{b} = \begin{bmatrix} \tilde{b}_n \\ \tilde{b}_{n-1} \\ \vdots \\ \tilde{b}_2 \\ \tilde{b}_1 \end{bmatrix}, \quad (11.60)$$

$$\tilde{U}_k = \{u_{k-n}, u_{k-n+1}, \dots, u_{k-1}\}^T, \quad b = \{b_n, b_{n-1}, \dots, b_1\}^T, \quad b = \mu\tilde{b}.$$

Assumption 11.4 The roots of the polynomial

$$\tilde{b}_1 z^{n-1} + \tilde{b}_2 z^{n-2} + \cdots + \tilde{b}_{n-1} z + \tilde{b}_n$$

lie outside some neighborhood of 1; that is,

$$\sum_{j=1}^{n} \tilde{b}_j \neq 0. \quad (11.61)$$

Remark 11.8 *If $\mu = 0$, then from (11.56) and (11.57) it follows that we have the system*

$$Y_{k+1} = A_n Y_k, \quad Y_0 = Y_0^0,$$

whose characteristic polynomial takes the form $(z-1)^n$ and $y_k = y_0^0 \ \forall\, k = 0, 1, \dots$.

Remark 11.9 *If u_k and w_k are bounded in (11.56), then, from Remark 11.8 and (11.57), we have*

$$\lim_{\mu\to 0}\{y_k(\mu) - y_{k-j}(\mu)\} = 0, \quad \forall\, j = 1, \dots, n \quad (11.62)$$

for the solutions of (11.56).

11.2.2 *Two-time-scale motion analysis*

Let us consider the output regulation problem (11.4) where, as above, the reference model in the form of the desired difference equation (11.9) is introduced; hence, the control problem (11.4) has been reformulated as the insensitivity condition (11.11). Then the control law is constructed as the difference equation

$$u_k = \sum_{j=1}^{q \geq n} d_j u_{k-j} + \lambda_0(\mu) e_k^F, \tag{11.63}$$

where

$$\lambda_0(\mu) = \mu^{-1}\tilde{\lambda}_0, \quad \tilde{\lambda}_0 \neq 0, \tag{11.64}$$

$$d_1 + d_2 + \cdots + d_q = 1, \quad q \geq n. \tag{11.65}$$

From (11.65) it follows that an equilibrium of (11.63) corresponds to the insensitivity condition (11.11).

Just as we represented (11.55) in the form (11.56), let us rewrite the desired difference equation (11.9) as the difference equation with small parameter

$$y_k = \sum_{j=1}^{n} a_{n,j} y_{k-j} + \mu \tilde{F}(Y_k, R_k) \tag{11.66}$$

and consider the closed-loop system equations

$$y_k = \sum_{j=1}^{n} a_{n,j} y_{k-j} + \mu \left\{ \tilde{f}(k, Y_k, W_k) + \sum_{j=1}^{n} \tilde{b}_j u_{k-j} \right\}, \quad Y_0 = Y^0(\mu), \tag{11.67}$$

$$u_k = \sum_{j=1}^{q \geq n} d_j u_{k-j} + \lambda_0(\mu)\{F(Y_k, R_k) - y_k\}, \quad U_0 = U^0. \tag{11.68}$$

By (11.66) we have that the substitution of (11.67) into (11.68) yields

$$y_k = \sum_{j=1}^{n} a_{n,j} y_{k-j} + \mu \left\{ \tilde{f}(k, Y_k, W_k) + \sum_{j=1}^{n} \tilde{b}_j u_{k-j} \right\}, Y_0 = Y^0(\mu), \tag{11.69}$$

$$\begin{aligned} u_k = {} & \sum_{j=1}^{n} [d_j - \mu\lambda_0(\mu)\tilde{b}_j] u_{k-j} + \sum_{j=n+1}^{q>n} d_j u_{k-j} \\ & + \mu\lambda_0(\mu)\{\tilde{F}(Y_k, R_k) - \tilde{f}(k, Y_k, W_k)\}, \quad U_0 = U^0. \end{aligned} \tag{11.70}$$

Denote

$$U_k = \{u_{k-q}, u_{k-q+1}, \ldots, u_{k-1}\}^T,$$

where the relationship between U_k and $\tilde{U}_k$ in (11.55) is explained by the following layout:

$$\{ \underbrace{u_{k-q}, \ldots, u_{k-n-1}, \overbrace{u_{k-n}, \ldots, u_{k-2}, u_{k-1}}^{\tilde{U}_k^T}}_{U_k^T} \}.$$

Hence $\tilde{U}_k = PU_k$, where

$$P = \mathrm{diag}\{p_1, \ldots, p_q\},$$
$$p_j = 0, \quad j = 1, \ldots, q-n,$$
$$p_j = 1, \quad j = q-n+1, \ldots, q,$$

for $q > n$, and $P = I$ for $q = n$.

In accordance with (11.59)–(11.60), the difference equations (11.69)–(11.70) can be rewritten in the form of the discrete-time state equations

$$Y_{k+1} = A_n Y_k + \mu B_n \{\tilde{f}(k, Y_k, W_k) + \tilde{b}^T P U_k\}, \; Y_0 = Y^0(\mu), \quad (11.71)$$
$$U_{k+1} = A_U \, U_k + B_U \, \{\tilde{F}(Y_k, R_k) - \tilde{f}(k, Y_k, W_k)\}, \; U_0 = U^0, \quad (11.72)$$

where

$$A_U = \begin{bmatrix} 0 & 1 & 0 & \cdots & 0 \\ 0 & 0 & 1 & \cdots & 0 \\ \cdots & \cdots & \cdots & \cdots & \cdots \\ 0 & 0 & 0 & \cdots & 1 \\ \beta_q & \beta_{q-1} & \beta_{q-2} & \cdots & \beta_1 \end{bmatrix}, \quad B_U = \begin{bmatrix} 0 \\ 0 \\ \cdots \\ 0 \\ \lambda_0 \end{bmatrix}.$$

By taking into account (11.64), we have

$$\beta_j = d_j - \lambda_0 \tilde{b}_j, \quad j = 1, \ldots, n, \quad (11.73)$$
$$\beta_j = d_j, \qquad j = n+1, \ldots, q. \quad (11.74)$$

Note that the matrices A_U, B_U do not depend on μ, while μ affects the rate of the transients of y_k. Property (11.62) implies that by decreasing μ we can induce two-time-scale transients in the closed-loop system (11.71)–(11.72), where the rate of the transients of y_k is much smaller than that of u_k. The desired degree of time-scale separation is provided as $\mu \to 0$. Then, as an asymptotic limit, from the closed-loop system equations (11.71)–(11.72) it follows that the FMS is governed by

$$U_{k+1} = A_U \, U_k + B_U \, \{\tilde{F}(Y_k, R_k) - \tilde{f}(k, Y_k, W_k)\}, \quad U_0 = U^0. \quad (11.75)$$

This is an LTI system where Y_k is the frozen vector during the transients in (11.75), i.e., $Y_k - Y_{k-1} \approx 0$.

Let us assume that the stability of the FMS subsystem (11.75) holds, and consider its steady state (or, more exactly, quasi-steady state). Then

$$U_{k+1} - U_k = 0$$

and, accordingly,

$$u_{k-q} = \cdots = u_{k-1} = u_k = u_k^{NID}, \tag{11.76}$$

where

$$u_k^{NID} = \left[\sum_{j=1}^{n} \tilde{b}_j\right]^{-1} \{\tilde{F}(Y_k, R_k) - \tilde{f}(k, Y_k, W_k)\}. \tag{11.77}$$

By taking into account (11.76) and, from (11.71)–(11.72) (or by substituting (11.76)–(11.77) into (11.71)), we obtain the SMS

$$Y_{k+1} = A_n Y_k + \mu B_n \tilde{F}(Y_k, R_k), \quad Y_0 = Y^0(\mu), \tag{11.78}$$

which is the same as (11.66).

So, if a sufficient time-scale separation between the fast and slow modes in the closed-loop system and stability of the FMS are provided, then after rapid decay of FMS transients we have in the closed-loop system the slow motions described by (11.66); hence, the output transient performance indices are as desired and are insensitive to parameter variations and external disturbances in (11.56).

The main result may be formulated as the following theorem.

Theorem 11.1 *If μ is sufficiently small, the two-time-scale transients are induced in the closed-loop system (11.71)–(11.72), where the FMS is governed by (11.75). The SMS is governed by (11.78) given that the FMS is stable.*

11.2.3 *Interrelationship with fixed point theorem*

The FMS (11.75) can be considered as a map

$$\Phi_U : U \to U, \tag{11.79}$$

where

$$U_{k+1} = \Phi_U(U_k, \Upsilon_k) \tag{11.80}$$

and

$$\Upsilon_k = B_U \{\tilde{F}(Y_k, R_k) - \tilde{f}(k, Y_k, W_k)\}.$$

Here Υ_k is considered as the vector of parameters and it has been assumed that $\Upsilon_k = \Upsilon_{k-j}$, $\forall\, j = 1, \dots, q$ and $\Upsilon_k = \Upsilon$.

Using the recursion (11.80) we can get the sequence $\{U_k\}$. If the difference equation (11.75) of the FMS is stable, then $\{U_k\}$ is convergent. This means that $U^\star = \{u^\star, u^\star, \dots, u^\star\}^T$ exists such that

$$\lim_{k\to\infty} \{U_k - U^\star\} = 0, \tag{11.81}$$

where $U^\star$ is the limit of $\{U_k\}$. Accordingly it is the fixed point of the map Φ_U, i.e.,

$$U^\star = \Phi_U(U^\star, \Upsilon). \tag{11.82}$$

By (11.65) we have $U^\star = U^{NID}$ where $U^{NID} = \{u^{NID}, \dots, u^{NID}\}^T$. Then

$$\lim_{k\to\infty} e^F(U_k) = 0. \tag{11.83}$$

Therefore if Φ_U is the contraction map and (11.65) holds, then the fixed point $U^\star$ of Φ_U corresponds to the control action that provides the control problem solution, i.e., $e^F(U^\star) = 0$.

Note that Φ_U is a contraction if all roots of the characteristic polynomial of the FMS (11.75) lie inside the unit disk.

Since there are time-varying parameters in (11.11), we have a time-varying quasi-fixed point $U^\star(k)$ of (11.82). Then, the condition

$$\lim_{k\to\infty} \{U_k - U^\star(k)\} \to 0 \tag{11.84}$$

and, accordingly, the requirement

$$\lim_{k\to\infty} e^F(U_k) \to 0 \tag{11.85}$$

are satisfied as $\mu \to 0$.

Similar to the above, we may consider the SMS (11.78) as a map

$$\Phi_Y : Y \to Y, \tag{11.86}$$

where

$$Y_{k+1} = \Phi_Y(Y_k, R_k). \tag{11.87}$$

Let us assume that

$$r_k = r_{k-j}, \quad \forall\, j = 1, \ldots, n \quad \text{and} \quad r_k = r.$$

Accordingly, we have

$$R_k = R_{k-j}, \quad \forall\, j = 1, \ldots, n, \quad \text{and} \quad R_k = \hat{R},$$

where, by definition, we put $\hat{R} = \{r, r, \ldots, r\}^T$.

Using the recursion (11.87) we can obtain the sequence $\{Y_k\}$. Since the desired difference equation is stable, $\{Y_k\}$ converges to a limit $Y^\star$:

$$\lim_{k\to\infty} \{Y_k - Y^\star\} = 0. \tag{11.88}$$

Accordingly, $Y^\star$ is the fixed point of Φ_Y:

$$Y^\star = \Phi_Y(Y^\star, \hat{R}). \tag{11.89}$$

From (11.6) it follows that $Y^\star = \hat{R}$.

So, from the stability of the desired difference equation (11.9) it follows that Φ_Y is a contraction map. If the condition (11.6) holds, then the fixed point $Y^\star$ of Φ_Y corresponds to the condition $r = y$ for $r = \text{const}$.

11.2.4 *Root placement of FMS characteristic polynomial*

Since the FMS (11.75) is an LTI system, similar to the continuous-time control systems with the highest derivative in feedback, we know that all conventional methods for linear discrete-time control systems (see, for instance, [Lindorff (1965); Chen (1993); Ogata (1994)]) can be applied in order to ensure stability and the allowable FMS transient performance indices for the discrete-time FMS (11.75).

In particular, in this section the design of controller parameters by root placement of the FMS characteristic polynomial is discussed.

The characteristic polynomial $A_{FMS}(z)$ of the FMS (11.75) has the following form:

$$A_{FMS}(z) = z^q - \beta_1 z^{q-1} - \cdots - \beta_{q-1} z - \beta_q. \tag{11.90}$$

Let us construct the desired characteristic polynomial $A^d_{FMS}(z)$ in the form

$$A^d_{FMS}(z) = z^q - \beta^d_1 z^{q-1} - \cdots - \beta^d_{q-1} z - \beta^d_q, \tag{11.91}$$

where the roots of (11.91) are selected in some small neighborhood of the origin in accordance with the requirements placed on the admissible transients in the FMS (11.75).

Theorem 11.2 *The condition*

$$A_{FMS}(z) = A^d_{FMS}(z) \tag{11.92}$$

holds if

$$\tilde{\lambda}_0 = \{1 - \beta^d_s\}\{\tilde{b}_s\}^{-1}, \tag{11.93}$$

$$d_j = \beta^d_j + \tilde{b}_j\tilde{\lambda}_0, \quad j = 1, \ldots, n, \tag{11.94}$$

$$d_j = \beta^d_j, \qquad j = n+1, \ldots, q, \tag{11.95}$$

where $\tilde{b}_s = \tilde{b}_1 + \tilde{b}_2 + \cdots + \tilde{b}_n$ *and* $\beta^d_s = \beta^d_1 + \beta^d_2 + \cdots + \beta^d_q$.

Proof. By substituting (11.93)–(11.95) into (11.73),(11.74), and (11.90), we get (11.91) as desired. ■

The deadbeat response of the FMS (11.75) can be obtained by choosing $A^d_{FMS}(z) = z^q$. Then from (11.93)–(11.95) we have

$$\tilde{\lambda}_0 = 1/\tilde{b}_s, \tag{11.96}$$

$$d_j = \tilde{b}_j\tilde{\lambda}_0, \quad j = 1, \ldots, n, \tag{11.97}$$

$$d_j = 0, \quad j = n+1, \ldots, q. \tag{11.98}$$

The deadbeat response of the FMS (11.75) has a finite settling time given by $t_{s,FMS} = qT_s$. Then the relationship

$$T_s \leq \frac{t_{s,SMS}}{q\eta} \tag{11.99}$$

may be used to estimate the sampling period in accordance with the required degree of time-scale separation between the fast and slow modes in the closed-loop system. Here $t_{s,SMS}$ is the settling time of the SMS and η is the degree of time-scale separation (typically $\eta \geq 10$).

11.2.5 *FMS design based on frequency-domain methods*

Let us consider a nonlinear time-varying discrete-time system

$$y_k = f(k, Y_k, W_k) + \sum_{j=1}^{n} b_j u_{k-j}$$

with control law given by

$$u_k = \sum_{j=1}^{q \geq n} d_j u_{k-j} + \lambda_0 \{F(Y_k, R_k) - y_k\}.$$

Let us represent this closed-loop system in the operator form

$$y_k = f_k(z^{-1}, y_k, w_k) + B(z^{-1})u_k, \tag{11.100}$$

$$D(z^{-1})u_k = \lambda_0 \{F_k(z^{-1}, y_k, w_k) - y_k\}, \tag{11.101}$$

where z^{-1} is considered as the backward shift operator and

$$B(z^{-1}) = b_1 z^{-1} + b_2 z^{-2} + \cdots + b_n z^{-n},$$
$$D(z^{-1}) = 1 - d_1 z^{-1} - d_2 z^{-2} - \cdots - d_q z^{-q}.$$

Accordingly, the desired difference equation (11.7) may be rewritten in the operator form

$$y_k = A^d(z^{-1})y_k + B^d(z^{-1})r_k, \tag{11.102}$$

where

$$A^d(z^{-1}) = a_1^d z^{-1} + a_2^d z^{-2} + \cdots + a_n^d z^{-n},$$
$$B^d(z^{-1}) = b_1^d z^{-1} + b_2^d z^{-2} + \cdots + b_n^d z^{-n}.$$

Substitution of (11.100) into (11.101) yields

$$y_k = f_k + B(z^{-1})u_k,$$
$$\{D(z^{-1}) + \lambda_0 B(z^{-1})\}u_k = \lambda_0 \{F_k - f_k\}.$$

As a result, the FMS equation in operator form is

$$u_k = H_{uf}(z^{-1})\{F_k - f_k\}, \tag{11.103}$$

where

$$H_{uf}(z^{-1}) = \frac{\lambda_0}{D(z^{-1}) + \lambda_0 B(z^{-1})} \tag{11.104}$$

and F_k, f_k are considered as the frozen variables during the transients in the FMS, i.e., $F_k \approx \text{const}$, $f_k \approx \text{const}$.

The block diagram representation of the discrete-time closed-loop system (11.100)–(11.101) is shown in Fig. 11.6. This may be regarded as the discrete-time counterpart of the continuous system (4.45)–(4.46) with the highest derivative in feedback loop shown in Fig. 5.3. Similar to Fig. 5.3, a portion of this diagram is highlighted by a circuit of dots, and this portion corresponds to the FMS (11.103).

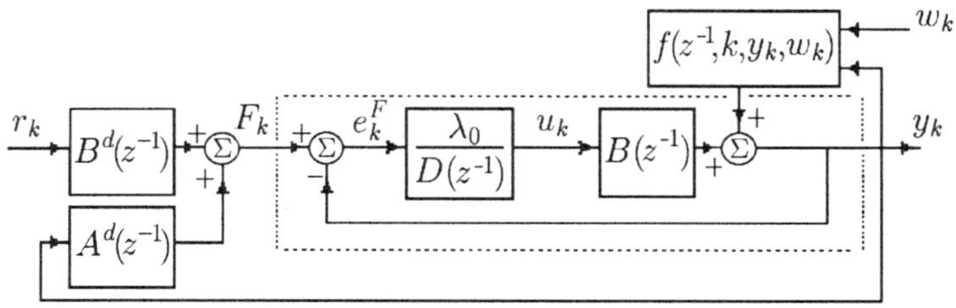

Fig. 11.6 Block diagram of the closed-loop system (11.100)–(11.101).

In order to avail ourselves of frequency-domain methods, let us transform the pulse transfer function $H_{uf}(z^{-1})$ in the z-plane over to the w-plane via the bilinear transformation

$$z = \frac{1 + \mathrm{w}\frac{T_s}{2}}{1 - \mathrm{w}\frac{T_s}{2}}. \tag{11.105}$$

The transfer function

$$G_{uf}(\mathrm{w}) = H_{uf}(z^{-1})\big|_{z=(1+\mathrm{w}T_s/2)(1-\mathrm{w}T_s/2)}$$

results. Here $G_{uf}(\mathrm{w})$ depends on the controller parameters, i.e., $G_{uf}(\mathrm{w}) = G_{uf}(\mathrm{w}, \lambda_0, d_1, \ldots, d_n)$.

The rational transfer function $G_{uf}(\mathrm{w})$ may be treated as $G_{uf}(s)$ in (5.26), with the same design procedure, to find the controller parameters. Note that by taking into account $z = e^{j\omega T_s}$ and $\mathrm{w} = j\nu$ we know that the relationship between ω and the fictitious frequency ν is defined by (see, e.g., [Chen (1993); Ogata (1994)])

$$\nu = \frac{2}{T_s}\tan\left(\omega\frac{T_s}{2}\right)$$

and by this the interval $\omega \in [0, \pi/T_s)$ is mapped into the interval $\nu \in [0, \infty)$.

The controller parameters $\lambda_0, d_1, \ldots, d_n$ can be found from the equation

$$G_{uf}(j\nu, \lambda_0, d_1, \ldots, d_n) = G_{uf}^d(j\nu),$$

where the desired sinusoidal transfer function $G_{uf}^d(j\nu)$ is constructed in accordance with the allowable transient performance indices for the FMS.

Alternatively, let us consider the block diagram of the open-loop discrete-time FMS shown in Fig. 11.7, where the feedback loop is broken.

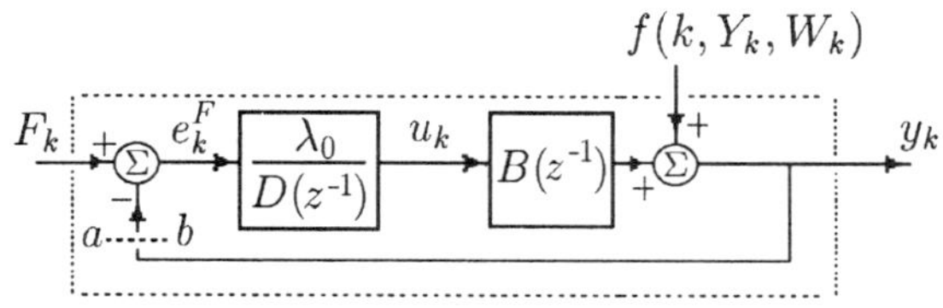

Fig. 11.7 Block diagram of the open-loop discrete-time FMS.

The pulse transfer function of the open-loop discrete-time FMS is

$$H_{zF}^o(z^{-1}) = \frac{\lambda_0}{D(z^{-1})} B(z^{-1}),$$

where $H_c(z^{-1}) = \lambda_0 / D(z^{-1})$ plays the role of a compensator in the FMS. Generalizing by analogy, we may design the compensator in the form

$$H_c(z^{-1}) = \frac{\Lambda(z^{-1})}{D(z^{-1})}, \tag{11.106}$$

where

$$\Lambda(z^{-1}) = \lambda_0 + \lambda_1 z^{-1} + \cdots + \lambda_l z^{-l}.$$

The expression

$$D(z^{-1}) u_k = \Lambda(z^{-1}) e_k^F \tag{11.107}$$

is the operator form of the control law corresponding to (11.106). From (11.107) the difference equation of the control law

$$u_k = \sum_{j=1}^{q \geq n} d_j u_{k-j} + \sum_{j=0}^{l} \lambda_j e_{k-j}^F \tag{11.108}$$

results. Note that (11.108) is the discrete counterpart of the continuous controller (5.73).

Using the bilinear transformation (11.105), the generalized compensator (11.106) can be designed via Bode plot methods [Lindorff (1965); Ogata (1994)] similar to the continuous-time linear compensator (5.72).

11.3 MIMO two-time-scale discrete-time control systems

11.3.1 *MIMO discrete-time systems*

The above design methodology for SISO discrete-time control systems by inducing two-time-scale motions in the closed-loop system can be easily extended for MIMO discrete-time systems. The needed relationships will be provided in this section in a simplified form, with the distinctive features caused by the sampling process relegated to the next chapter.

Let us consider a discrete-time MIMO system described by the following difference equation:

$$y_k = \sum_{j=1}^{n} A_j y_{k-j} + \sum_{j=1}^{n} B_j u_{k-j} + \sum_{j=1}^{n} \hat{B}_j w_{k-j}, \qquad (11.109)$$

where $k = 0, 1, \ldots$ is discrete time; $y_k \in \mathbb{R}^p$ is the output, available for measurement; $u_k \in \mathbb{R}^p$ is the control; $w_k \in \mathbb{R}^p$ is the external disturbance, unavailable for measurement.

Assumption 11.5 The roots of the characteristic polynomial of (11.109) lie in some small neighborhood of 1; in other words, the unforced system (11.109) has a sufficiently small rate of transients in comparison with the deadbeat response of the discrete-time system having the same order.

Assumption 11.6 The condition

$$\det \left[\sum_{j=1}^{n} B_j \right] \neq 0 \qquad (11.110)$$

holds.

Assumptions 11.5 and 11.6 correspond to Assumptions 11.1 and 11.2 for SISO discrete-time systems, respectively.

11.3.2 *Control law*

Control problem

We wish to design a control system for which

$$\lim_{k\to\infty} e_k = 0, \qquad (11.111)$$

where $e_k = r_k - y_k$ is the error of the reference input realization, $r_k \in \mathbb{R}^p$ being the reference input.

Moreover, the controlled transients of the output vector y_k should have desired performance indices such as overshoot, settling time, and system type assigned independently for each of its components. So, the design goal for MIMO control systems is to provide decoupling of output transients and, at the same time, independence of these transients from external disturbances and varying parameters of the system (11.109).

Desired difference equation and insensitivity condition

The control problem (11.111) can be solved if the behavior of the output transients of y_k fulfills a desired stable difference equation

$$y_k = F(y_{k-1}, \ldots, y_{k-n}, r_{k-1}, \ldots, r_{k-n}). \tag{11.112}$$

For instance, the reference model (11.112) may be defined by the stable linear difference equation

$$y_k = \sum_{j=1}^{n} A_j^d y_{k-j} + \sum_{j=1}^{n} B_j^d r_{k-j}, \tag{11.113}$$

where

$$I_p - \sum_{j=1}^{n} A_j^d = \sum_{j=1}^{n} B_j^d \quad \text{and} \quad \det\left[\sum_{j=1}^{n} B_j^d\right] \neq 0. \tag{11.114}$$

Parameters of (11.113) are selected based on the required output transient performance indices. In order to provide decoupling of the control channels, let us assume that A_j^d and B_j^d are diagonal matrices for all $j = 1, \ldots, n$.

Assumption 11.7 The roots of the characteristic polynomial of (11.113) reside in the unit disk in some small neighborhood of 1.

Let

$$e_k^F = F_k - y_k \tag{11.115}$$

be the error of the desired dynamics realization, where F_k is the desired dynamics vector assigned by $F_k = F(y_{k-1}, \ldots, y_{k-n}, r_{k-1}, \ldots, r_{k-n})$. Then equation (11.112), defining the desired behavior of y_k, is fulfilled if and only if

$$e_k^F = 0 \tag{11.116}$$

for all $k = 0, 1, \ldots$.

If (11.116) holds, then the output behavior is insensitive to parameter variations and external disturbances in the system (11.109); i.e., (11.116) is the insensitivity condition.

From (11.109), (11.112), and (11.113), it follows that (11.116) can also be viewed as the difference equation

$$e^F(u_{k-n}, \ldots, u_{k-2}, u_{k-1}) = 0 \tag{11.117}$$

with varying parameters, where (11.117) has the form

$$\sum_{j=1}^{n} \{[A_j^d - A_j]y_{k-j} + B_j^d r_{k-j} - B_j u_{k-j} - \hat{B}_j w_{k-j}\} = 0. \tag{11.118}$$

So the control problem (11.111) has been reformulated as a problem of finding the solution to (11.117) when its varying parameters are unknown.

Discrete-time control law

To fulfill the requirement of equation (11.117), let us construct the control law in the form

$$u_k = \sum_{j=1}^{n} D_j u_{k-j} + \Lambda e_k^F, \tag{11.119}$$

where

$$\sum_{j=1}^{n} D_j = I_p \quad \text{and} \quad \det \Lambda \neq 0. \tag{11.120}$$

From (11.120) it follows that an equilibrium of (11.119) is the solution of (11.117).

11.3.3 Two-time-scale motion analysis

Fast motion subsystem

The equations of the discussed closed-loop system have the following form:

$$y_k = \sum_{j=1}^{n} A_j y_{k-j} + \sum_{j=1}^{n} B_j u_{k-j} + \sum_{j=1}^{n} \hat{B}_j w_{k-j}, \tag{11.121}$$

$$u_k = \sum_{j=1}^{n} D_j u_{k-j} + \Lambda\{F_k - y_k\}. \tag{11.122}$$

From (11.112)–(11.113) it follows that the closed-loop system equations (11.121)–(11.122) can be rewritten as

$$y_k = \sum_{j=1}^{n} A_j y_{k-j} + \sum_{j=1}^{n} B_j u_{k-j} + \sum_{j=1}^{n} \hat{B}_j w_{k-j}, \tag{11.123}$$

$$u_k = \sum_{j=1}^{n} \mathcal{B}_j u_{k-j} + \Lambda \sum_{j=1}^{n} \{[A_j^d - A_j] y_{k-j} + B_j^d r_{k-j} - \hat{B}_j w_{k-j}\}, \tag{11.124}$$

where $\mathcal{B}_j = D_j - \Lambda B_j$ for all $j = 1, \ldots, n$.

Let us consider how the anticipated multi-time-scale process formation is used to provide the desired output transients of y_k under uncertainty. First, we assume that the rate of the transients of y_k in (11.123)–(11.124) is much smaller than that of the transients of u_k.

Then, as an asymptotic limit, from the closed-loop system (11.123)–(11.124) the FMS

$$u_k = \sum_{j=1}^{n} \mathcal{B}_j u_{k-j} + \Lambda \sum_{j=1}^{n} \{[A_j^d - A_j] y_{k-j} + B_j^d r_{k-j} - \hat{B}_j w_{k-j}\} \tag{11.125}$$

results, where $y_k - y_{k-j} = 0, \ \forall\, j = 1, \ldots, n$, i.e., during the transients in the system (11.125).

The above root placement approach can be used to ensure stability and desired performance indices of the dynamical behavior of u_k by selection of the matrices Λ and $D_1, \ldots, D_n$. For instance, all roots of the characteristic polynomial of the FMS (11.125) are placed at the origin, hence the deadbeat response of the FMS (11.125) is provided if the control law parameters Λ, D_j

are selected as follows:

$$\Lambda = \left[\sum_{j=1}^{n} B_j\right]^{-1} \quad \text{and} \quad D_j = \Lambda B_j \,, \quad \forall\, j = 1, \ldots, n. \tag{11.126}$$

Slow motion subsystem

Second, we assume that the FMS subsystem (11.125) is stable and consider its (quasi-) steady state, i.e.,

$$u_k - u_{k-j} = 0, \quad \forall\, j = 1, \ldots, n. \tag{11.127}$$

Then, from (11.123)–(11.124) and (11.127) we get

$$u_k = u_k^{LID},$$

where

$$u_k^{LID} = \left[\sum_{j=1}^{n} B_j\right]^{-1} \sum_{j=1}^{n} \{[A_j^d - A_j] y_{k-j} + B_j^d r_{k-j} - \hat{B}_j w_{k-j}\}. \tag{11.128}$$

Remark 11.10 *Note that the discrete-time control function u_k^{LID} given by (11.128) corresponds to the insensitivity condition (11.116), that is, u_k^{LID} is the discrete-time counterpart of the nonlinear inverse dynamics solution (8.18).*

Remark 11.11 *From (11.128), we see that u_k^{LID} exists if (11.110) is satisfied. So (11.110) is the discrete-time counterpart of (8.4).*

By substituting (11.127) into (11.123)–(11.124) (or (11.127)–(11.128) into (11.123)), we obtain the SMS of (11.123)–(11.124), which is the same as the desired difference equation (11.113). By this, the solution of the above-stated control problem (11.111) is provided.

Note that if the fast and slow motions are sufficiently separated, then the required accuracy of realization for the desired dynamics given by (11.112) is provided; hence, we get the solution of the control problem (11.111) despite the existence of unknown external disturbances and varying parameters of the system (11.109).

The variations of the matrices B_j in the closed-loop system (11.123), (11.124) are possible within the domain where the FMS (11.125) is stable and fast and slow motion separation is maintained.

11.3.4 *Example*

Let us consider the discrete-time 2-input 2-output system given by

$$y_k = \sum_{j=1}^{2} A_j y_{k-j} + \mu \left[\sum_{j=1}^{2} \tilde{A}_j y_{k-j} + \sum_{j=1}^{2} \tilde{B}_j u_{k-j} + \sum_{j=1}^{2} \hat{B}_j w_{k-j} \right], \quad (11.129)$$

where $y_k = \{y_{1,k}, y_{2,k}\}^T$, $u_k = \{y_{1,k}, u_{2,k}\}^T$, $w_k = \{w_{1,k}, w_{2,k}\}^T$, $\mu = 0.1$, and

$$A_1 = \begin{bmatrix} 1 & 0 \\ 0 & 2 \end{bmatrix}, \quad A_2 = \begin{bmatrix} 0 & 0 \\ 0 & -1 \end{bmatrix}, \quad \tilde{A}_1 = \begin{bmatrix} -0.2 & 0.1 \\ 0.2 & 0.2 \end{bmatrix}, \quad \tilde{A}_2 = \begin{bmatrix} 0.1 & 0.2 \\ -0.3 & 0.1 \end{bmatrix},$$

$$\tilde{B}_1 = \begin{bmatrix} 1 & -0.5 \\ 1 & 0 \end{bmatrix}, \quad \tilde{B}_2 = \begin{bmatrix} -1 & 1 \\ 1 & 2 \end{bmatrix}, \quad \hat{B}_1 = \begin{bmatrix} 1 & 0 \\ 0 & 1 \end{bmatrix}, \quad \hat{B}_2 = 0.$$

All roots of the characteristic polynomial of (11.129) are 1 if $\mu = 0$.

Require that the controlled outputs $y_1(t)$, $y_2(t)$ behave as step responses of the transfer functions

$$G_1^d(s) = \frac{1}{\tau_1 s + 1}, \quad G_2^d(s) = \frac{1}{(\tau_2 s + 1)^2}. \quad (11.130)$$

Then the pulse transfer functions $H_1^d(z)$, $H_2^d(z)$ of series connections of a ZOH and the continuous-time systems (11.130) are the functions

$$H_1^d(z) = \frac{1 - d_1}{z - d_1}, \quad (11.131)$$

$$H_2^d(z) = \frac{(1 - d_2 - \tau_2^{-1} T_s d_2) z + d_2 (d_2 - 1 + \tau_2^{-1} T_s)}{z^2 - 2 d_2 z + d_2^2}, \quad (11.132)$$

where $d_1 = \exp(-T_s/\tau_1)$, $d_2 = \exp(-T_s/\tau_2)$. As a result, from (11.131)–(11.132) the reference model of the form (11.112) results:

$$\begin{aligned} y_{1,k} &= d_1 y_{1,k-1} + (1 - d_1) r_{1,k-1}, \\ y_{2,k} &= 2 d_2 y_{2,k-1} - d_2^2 y_{2,k-2} + (1 - d_2 - \tau_2^{-1} T_s d_2) r_{2,k-1} \\ &\quad + d_2 (d_2 - 1 + \tau_2^{-1} T_s) r_{2,k-2}. \end{aligned} \quad (11.133)$$

In accordance with (11.64) and (11.119), by taking into account (11.133) calculated for $T_s = 1$ s, $\tau_1 = 10$ s, and $\tau_2 = 4$ s, we obtain the control law

$$u_k = \sum_{j=1}^{2} D_j u_{k-j} + \mu^{-1} \Lambda \{ F(y_{k-1}, y_{k-2}, r_{k-1}, r_{k-2}) - y_k \}, \quad (11.134)$$

where from (11.126) we have

$$D_1 = \begin{bmatrix} -1.5 & 1 \\ 2 & -1 \end{bmatrix}, \quad D_2 = \begin{bmatrix} 2.5 & -1 \\ -2 & 2 \end{bmatrix}, \quad \Lambda = \begin{bmatrix} -2 & 0.5 \\ 2 & 0 \end{bmatrix}. \tag{11.135}$$

Simulation results for the system (11.129) and (11.134) are displayed in Figs. 11.8–11.10, where $t \in [0, 100]$ s.

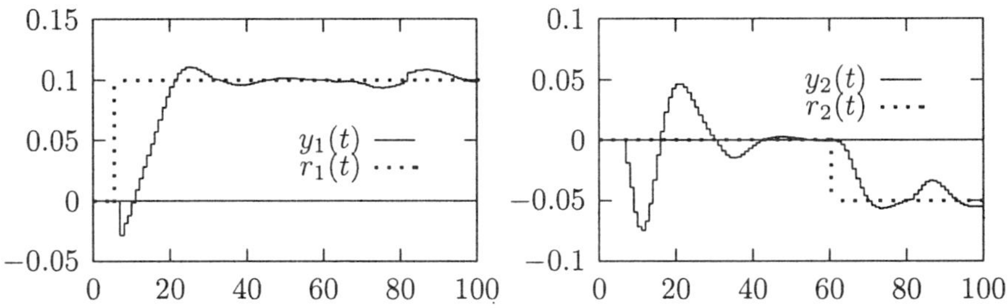

Fig. 11.8 Simulation results for $r_1(t), r_2(t), y_1(t), y_2(t)$ in the system (11.129) and (11.134).

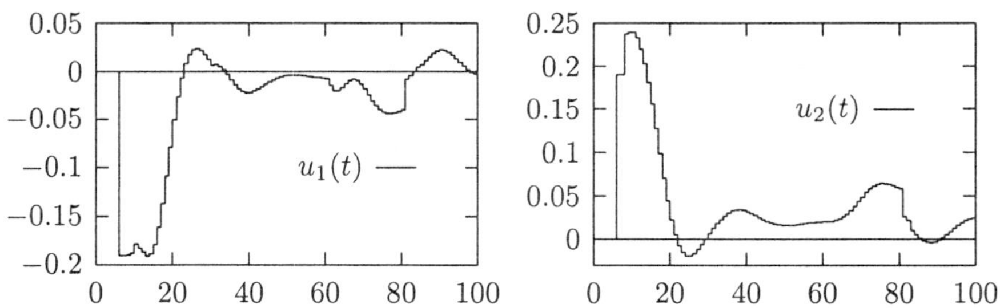

Fig. 11.9 Simulation results for $u_1(t), u_2(t)$ in the system (11.129) and (11.134).

11.4 Notes

In this chapter, the procedure for control law design and analysis of fast and slow motions in the discussed discrete-time control system have been given. This has been based on results published in [Yurkevich (1993a); Yurkevich (1996)]. It has been shown that under some additional conditions, the SMS is the same as the constructed desired stable difference equation and, accordingly, the desired output transients are guaranteed

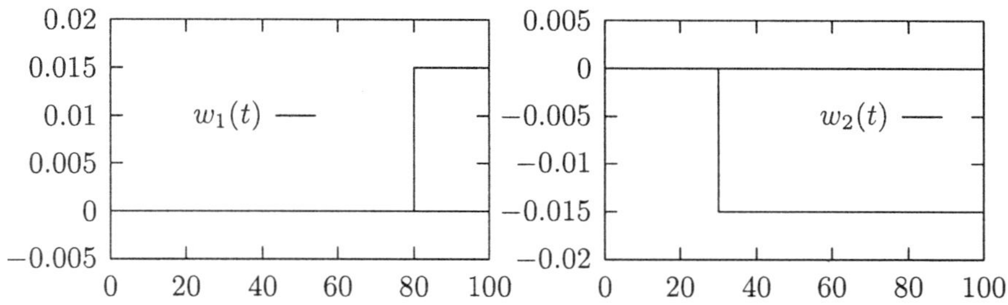

Fig. 11.10 Simulation results for $w_1(t), w_2(t)$ in the system (11.129) and (11.134).

fully in the discrete-time closed-loop system after damping of the discrete-time FMS transients. So, if a sufficient time-scale separation between the fast and slow modes in the discrete-time closed-loop system and stability of the FMS are provided, then the output transient performance indices are insensitive to parameter variations and external disturbances of the system.

Design of discrete-time control system for a reactive ion etching (RIE) system based on the presented above approach was discussed in [Tudoroiu *et al.* (2003a)].

The presented design methodology is the discrete-time counterpart of the above approach to the continuous-time control system design with the highest derivative in feedback. The results of this chapter can be extended for sampled-data control system design by taking into consideration the particularities of the model of a series connection of a ZOH and a continuous-time system with high sampling rate. This will be shown in the next chapter.

11.5 Exercises

11.1 Construct the reference model in the form of the 2nd-order difference equation (11.7) based on the $\mathcal{Z}$-transform in such a way that the step output response with zero initial conditions meets the requirements $t_s^d \approx 1$ s, $\sigma^d \approx 0\,\%$ given that the sampling period is $T_s = 0.05$ s. Compare simulation results with the assignment.

11.2 Construct the reference model (11.7) in the form of a type-1 system given that the characteristic polynomial of the system has the following roots: $z_1 = 0.9$; $z_2 = 0.8 + j0.1$; $z_3 = 0.8 - j0.1$. Simulate the output response of the reference model with zero initial conditions for the reference input signals (a) $r_k = 1(k)$; (b) $r_k = k$; (c) $r_k = k^2$, where

$k = 0, 1, \ldots$.

11.3 Construct the reference model (11.7) in the form of a type-2 system given that the characteristic polynomial of the system has the following roots: $z_1 = 0.8$; $z_2 = 0.7 + j0.1$; $z_3 = 0.7 - j0.1$. Simulate the output response of the reference model with zero initial conditions for the reference input signals (a) $r_k = 1(k)$; (b) $r_k = k$; (c) $r_k = k^2$; (d) $r_k = k^3$, where $k = 0, 1, \ldots$.

11.4 Determine by the $\mathcal{Z}$-transform the discrete-time model of the system

$$y^{(2)} - y = 2u \tag{11.136}$$

preceded by ZOH. Design the control law (11.14) to meet the following specifications: $t_s^d \approx 2$ s; $\sigma^d \approx 0\%$; $\eta \geq 10$. Compare simulation results of the step output response of the closed-loop control system with the assignment. Note that η is the degree of time-scale separation between stable fast and slow motions defined by (1.73).

11.5 Determine by the $\mathcal{Z}$-transform the discrete-time model of the system

$$y^{(2)} - y^{(1)} = 2u \tag{11.137}$$

preceded by ZOH. Design the control law (11.14) to meet the following specifications: $t_s^d \approx 4$ s; $\sigma^d \approx 0\%$; $\eta \geq 10$. Compare simulation results of the step output response of the closed-loop control system with the assignment.

11.6 Determine by the $\mathcal{Z}$-transform the discrete-time model of the system

$$y^{(2)} - 2y^{(1)} + 2y = u \tag{11.138}$$

preceded by ZOH. Design the control law (11.14) to meet the following specifications: $t_s^d \approx 3$ s; $\sigma^d \approx 0\%$; $\eta \geq 10$. Compare simulation results of the step output response of the closed-loop control system with the assignment.

11.7 Determine the discrete-time model of the system

$$y^{(2)} - 2y^{(1)} - 3y = 4u \tag{11.139}$$

preceded by ZOH. Design the discrete-time control law (11.14) to meet the following specifications: $t_s^d \approx 1$ s; $\sigma^d \approx 0\%$; $\eta \geq 10$. Compare simulation results of the step output response of the closed-loop control system with the assignment.

Chapter 12

Design of sampled-data control systems

In this chapter the design methodology for a discrete-time control system with two-time-scale motions is extended for the purpose of sampled-data control system design. This is done by taking into account the particularities of the model of a series connection of a ZOH and a continuous-time system on the condition of high sampling rate. We derive an approximate discrete-time model for a nonlinear time-varying system preceded by a ZOH. The model takes the form of a difference equation having a small parameter, where the parameter depends on the sampling period. Both SISO and MIMO sampled-data control systems are treated.

12.1 SISO sampled-data control systems

12.1.1 *Reduced order pulse transfer function*

As a preliminary, let us consider the asymptotic properties of a pulse transfer function for a series connection of a ZOH and a continuous-time linear system with high sampling rate. We will show that the approximate model of the sampled-data system is a difference equation with a small parameter that depends on the sampling period. Hence, the previous discrete-time control system design methodology can be extended to sampled-data control systems.

Let $G(s)$ be a rational, strictly proper, continuous-time transfer function

$$G(s) = \frac{b_m s^m + \cdots + b_1 s + b_0}{a_n s^n + \cdots + a_1 s + a_0} \tag{12.1}$$

with relative degree $\alpha = n - m > 0$, where $a_n \neq 0$ and $b_m \neq 0$.

Assumption 12.1 The roots of the polynomial

$$b_m s^m + b_{m-1} s^{m-1} + \cdots + b_1 s + b_0$$

lie in the stable half-plane $\operatorname{Re} s < 0$ or, in other words, the internal stability requirement is satisfied for the system given by (12.1).

Consider the Laurent series expansion of $G(s)$ about $s = 0$:

$$G(s) = \sum_{j=\alpha}^{\infty} \frac{m_j}{s^j},$$

where m_j is the jth Markov parameter of (12.1).

Then, based on the $\mathcal{Z}$-transform, it is easy to find the pulse transfer function $H(z)$ of a series connection of a ZOH and a continuous-time system with the transfer function $G(s)$:

$$H(z) = \sum_{j=\alpha}^{\infty} T_s^j \frac{m_j}{j!} \frac{\mathcal{E}_j(z)}{(z-1)^j}. \tag{12.2}$$

Here T_s is the sampling period and the $\mathcal{E}_j(z)$ are known polynomials recently termed the Euler polynomials [Sobolev (1977); Åström *et al.* (1984); Weller *et al.* (1997); Błachuta (1999a)]. They are calculated via the recursion

$$\mathcal{E}_{l+1}(z) = (1 + lz)\mathcal{E}_l(z) + z(1-z)\frac{d\mathcal{E}_l(z)}{dz}$$

with $\mathcal{E}_1(z) = 1$, and $l = 1, 2, \ldots$, where

$$\mathcal{E}_l(z) = \epsilon_{l,1} z^{l-1} + \epsilon_{l,2} z^{l-2} + \cdots + \epsilon_{l,l}, \tag{12.3}$$

$$\epsilon_{l,i} = \sum_{j=1}^{i} (-1)^{i-j} j^l \binom{l+1}{i-j}, \quad i = 1, \ldots, l, \tag{12.4}$$

$$\mathcal{E}_l(1) = l\,!, \tag{12.5}$$

$$\left.\frac{d\mathcal{E}_l(z)}{dz}\right|_{z=1} = \frac{l!\,(l-1)}{2}. \tag{12.6}$$

Alternatively, they can be obtained from the following expression [Tsypkin

(1964)]:

$$\mathcal{E}_l(z) = l! \ \det \begin{bmatrix} 1 & 1-z & 0 & \cdots & 0 \\ \frac{1}{2!} & 1 & 1-z & \cdots & 0 \\ \frac{1}{3!} & \frac{1}{2!} & 1 & \cdots & 0 \\ \cdots & \cdots & \cdots & \cdots & \cdots \\ \frac{1}{l!} & \frac{1}{(l-1)!} & \frac{1}{(l-2)!} & \cdots & 1 \end{bmatrix}.$$

In particular, we have

$$\begin{aligned} \mathcal{E}_1(z) &= 1, \\ \mathcal{E}_2(z) &= z+1, \\ \mathcal{E}_3(z) &= z^2+4z+1, \\ \mathcal{E}_4(z) &= z^3+11z^2+11z+1, \\ \mathcal{E}_5(z) &= z^4+26z^3+66z^2+26z+1. \end{aligned}$$

The properties of the roots of the Euler polynomials $\mathcal{E}_l(z)$ are as follows [Sobolev (1977); Błachuta (1999b)]:

(i) All roots of $\mathcal{E}_l(z)$ are coprime and real negative numbers.
(ii) If $z = \bar{z}$ is the root of $\mathcal{E}_l(z)$ so that $\mathcal{E}_l(\bar{z}) = 0$, then $\mathcal{E}_l(1/\bar{z}) = 0$.
(iii) $\mathcal{E}_l(-1) = 0$ if l is even number, $l > 0$.

Note that the pulse transfer function $H(z)$ has the following property [Åström *et al.* (1984)]:

$$\lim_{T_s \to 0} T_s^{m-n} H(z) = \frac{m_\alpha}{\mathcal{E}_\alpha(1)} \frac{\mathcal{E}_\alpha(z)(z-1)^m}{(z-1)^n}. \tag{12.7}$$

So, if $m > 0$, then from (12.7) it follows that there exists some small sampling period T_s such that m zeros of $H(z)$ fall in any small neighborhood of 1. As a result of (12.7), Assumption 11.2 is not satisfied for the difference equation (11.1), which corresponds to the pulse transfer function $H(z)$ given by (12.2).

In accordance with (12.2) and by following through [Błachuta (1997)], we can see that if $T_s \to 0$, then for any finite $z \in \mathcal{C}$, $z \neq 1$ the pulse transfer function $H(z)$ admits the following approximation:

$$H(z) \approx H_\alpha(z), \tag{12.8}$$

where

$$H_\alpha(z) = T_s^\alpha \frac{m_\alpha}{\alpha\,!} \frac{\mathcal{E}_\alpha(z)}{(z-1)^\alpha}, \quad m_\alpha = \frac{b_m}{a_n}. \tag{12.9}$$

Then, in order to base controller design on the above method, the reduced-order pulse transfer function $H_\alpha(z)$ may be used as the approximate model of the continuous-time linear system (12.1) preceded by a ZOH with high sampling rate. From (12.9) the difference equation

$$y_k = \sum_{j=1}^{\alpha} a_{\alpha,j} y_{k-j} + \mu \sum_{j=1}^{\alpha} m_\alpha \frac{\epsilon_{\alpha,j}}{\alpha\,!} u_{k-j} \tag{12.10}$$

follows, where μ is the small parameter and

$$\mu = T_s^\alpha, \quad y_k = y(t)|_{t=kT_s}, \quad u_k = u(t)|_{t=kT_s},$$

$$(z-1)^\alpha = z^\alpha - a_{\alpha,1} z^{\alpha-1} - a_{\alpha,2} z^{\alpha-2} - \cdots - a_{\alpha,\alpha}, \tag{12.11}$$

$$a_{\alpha,j} = (-1)^{j+1} \frac{\alpha\,!}{(\alpha-j)\,!\,j\,!}, \tag{12.12}$$

Note that (12.10) has the same properties as (11.56), which were discussed above. Hence, the two-time scale technique for controller design can be applied. The approach involving approximate model construction can be extended to nonlinear control systems as will be shown later in this chapter.

12.1.2 *Input-output approximate model of linear system*

Let us consider the state-space model of (12.1) preceded by ZOH:

$$dX/dt = AX + Bu, \quad y = CX, \tag{12.13}$$

where $X \in \mathbb{R}^n$, $y \in \mathbb{R}^1$, $u \in \mathbb{R}^1$. From (12.13) it follows that

$$d^\alpha y/dt^\alpha = CA^\alpha X + CA^{\alpha-1} Bu, \tag{12.14}$$

where

$$CA^{\alpha-1} B = m_\alpha.$$

Let us introduce a new time scale $t_0 = t/T_s$ in (12.13). Then

$$dX/dt_0 = T_s\{AX + Bu\}, \quad y = CX. \tag{12.15}$$

Accordingly, from (12.14) it follows that

$$d^\alpha y/dt_0^\alpha = T_s^\alpha\{CA^\alpha X + m_\alpha u\}. \tag{12.16}$$

It is easy to see that if $T_s \to 0$, then $dX/dt_0 \to 0$ and, accordingly, $X(t) \to$ const in the new time scale. So, if the sampling period T_s is sufficiently small, then it may be assumed that, at least during the sampling period T_s, the condition $X(t) =$ const is satisfied for $kT_s \leq t < (k+1)T_s$. Then, as a result of the $\mathcal{Z}$-transform of (12.16), it follows that

$$y(z) = \frac{\mathcal{E}_\alpha(z)}{\alpha\,!\,(z-1)^\alpha} T_s^\alpha \left\{CA^\alpha X(z) + m_\alpha u(z)\right\}. \tag{12.17}$$

From (12.17) the difference equation

$$y_k = \sum_{j=1}^{\alpha} a_{\alpha,j} y_{k-j} + T_s^\alpha \sum_{j=1}^{\alpha} \frac{\epsilon_{\alpha,j}}{\alpha\,!}\{CA^\alpha X_{k-j} + m_\alpha u_{k-j}\} \tag{12.18}$$

results given that the high sampling rate takes place, where

$$X_k = X(t)|_{t=kT_s}, \quad g_k = g(t, X(t))|_{t=kT_s}.$$

Difference equation (12.18) will be used below in order to approximately describe the behavior of a discrete output sequence $\{y_k\}_{k=0}^\infty$ in a new time scale t_0 or, in other words, in the discrete time scale $k = 0, 1, 2, \ldots$.

12.1.3 *Control law*

Desired difference equation

Let us construct a stable differential equation

$$y^{(\alpha)} = F(y^{(\alpha-1)}, \ldots, y^{(1)}, y, r^{(\rho)}, \ldots, r), \tag{12.19}$$

which follows from the transfer function

$$G_{yr}^d(s) = \frac{b_\rho^d s^\rho + b_{\rho-1}^d s^{\rho-1} + \cdots + b_1^d s + b_0^d}{s^\alpha + a_{\alpha-1}^d s^{\alpha-1} + \cdots + a_1^d s + a_0^d},$$

where $a_0^d = b_0^d$, $\rho < \alpha$, and the parameters of $G_{yr}^d(s)$ are selected based on the required output transient performance indices.

Similar to (11.5), by the $\mathcal{Z}$-transform of a series connection of ZOH and a continuous-time system with the transfer function $G_{yr}^d(s)$, we can obtain the desired stable difference equation

$$y_k = F(Y_k, R_k), \tag{12.20}$$

where $Y_k = \{y_{k-\alpha}, \dots, y_{k-1}\}^T$ and $R_k = \{r_{k-\alpha}, \dots, r_{k-1}\}^T$. In particular, (12.20) has the following form:

$$y_k = \sum_{j=1}^{\alpha} a_j^d y_{k-j} + \sum_{j=1}^{\alpha} b_j^d r_{k-j}. \tag{12.21}$$

By property (12.7), let us assume that (12.21) may be rewritten in the form

$$y_k = \sum_{j=1}^{\alpha} a_{\alpha,j} y_{k-j} + T_s^{\alpha} \tilde{F}(Y_k, R_k, T_s). \tag{12.22}$$

Theorem 12.1 *From (12.19), (12.22) it follows that*

$$\lim_{T_s \to 0} \tilde{F}(Y_k, R_k, T_s) = F(y^{(\alpha-1)}, \dots, y^{(1)}, y, r^{(\rho)}, \dots, r^{(1)}, r)\Big|_{t=kT_s}. \tag{12.23}$$

Proof. We have

$$\lim_{T_s \to 0} \frac{y_k - y_{k-1}}{T_s} = y^{(1)}(t)\Big|_{t=kT_s},$$

$$\lim_{T_s \to 0} \frac{y_k - 2y_{k-1} + y_{k-2}}{T_s^2} = y^{(2)}(t)\Big|_{t=kT_s}.$$

Similarly, from (12.11) and (12.12), the equation

$$\lim_{T_s \to 0} \frac{y_k - \sum_{j=1}^{\alpha} a_{\alpha,j} y_{k-j}}{T_s^{\alpha}} = y^{(\alpha)}(t)\Big|_{t=kT_s}$$

follows. Hence the proof is complete. ∎

Insensitivity condition

Denote

$$e_k^F = F_k - y_k, \tag{12.24}$$

where e_k^F is the error of the desired dynamics realization and $F_k = F(Y_k, R_k)$ is defined by (12.20). As a result, the discussed output regulation problem has been reformulated as the requirement

$$e_k^F = 0, \quad \forall\, k = 0, 1, \dots.$$

This is the insensitivity condition for the output transient performance indices with respect to inherent dynamical properties and parameter variations in the system (12.13).

Control law structure

To fulfill the requirement $e_k^F = 0$ let us consider the control law

$$u_k = \sum_{j=1}^{q\geq\alpha} d_j u_{k-j} + \lambda(T_s)\, e_k^F, \tag{12.25}$$

where

$$\lambda(T_s) = T_s^{-\alpha}\tilde{\lambda}, \qquad \tilde{\lambda} \neq 0, \tag{12.26}$$

$$d_1 + d_2 + \cdots + d_\alpha = 1. \tag{12.27}$$

From (12.27) it follows that an equilibrium of (12.25) corresponds to the insensitivity condition.

The next sections are devoted to the problem of closed-loop system analysis and selection of controller parameters.

12.1.4 *Closed-loop system analysis*

Fast motions

Let us consider the difference equation (12.18) under control in the form of (12.25)–(12.27), that is

$$y_k = \sum_{j=1}^{\alpha} a_{\alpha,j} y_{k-j} + T_s^{\alpha} \sum_{j=1}^{\alpha} \frac{\epsilon_{\alpha,j}}{\alpha!}\{CA^\alpha X_{k-j} + m_\alpha u_{k-j}\}, \tag{12.28}$$

$$u_k = \sum_{j=1}^{q\geq\alpha} d_j u_{k-j} + T_s^{-\alpha}\tilde{\lambda}\{F(Y_k, R_k) - y_k\}. \tag{12.29}$$

From (12.18), (12.22), and (12.24), and by substituting (12.28) into (12.29), we find that (12.28) and (12.29) may be rewritten in the form

$$y_k = \sum_{j=1}^{\alpha} a_{\alpha,j} y_{k-j} + T_s^{\alpha} \sum_{j=1}^{\alpha} \frac{\epsilon_{\alpha,j}}{\alpha!} \{CA^{\alpha} X_{k-j} + m_{\alpha} u_{k-j}\}, \tag{12.30}$$

$$u_k = \sum_{j=1}^{\alpha} \left\{ d_j - \tilde{\lambda} m_{\alpha} \frac{\epsilon_{\alpha,j}}{\alpha!} \right\} u_{k-j} + \sum_{j=\alpha+1}^{q>\alpha} d_j u_{k-j}$$

$$+\tilde{\lambda} \left\{ \tilde{F}(Y_k, R_k, T_s) - \sum_{j=1}^{\alpha} \frac{\epsilon_{\alpha,j}}{\alpha!} CA^{\alpha} X_{k-j} \right\}. \tag{12.31}$$

In accordance with (12.15), it is easy to see that in a new time scale t_0 we have the following fact. If $T_s \to 0$, then the rate of output transients of (12.30) is decreased. Accordingly, the fast and slow modes are induced in the closed-loop input-output system (12.30)–(12.31) as $T_s \to 0$, where a time-scale separation between these modes is represented explicitly by the small parameter T_s.

Theorem 12.2 *If $T_s \to 0$, then from (12.30)–(12.31) it follows that*

$$u_k = \sum_{j=1}^{q \geq \alpha} \beta_j u_{k-j} + \tilde{\lambda} \left\{ \tilde{F}(Y_k, R_k, T_s) - \sum_{j=1}^{\alpha} \frac{\epsilon_{\alpha,j}}{\alpha!} CA^{\alpha} X_{k-j} \right\} \tag{12.32}$$

is the FMS equation, where

$$\beta_j = d_j - \tilde{\lambda} m_{\alpha} \epsilon_{\alpha,j} \{\alpha!\}^{-1}, \quad \forall\, j = 1, \ldots, \alpha, \tag{12.33}$$

$$\beta_j = d_j, \quad \forall\, j = \alpha+1, \ldots, q, \tag{12.34}$$

and the state vector $X(t)$ and the output $y(t)$ of the system of (12.15) are constants during the transients in the FMS (12.32), i.e., $X_k - X_{k-j} \approx 0$ and $y_k - y_{k-j} \approx 0$, $\forall\, j = 1, \ldots, q$.

Proof. By taking into account (12.15) and (12.16), we have that from (12.30)–(12.31) the FMS equation (12.32) follows as $T_s \to 0$. ■

Slow motions of output response

Theorem 12.3 *If the FMS (12.32) is asymptotically stable and its (quasi-) steady state takes place, i.e.,*

$$u_k - u_{k-j} = 0, \quad \forall\, j = 1, \ldots, q, \tag{12.35}$$

then $u_k = u_k^{LID}$, *where*

$$u_k^{LID} = m_\alpha^{-1}\left\{\tilde{F}(Y_k, R_k, T_s) - \sum_{j=1}^{\alpha}\frac{\epsilon_{\alpha,j}}{\alpha!}CA^\alpha x_{k-j}\right\}. \tag{12.36}$$

Proof. Let the FMS (12.32) be asymptotically stable. Hence, by taking into account (12.5), (12.27), and (12.35), we find that from the FMS (12.32) the expression (12.36) results. ■

Theorem 12.4 *If $T_s \to 0$, then the SMS equation, which describes the behaviour of y_k in the closed-loop input-output system (12.30)–(12.31), is the same as (12.22).*

Proof. Let the FMS (12.32) be asymptotically stable and $T_s \to 0$. Then after damping of FMS transients in (12.30)–(12.31) we find that (12.35) and (12.36) are fulfilled. Substitution of (12.35) and (12.36) into (12.30) with due regard for (12.5) yields the SMS equation, which is the same as (12.22). ■

Theorem 12.5 *Let the FMS (12.32) be asymptotically stable. In the closed-loop system (12.30)–(12.31) we have*

$$\{u_k^{LID} - u^{LID}(t)\big|_{t=kT_s}\} \to 0 \tag{12.37}$$

as $T_s \to 0$, where

$$u^{LID}(t) = \frac{1}{m_\alpha}\{F(y^{(\alpha-1)}(t), \ldots, y(t), r^{(\rho)}(t), \ldots, r(t)) - CA^\alpha X(t)\} \tag{12.38}$$

is the linear inverse dynamics solution.

Proof. If $T_s \to 0$, then from (12.15) it follows that $X_k - X_{k-j} \to 0,\ \forall j = 1, \ldots, q$. From (12.5), (12.23), and (12.36), we obtain (12.38). ■

Corollary 12.1 *If $T_s \to 0$, then from (12.37) and (12.38) it follows that the behavior of $y(t)$ tends to the solution of (12.19); hence, there are no hidden output dynamics (oscillations) between sampling instants in the discussed discrete-time control system.*

Note that the system of equations (12.28)–(12.29) is not exactly a closed-loop system, since the state space equation for X_k is not taken into account. The system (12.28)–(12.29) reflects only input-output behavior in the closed-loop system. Hence, the above expressions are valid given that Assumption 12.1 is satisfied. We must emphasize that internal stability of the system (12.13) is the essential point in obtaining the desired input-output mapping, because the effect of the observable internal dynamics can be canceled only if the internal dynamics are smooth and bounded in a specified region of state space. This inference corresponds fully to Remarks 7.11 and 8.12, concerned with the problem of continuous-time control system design.

12.1.5 *Selection of controller parameters*

The above root placement approach or frequency-domain methods can be used to provide the stability and allowable FMS transient performance indices for the FMS (12.32). For instance, the root placement approach can be easily applied to obtain the controller parameters $\tilde{\lambda}, d_1, \ldots, d_q$.

Consider the characteristic polynomial of the FMS (12.32)

$$A_{FMS}(z) = z^q - \beta_1 z^{q-1} - \cdots - \beta_{q-1} z - \beta_q \tag{12.39}$$

and its desired characteristic polynomial

$$A^d_{FMS}(z) = z^q - \beta^d_1 z^{q-1} - \cdots - \beta^d_{q-1} z - \beta^d_q, \tag{12.40}$$

where the roots of (12.40) are selected in some small neighborhood of zero in accordance with the requirements placed on the admissible transients in the FMS.

Theorem 12.6 *The condition*

$$A_{FMS}(z) = A^d_{FMS}(z) \tag{12.41}$$

is fulfilled if and only if the following holds:

$$\tilde{\lambda} = \{1 - \beta^d_s\} m_\alpha^{-1}, \tag{12.42}$$

$$d_j = \beta^d_j + \tilde{\lambda} m_\alpha \epsilon_{\alpha,j} \{\alpha!\}^{-1}, \quad \forall j = 1, \ldots, \alpha, \tag{12.43}$$

$$d_j = \beta^d_j, \quad \forall\, j = \alpha + 1, \ldots, q, \tag{12.44}$$

where $\beta^d_s = \beta^d_1 + \beta^d_2 + \cdots + \beta^d_q$.

Proof. Substitution of (12.42), (12.43), and (12.44) into (12.33), (12.34), and (12.39) yields (12.40). ∎

Corollary 12.2 *Let* $A^d_{FMS}(z) = z^q$. *Then from (12.42), (12.43), and*

(12.44) it follows that the parameters

$$\tilde{\lambda} = m_{\alpha}^{-1}, \tag{12.45}$$

$$d_j = \epsilon_{\alpha,j}\{\alpha\,!\}^{-1}, \quad \forall j = 1, \dots, \alpha, \tag{12.46}$$

$$d_j = 0, \quad \forall\, j = \alpha + 1, \dots, q, \tag{12.47}$$

provide the deadbeat response of the FMS (12.32).

Corollary 12.3 *Let $q = \alpha$. Then from (12.25), (12.26), (12.45), and (12.46), we obtain the following universal digital control law:*

$$u_k = \sum_{j=1}^{q=\alpha} \frac{\epsilon_{\alpha,j}}{\alpha\,!} u_{k-j} + \{T_s^{\alpha} m_{\alpha}\}^{-1} e_k^F, \tag{12.48}$$

where its parameters depend on the relative degree α of the continuous-time system (12.13), the αth Markov parameter, the coefficients of the Euler polynomial $\mathcal{E}_{\alpha}(z)$, and the sampling period T_s.

Remark 12.1 *If $q = \alpha$ and $A_{FMS}(z) = z^{\alpha}$, then, similar to (11.99), the inequality $T_s \leq t_{s,SMS}\{\alpha\,\eta\}^{-1}$ can be used to estimate the sampling period in accordance with the required degree η of time-scale separation between the fast and slow modes in the closed-loop system, where $t_{s,SMS}$ is the settling time in the SMS.*

Remark 12.2 *Note that the internal dynamics are included in the SMS of the closed-loop system given that the desired reference input-controlled output map is satisfied. Therefore, the rate of transients in the internal subsystem should be taken into account in order to make a proper selection of the required sampling rate.*

12.1.6 *Nonlinear sampled-data systems*

Approximate model

The above approach to approximate model derivation can also be used for nonlinear systems preceded by ZOH with high sampling rate. For instance, let us consider the nonlinear system (4.27)

$$x^{(n)} = f(t, X) + g(t, X)u + w, \quad y = x, \quad X(0) = X^0,$$

preceded by ZOH, where $y \in \mathbb{R}^1$ is the output, available for measurement; $u \in \mathbb{R}^1$ is the control; w is the external disturbance, unavailable for measurement; $X = \{x, x^{(1)}, \dots, x^{(n-1)}\}^T$; $X(0)$ is the initial state at $t = 0$.

We can obtain the state-space equations of (4.27) given by

$$\begin{aligned} \dot{x}_i &= x_{i+1}, \quad i = 1, \ldots, n-1, \\ \dot{x}_n &= f(\cdot) + g(\cdot)u + w, \\ y &= x_1. \end{aligned}$$

Let us introduce the new time scale $t_0 = t/T_s$. We obtain

$$\begin{aligned} \frac{d}{dt_0} x_i &= T_s x_{i+1}, \quad i = 1, \ldots, n-1, \\ \frac{d}{dt_0} x_n &= T_s\{f(\cdot) + g(\cdot)u + w\}, \\ y &= x_1, \end{aligned} \tag{12.49}$$

where $dX/dt_0 \to 0$ as $T_s \to 0$. From (12.49) it follows that

$$\frac{d^n y}{dt_0^n} = T_s^n \{f(\cdot) + g(\cdot)u + w\}. \tag{12.50}$$

Assume that the sampling period T_s is sufficiently small such that the conditions $X(t) = \text{const}$, $g(t, X) = \text{const}$ hold for $kT_s \leq t < (k+1)T_s$. Then, by taking the $\mathcal{Z}$-transform of (12.50), we get

$$y(z) = \frac{\mathcal{E}_n(z)}{n!\,(z-1)^n} T_s^n \left\{ f(z) + \{gu\}(z) + w(z) \right\}. \tag{12.51}$$

From (12.51) we get the difference equation

$$\begin{aligned} y_k &= \sum_{j=1}^{n} a_{n,j} y_{k-j} \\ &\quad + T_s^n \sum_{j=1}^{n} \frac{\epsilon_{n,j}}{n!} \left\{ f_{k-j} + g_{k-j} u_{k-j} + w_{k-j} \right\}, \ Y_0 = Y^0(T_s), \end{aligned} \tag{12.52}$$

given that the high sampling rate takes place, where $Y_0 = Y^0(T_s)$ is the vector of the initial conditions, $w_k = w(t)|_{t=kT_s}$, $f_k = f(t)|_{t=kT_s}$.

Control law

Similar to (12.25) and in accordance with (12.52), we get the following control law:

$$u_k = \sum_{j=1}^{q \geq n} d_j u_{k-j} + \lambda(T_s) \left\{ -y_k + \sum_{j=1}^{n} a_j^d y_{k-j} + \sum_{j=1}^{n} b_j^d r_{k-j} \right\}, \tag{12.53}$$

where

$$\lambda(T_s) = T_s^{-n}\tilde{\lambda}, \qquad \tilde{\lambda} \neq 0,$$
$$d_1 + d_2 + \cdots + d_n = 1,$$

and the reference model of the desired output behavior is given by (11.7).

The analysis of the closed-loop system (12.52)–(12.53) properties is the same as was represented above.

Limitations of control accuracy

Limitations of the control accuracy due to external disturbance signal w_k for the discussed control system can be found based on (11.41) and (11.47).

Let $g = \text{const}$, $q = n$, and the control law parameters of (12.53) be chosen in accordance with the requirement of the deadbeat response of FMS. Then, by taking into account the errors of the implementation of the calculated controller parameters represented by $\Delta d_1, \ldots, \Delta d_n$, we get

$$\lambda(T_s) = \{T_s^n g\}^{-1}, \tag{12.54}$$
$$d_j = \frac{\epsilon_{n,j}}{n!} + \Delta d_j, \qquad \forall\, i = 1, \ldots, n, \tag{12.55}$$

where g plays the role of the nth Markov parameter and is assumed known.

Then, from (11.41) and (12.52)–(12.53), we obtain the limit property of the steady-state error e_w^s due to the constant disturbance signal $w_k = w^s$:

$$\lim_{T_s \to 0} \frac{e_w^s}{T_s^n} = \Delta d_s \left\{ \left[1 - \sum_{j=1}^{n} a_j^d\right] [1 - \Delta d_s] - \Delta d_s \sum_{j=1}^{n} (a_j^d - a_{n,j}) \right\}^{-1} w^s. \tag{12.56}$$

Proof of (12.56) see in Appendix A.8.

By definition, Δd_s is given by (11.38). Assume that

$$\Delta d_s = 0.$$

Consider the ramp disturbance signal given by (11.44)

$$w_k = w^v T_s k, \quad k = 0, 1, \ldots.$$

Then, from (11.47), we get

$$\lim_{T_s \to 0} \frac{e_w^v}{T_s^{n+1}} = -\left[1 - \sum_{j=1}^{n} a_j^d\right]^{-1} \left[\frac{n+1}{2} + \sum_{j=1}^{n} j \Delta d_j\right] w^v. \tag{12.57}$$

Proof of (12.57) see in Appendix A.9.

The sampling period T_s can be found based on the relationships (12.56)–(12.57) in addition to (11.99).

12.1.7 *Example*

Let us consider a control problem for the system

$$G(s) = \frac{2.2(s+1)}{(s^3 - 2.5s^2 + 3s - 1)} \tag{12.58}$$

preceded by ZOH, where $m_\alpha = 2.2$ and $\alpha = 2$. In accordance with (12.3) we have $\mathcal{E}_2(z) = z + 1$. Require that the controlled output $y(t)$ behave as a solution to the desired stable difference equation

$$G^d(s) = \frac{1}{(\tau_d s + 1)^2}. \tag{12.59}$$

Then the pulse transfer function $H^d(z)$ of a series connection of ZOH and the continuous-time system of (12.59) is the function

$$H^d(z) = \frac{(1 - d - \tau_d^{-1}T_s d)z + d(d - 1 + \tau_d^{-1}T_s)}{z^2 - 2dz + d^2},$$

where $d = \exp(-T_s/\tau_d)$. As a result, the discrete-time controller (12.48) becomes

$$\begin{aligned} u_k = {} & 0.5u_{k-1} + 0.5u_{k-2} + \{T_s^2 m_2\}^{-1}\{-y_k + 2dy_{k-1} - d^2 y_{k-2} \\ & + (1 - d - \tau_d^{-1}T_s d)r_{k-1} + d(d - 1 + \tau_d^{-1}T_s)r_{k-2}\}. \end{aligned} \tag{12.60}$$

Simulation results for (12.58) controlled by the algorithm (12.60) are displayed in Fig. 12.1 for the time interval $t \in [0, 4]$ s, where $u(t) = u_k$, $\forall\, kT_s \le t < (k+1)T_s$, $T_s = 0.05$ s, and $\tau_d = 0.6$ s.

12.2 MIMO sampled-data control systems

12.2.1 *Control problem*

This section deals with the control system shown in Fig. 10.1 (see p. 234), where the output signals of the DAC and ADC are assumed to be sampled-data signals and amplitude quantization of signals is not taken into account.

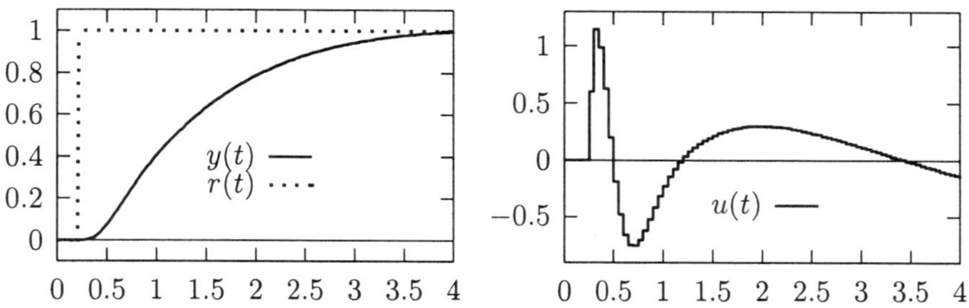

Fig. 12.1 Output response of the system (12.58) and (12.60) for a step reference input $r(t)$.

Assume that the continuous-time system is the nonlinear time-varying system given by

$$\dot{X} = f(t, X) + G(t, X)u, \;\; X(0) = X^0, \tag{12.61}$$

$$y = h(t, X). \tag{12.62}$$

Here t denotes time, $t > 0$; $y(t)$ is the output, $y \in \mathbb{R}^p$; $X(t)$ is the state, $X \in \mathbb{R}^n$; $X(0) = X^0$ is the initial state, $X^0 \in \Omega_X$; Ω_X is a bounded set, $\Omega_X \subset \mathbb{R}^n$; $u(t)$ is the control, $u \in \Omega_u \subset \mathbb{R}^p$; $p \leq n$; $f(t, X)$, $G(t, X)$, $h(t, X)$ are smooth $\forall\ (t, X) \in \Omega_{t,X} = \Omega_X \times [t, \infty)$.

We wish to design a control system for which

$$\lim_{k \to \infty} e_k = 0, \tag{12.63}$$

where $e_k = e(t)|_{t=kT_s}$ is the error of the reference input realization, $e_k = r_k - y_k$; $y_k = y(t)|_{t=kT_s}$ is the sample point of the output $y(t)$; $r_k = r(t)|_{t=kT_s}$ is the sample point of the reference input.

Moreover, the controlled transients $e_k \to 0$ should have desired performance indices. These transients should not depend on external disturbances and varying parameters of the system (12.61)–(12.62), where the influence of all external disturbances $w(t)$ and varying parameters is represented by the dependence of $f(t, X)$, $G(t, X)$, and $h(t, X)$ on t.

Note that throughout this section the problem of output regulation is discussed on the assumption that the above realizability of the desired output behavior is satisfied, that is, the invertibility (Assumption 10.1) and the internal stability of the continuous-time system (Assumption 10.2) both hold. It is also assumed that the DAC acts as a ZOH (Assumption 10.3).

12.2.2 *MIMO continuous-time system preceded by ZOH*

Following [Porter (1970)], by differentiating each component of the output vector in (12.62) along the trajectories of (12.61), we obtain

$$y_* = H^*(t, X) + G^*(t, X)u, \tag{12.64}$$

where

$$y_* = \left\{ \frac{d^{\alpha_1} y_1}{dt^{\alpha_1}}, \frac{d^{\alpha_2} y_2}{dt^{\alpha_2}}, \dots, \frac{d^{\alpha_p} y_p}{dt^{\alpha_p}} \right\}^T, \quad H^* = \{h_1^*, h_2^*, \dots, h_p^*\}^T,$$

and α_i is the relative degree of the system (12.61)–(12.62) with respect to the output $y_i(t)$, where $i = 1, \dots, p$. By Assumption 10.1, the condition

$$\det G^*(t, X) \neq 0, \quad \forall\ (t, X) \in \Omega_{t,X}$$

is satisfied.

Let us introduce a new time scale $t_0 = t/T_s$ in (12.61). Then we get

$$dX/dt_0 = T_s\{f(\cdot) + G(\cdot)u\}, \quad X(0) = X^0, \tag{12.65}$$

where $dX/dt_0 \to 0$ as $T_s \to 0$.

Similar to (8.36), let us introduce the auxiliary control vector $\tilde{u}(t)$, where the actual control vector $u(t)$ depends on $\tilde{u}(t)$ through the matching matrix K_0, that is

$$u = K_0 \tilde{u}. \tag{12.66}$$

Moreover, let us assume that

$$K_0 = \{G^*\}^{-1}.$$

As a result, the mapping of $\tilde{u}$ to $y(t)$ in the new time scale t_0 is given by

$$\left\{ \frac{d^{\alpha_1} y_1}{dt_0^{\alpha_1}}, \frac{d^{\alpha_2} y_2}{dt_0^{\alpha_2}}, \dots, \frac{d^{\alpha_p} y_p}{dt_0^{\alpha_p}} \right\}^T = \boldsymbol{\mathcal{T}}\{H^*(\cdot) + \tilde{u}\}, \tag{12.67}$$

where $\boldsymbol{\mathcal{T}} = \operatorname{diag}\{T_s^{\alpha_1}, T_s^{\alpha_2}, \dots, T_s^{\alpha_p}\}$.

Assume that the sampling period T_s is sufficiently small that we can take $H^*(t, X) = \text{const}$. Then, similar to (12.17), by taking the $\mathcal{Z}$-transform of (12.67) we obtain

$$y_i(z) = \frac{\mathcal{E}_{\alpha_i}(z)}{\alpha_i!\ (z-1)^{\alpha_i}} T_s^{\alpha_i} \left\{ h_i^*(z) + \tilde{u}_i(z) \right\}, \quad i = 1, \dots, p, \tag{12.68}$$

where $y_{i,k} = y_i(t)|_{t=kT_s}$, $\tilde{u}_{i,k} = \tilde{u}_i(t)|_{t=kT_s}$, $h^*_{i,k} = h^*_i(t, X(t))|_{t=kT_s}$, and $\mathcal{E}_i(z)$ are the Euler polynomials (12.3).

From (12.68) we obtain the difference equations

$$\begin{aligned} y_{i,k} &= \sum_{j=1}^{\alpha_i} a_{\alpha_i,j} y_{i,k-j} \\ &+ T_s^{\alpha_i} \sum_{j=1}^{\alpha_i} \frac{\epsilon_{\alpha_i,j}}{\alpha_i\,!} \left\{ h^*_{i,k-j} + \tilde{u}_{i,k-j} \right\}, \quad i = 1, \ldots, p, \end{aligned} \tag{12.69}$$

as an approximate model of the mapping of $\tilde{u}$ to y in the discrete time given that the high sampling rate takes place.

Remark 12.3 *By taking $T_s = 0$, from (12.69) we get the difference equations*

$$y_{i,k} = \sum_{j=1}^{\alpha_i} a_{\alpha_i,j} y_{i,k-j}, \quad i = 1, \ldots, p, \tag{12.70}$$

having characteristic polynomials $(z-1)^{\alpha_i}$, $\forall\, i = 1, \ldots, p$.

12.2.3 *Control law*

Desired difference equations

Consider the stable continuous-time transfer function

$$G_i^d(s) = \frac{b^d_{i,\rho_i} s^{\rho_i} + b^d_{i,\rho_i-1} s^{\rho_i-1} + \cdots + b^d_{i,0}}{s^{\alpha_i} + a^d_{i,\alpha_i-1} s^{\alpha_i-1} + \cdots + a^d_{i,0}}, \tag{12.71}$$

where $b^d_{i,0} = a^d_{i,0}$ and the parameters of $G_i^d(s)$ are selected based on the required output transient performance indices of $y_i(t)$. From (12.71) the differential equation

$$y_i^{(\alpha_i)} = F_i(Y_i, R_i) \tag{12.72}$$

can be obtained, where $Y_i = [y_i, \ldots, y_i^{(\alpha_i-1)}]^T$, $R_i = [r_i, \ldots, r_i^{(\rho_i)}]^T$, $\rho_i < \alpha_i$, and $r_i = y_i$ at the equilibrium of (12.72) for $r_i = \text{const}$. As a result, the set of p equations (12.71) gives the reference model for the desired behavior of the output vector $y(t)$ in the form

$$y_* = F(Y, R), \tag{12.73}$$

where

$$Y = \{y_1, \ldots, y_1^{(\alpha_1 - 1)}, y_2, \ldots, y_p^{(\alpha_p - 1)}\}^T, \ R = \{r_1, \ldots, r_1^{(\rho_1)}, r_2, \ldots, r_p^{(\rho_p)}\}^T.$$

Similar to (11.5), by the $\mathcal{Z}$-transform from (12.71), the pulse transfer function $H_i^d(z)$ of a series connection of ZOH and a continuous-time system with the transfer function $G_i^d(s)$ can be obtained, where

$$H_i^d(z)\big|_{z=1} = 1. \tag{12.74}$$

From $H_i^d(z)$ the desired stable difference equation (reference model)

$$y_{i,k} = F_i(Y_{i,k}, R_{i,k})$$

follows, which, similar to (12.22), can be represented in the form

$$y_{i,k} = \sum_{j=1}^{\alpha_i} a_{\alpha_i,j} y_{i,k-j} + T_s^{\alpha_i} \tilde{F}_i(Y_{i,k}, R_{i,k}, T_s), \tag{12.75}$$

where $r_{i,k} = y_{i,k}$ at the equilibrium for all $i = 1, \ldots, p$.

Note that, in accordance with (12.23), the asymptotic limit

$$\lim_{T_s \to 0} \tilde{F}_i(Y_{i,k}, R_{i,k}, T_s) = F_i(y_i^{(\alpha_i - 1)}, \ldots, y_i, r_i^{(\rho_i)}, \ldots, r_i^{(1)}, r_i)\Big|_{t=kT_s} \tag{12.76}$$

holds.

Insensitivity condition

Let us denote $e^F = F(Y, R) - y_*$. Accordingly, if

$$\mathbf{e}^F = 0, \tag{12.77}$$

then the desired behavior of $y(t)$ with prescribed dynamics of (12.73) is fulfilled, where (12.77) is the insensitivity condition for the output transient performance indices with respect to the external disturbances and varying parameters of the system (12.61)–(12.62). At the same time, since all p transfer functions (12.71) are mutually independent, channel decoupling is provided if (12.77) holds.

Note, in accordance with Assumption 10.2, that the stability or at least boundedness of the internal behavior of (12.61)–(12.62) occurs if (12.77) is satisfied.

Define $F_{i,k} = F_i(Y_{i,k}, R_{i,k})$ and denote

$$e_{i,k}^F = F_{i,k} - y_{i,k}. \tag{12.78}$$

Then the desired behavior of $y_{i,k}$ is fulfilled if and only if

$$e_{i,k}^F = 0 \tag{12.79}$$

for all $k = 0, 1, \ldots$. If (12.79) holds, then the output transient performance indices of $y_{i,k}$ are insensitive to external disturbances and parameter variations in the system (12.61)–(12.62).

Control law structure

In order to fulfill the requirement of (12.79), let us construct the control law in the form of the difference equation

$$\tilde{u}_{i,k} = \sum_{j=1}^{q_i \geq \alpha_i} d_{i,j} \tilde{u}_{i,k-j} + \lambda_i(T_s)\, e_{i,k}^F, \tag{12.80}$$

where $i = 1, \ldots, p$ and

$$\lambda_i(T_s) = T_s^{-\alpha_i} \tilde{\lambda}_i, \qquad \tilde{\lambda}_i \neq 0, \tag{12.81}$$

$$d_{i,1} + d_{i,2} + \cdots + d_{i,q_i} = 1. \tag{12.82}$$

From (12.82) it follows that an equilibrium of (12.80) is the solution of equation (12.79).

12.2.4 *Fast-motion subsystem*

Consider the system of the difference equations (12.69) with the discussed control law (12.80), that is, the closed-loop input-output system equations

$$y_{i,k} = \sum_{j=1}^{\alpha_i} a_{\alpha_i,j} y_{i,k-j} + T_s^{\alpha_i} \sum_{j=1}^{\alpha_i} \frac{\epsilon_{\alpha_i,j}}{\alpha_i\,!} \{h_{i,k-j}^* + \tilde{u}_{i,k-j}\}, \tag{12.83}$$

$$\tilde{u}_{i,k} = \sum_{j=1}^{q_i \geq \alpha_i} d_{i,j} \tilde{u}_{i,k-j} + T_s^{-\alpha_i} \tilde{\lambda}_i\, e_{i,k}^F, \quad i = 1, \ldots, p. \tag{12.84}$$

Theorem 12.7 *If $T_s \to 0$, then fast and slow modes are induced in the closed-loop system (12.83)–(12.84). Here*

$$\tilde{u}_{i,k} = \sum_{j=1}^{q_i \geq \alpha_i} \beta_{i,j} \tilde{u}_{i,k-j} + \tilde{\lambda}_i \left\{ \tilde{F}_i - \sum_{j=1}^{\alpha_i} \frac{\epsilon_{\alpha_i,j}}{\alpha_i\,!} h_{i,k-j}^* \right\} \tag{12.85}$$

is the FMS equation of the ith channel, where $h^*_{i,k} - h^*_{i,k-j} \approx 0$, $y_{i,k} - y_{i,k-j} \approx 0$, $\forall\, j = 1, \ldots, q_i$ *and*

$$\beta_{i,j} = d_{i,j} - \tilde{\lambda}_i \epsilon_{\alpha_i,j} \{\alpha_i\,!\}^{-1}, \quad \forall\, j = 1, \ldots, \alpha_i, \tag{12.86}$$

$$\beta_{i,j} = d_{i,j}, \qquad \forall\, j = \alpha_i + 1, \ldots, q_i. \tag{12.87}$$

Proof. By taking into account (12.78) and by substituting (12.83) into (12.84), we find that the closed-loop system equations (12.83)–(12.84) may be rewritten in the form

$$y_{i,k} = \sum_{j=1}^{\alpha_i} a_{\alpha_i,j} y_{i,k-j} + T_s^{\alpha_i} \sum_{j=1}^{\alpha_i} \frac{\epsilon_{\alpha_i,j}}{\alpha_i\,!} \{h^*_{i,k-j} + \tilde{u}_{i,k-j}\}, \tag{12.88}$$

$$\begin{aligned} \tilde{u}_{i,k} = {} & \sum_{j=1}^{\alpha_i} \left\{ d_{i,j} - \tilde{\lambda}_i \frac{\epsilon_{\alpha_i,j}}{\alpha_i\,!} \right\} \tilde{u}_{i,k-j} + \sum_{j=\alpha_i+1}^{q_i} d_{i,j} \tilde{u}_{i,k-j} \\ & + \tilde{\lambda}_i \left\{ \tilde{F}_i(Y_{i,k}, R_{i,k}, T_s) - \sum_{j=1}^{\alpha_i} \frac{\epsilon_{\alpha_i,j}}{\alpha_i\,!} h^*_{i,k-j} \right\}. \end{aligned} \tag{12.89}$$

From (12.65) and (12.67) we see that the rate of output transients of (12.88) is decreased in a discrete-time scale k as $T_s \to 0$. Accordingly, fast and slow modes are induced in the closed-loop system (12.88)–(12.89) as $T_s \to 0$, where a time-scale separation between these modes is represented by the parameter T_s. If T_s is sufficiently small, then

$$h^*_{i,k} - h^*_{i,k-j} \approx 0, \quad y_{i,k} - y_{i,k-j} \approx 0, \quad \forall\, j = 1, \ldots, q_i. \tag{12.90}$$

Thus, as an asymptotic limit, from (12.88)–(12.89) and (12.90) the equation (12.85) of the FMS follows. ∎

12.2.5 *Selection of controller parameters*

Asymptotic stability of the FMS for the ith channel, desired transient performance indices of $\tilde{u}_{i,k}$, and the desired settling time of the FMS can be achieved by selection of the control law parameters. For instance, the above root placement approach or frequency-domain methods may be used to choose the control law parameters in accordance with the requirements placed on the admissible transients in FMS (12.85).

Let $q_i = \alpha_i$. Then from (12.85) we obtain the characteristic polynomial $A_i^{FMS}(z)$ of the FMS equation of the ith channel in the form

$$A_i^{FMS}(z) = z^{\alpha_i} - \beta_{i,1} z^{\alpha_i - 1} - \cdots - \beta_{i,\alpha_i}. \tag{12.91}$$

The deadbeat response of the FMS (12.85) is provided and, accordingly, the settling time t_s^{FMS} of the FMS of the ith channel is equal to $\alpha_i T_s$ if the requirement

$$A_i^{FMS}(z) = z^{\alpha_i} \tag{12.92}$$

is satisfied. Similar to (12.45), (12.46), and (12.47), from (12.92) and expressions (12.82) and (12.86) we obtain the controller parameters

$$d_{i,j} = \epsilon_{\alpha_i,j}\{\alpha_i\,!\}^{-1}, \quad \forall\, j = 1, \ldots, \alpha_i, \tag{12.93}$$

$$\tilde{\lambda}_i = 1, \qquad i = 1, \ldots, p. \tag{12.94}$$

We can see that the parameters $d_{i,j}$ of the controller with deadbeat response of the FMS depend on the relative degrees $\{\alpha_i\}_{i=1}^{p}$ of continuous-time system (12.61)–(12.62) and the coefficients of the Euler polynomials.

12.2.6 *Slow-motion subsystem*

Theorem 12.8 *If a (quasi-) steady state for the FMS (12.85) takes place, i.e.,*

$$\tilde{u}_{i,k} - \tilde{u}_{i,k-j} = 0, \quad \forall\, j = 1, \ldots, q_i\,, \tag{12.95}$$

then $\tilde{u}_{i,k} = \tilde{u}_{i,k}^{NID}$, *where*

$$\tilde{u}_{i,k}^{NID} = \tilde{F}_i(Y_{i,k}, R_{i,k}, T_s) - \sum_{j=1}^{\alpha_i} \frac{\epsilon_{\alpha_i,j}}{\alpha_i\,!} h_{i,k-j}^{*}\,. \tag{12.96}$$

Proof. From (12.82), (12.85), (12.86), (12.87), and (12.95), the expression (12.96) results. ∎

Theorem 12.9 *If $T_s \to 0$ and the FMS of (12.85) is asymptotically stable, then the SMS equation of $y_{i,k}$ in the closed-loop system (12.88)–(12.89) is the same as (12.75).*

Proof. Let the FMS be asymptotically stable. If $T_s \to 0$, then after the FMS transients in (12.88)–(12.89), we find that (12.95) and (12.96) are fulfilled. By substituting (12.95) and (12.96) into (12.88), we obtain the SMS equation, which is the same as (12.75). ■

Theorem 12.10 *If $T_s \to 0$, then $\tilde{u}_{i,k}^{NID} - \tilde{u}_i^{NID}(t)\big|_{t=kT_s} \to 0$, where*

$$\tilde{u}_i^{NID}(t) = F_i(Y_i(t), R_i(t)) - h_i^*(t, X(t)) \tag{12.97}$$

is the nonlinear inverse dynamics solution.

Proof. If $T_s \to 0$, then from (12.65) it follows that $X_k - X_{k-j} \to 0, \ \forall\, j = 1, \ldots, q_i$. In accordance with (12.5) and (12.76) we find that from (12.96), the expression (12.97) follows. ■

Corollary 12.4 *If $T_s \to 0$, then from (12.97) it follows that the behavior of $y_i(t)$ tends to the solution of (12.72). Accordingly, the controlled output transients in the closed-loop system have desired performance indices after fast ending of FMS transients. Therefore, no hidden oscillations exist between sampling instants in the discussed sampled-data control system.*

Note that the system of equations (12.88)–(12.89) is not exactly a closed-loop system, since the state space equation for X is not taken into account. The system (12.88)–(12.89) reflects only input-output behavior in the closed-loop system and the above expressions are valid given that Assumption 10.2 is satisfied. We emphasize that the internal stability of the system (12.61)–(12.62) is the essential point in obtaining the desired input-output mapping, since the effect of the observable internal dynamics can be canceled only if the internal dynamics are stable or at least bounded in a specified region of the system state space.

12.2.7 *Example*

Consider the following nonlinear time-varying system given by (7.73) (see p. 170):

$$\begin{aligned}
\dot{x}_1 &= x_2 + x_3(x_1 - x_3)(x_3 + x_4 - x_1) + [2 + \sin(x_4)]u_1 + u_2, \\
\dot{x}_2 &= -(x_1 - x_3)(x_3 + x_4 - x_1) + w_2(t) - u_1 + [1 + 0.5\sin(x_3)]u_2, \\
\dot{x}_3 &= x_3(x_1 - x_3)(x_3 + x_4 - x_1) + [2 + \sin(x_4)]u_1 + u_2, \\
\dot{x}_4 &= x_2 - 12(x_3 + x_4 - x_1) + w_4(t) + u_1 + u_2, \\
y_1 &= x_1 - x_3, \qquad y_2 = x_3.
\end{aligned}$$

From (7.73) it follows that $\alpha_1 = 2, \alpha_2 = 1$ and

$$G^* = \begin{bmatrix} -1 & 1 + 0.5\sin(x_5) \\ 2 + \sin(x_4) & 1 \end{bmatrix}, \tag{12.98}$$

where Assumptions 10.1, 10.2 are satisfied (see p. 170). Assume that $K_0 = \{k_{ij}\} \approx G^*$, where

$$K_0 = \begin{bmatrix} -1/3 \; 1/3 \\ 2/3 \; 1/3 \end{bmatrix}.$$

Require that the controlled outputs $y_1(t)$, $y_2(t)$ behave as step responses of the transfer functions

$$G_1^d(s) = \frac{1}{(\tau_1 s + 1)^2}, \quad G_2^d(s) = \frac{1}{\tau_2 s + 1}. \tag{12.99}$$

Then pulse transfer functions $H_1^d(z)$, $H_2^d(z)$ of series connections of ZOH and continuous-time systems of (12.99) are the functions

$$H_1^d(z) = \frac{(1 - d_1 - \tau_1^{-1}T_s d_1)z + d_1(d_1 - 1 + \tau_1^{-1}T_s)}{z^2 - 2d_1 z + d_1^2}, \quad H_2^d(z) = \frac{1 - d_2}{z - d_2},$$

where $d_1 = \exp(-T_s/\tau_1)$, $d_2 = \exp(-T_s/\tau_2)$. As a result, the discrete-time controller has the form

$$\begin{aligned} \tilde{u}_{1,k} &= 0.5\tilde{u}_{1,k-1} + 0.5\tilde{u}_{1,k-2} \\ &\quad + T_s^{-2}\{-y_{1,k} + 2d_1 y_{1,k-1} - d_1^2 y_{1,k-2} \\ &\quad + (1 - d_1 - \tau_1^{-1}T_s d_1) r_{1,k-1} \\ &\quad + d_1(d_1 - 1 + \tau_1^{-1}T_s) r_{1,k-2}\}, \\ \tilde{u}_{2,k} &= \tilde{u}_{2,k-1} + T_s^{-1}\{-y_{2,k} + d_2 y_{2,k-1} + (1 - d_2) r_{2,k-1}\}, \end{aligned} \tag{12.100}$$

where $\{u_{1,k}, u_{2,k}\}^T = K_0\{\tilde{u}_{1,k}, \tilde{u}_{2,k}\}^T$.

Simulation results for (7.73) controlled by the algorithm (12.100) are displayed in Fig. 12.2 for the time interval $t \in [0, 3]$ s, where $u(t) = u_k$, $\forall\, kT_s \le t < (k+1)T_s$, $T_s = 0.05$ s, $\tau_1 = 0.5$ s, $\tau_2 = 0.4$ s, and $w_4(t) = 0$, $\forall\, t$.

12.3 Notes

Methods for digital nonlinear control system design have proliferated in recent times, with attention given to such problems as digital implementation of nonlinear systems with input-to-state stabilizing controller [Teel

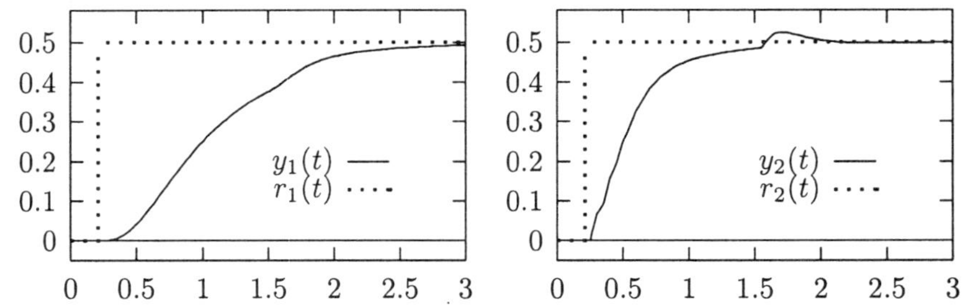

Fig. 12.2 Output step responses in the closed-loop system (7.73) and (12.100).

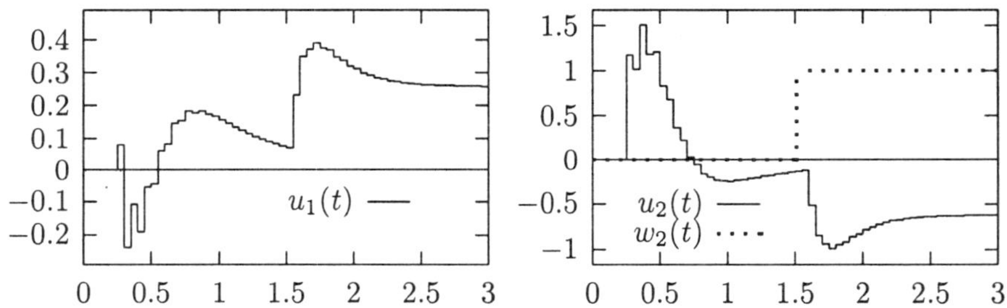

Fig. 12.3 Output step responses in the closed-loop system (7.73) and (12.100).

et al. (1998)], multirate composite control strategies [Djemai *et al.* (1999)], discrete-time approximated linearization [Barbot *et al.* (1999)], performance of systems with state feedback controller and high-gain observer under sampled data [Dabroom and Khalil (2001)]. A digital implementation of the acceleration feedback control law for robot manipulator was discussed in [Studenny *et al.* (1991)].

The approach of this chapter is related to control systems with the highest output derivative in feedback [Vostrikov (1977a)] and its digital implementation is discussed in [Mutschkin (1988); Fehrmann *et al.* (1989)]. In this chapter, the results published in [Yurkevich (1993a); Yurkevich (1999a); Yurkevich (2000b); Yurkevich (2002)] were extended and a systematic design procedure for sampled-data control systems with high sampling rate has been represented. The distinctive feature of the discussed control systems is that two-time scale motions are induced in the closed-loop system with increased sampling period. The stability of the fast transients and a sufficient degree of time-scale separation between the fast and slow modes in the closed-loop system are provided by selection of the controller parameters and sampling rate. As a result, the slow modes have the desired output transient performance indices and are insensitive to external disturbances and parameter variations of the system. The design methodology allows us to find simple analytical expressions for digital controller parameters for nonlinear time-varying systems. Design of a digital flight controller for an aircraft based on the presented approach was discussed in [Blachuta *et al.* (2000); Yurkevich *et al.* (2000)], while a digital controller for robot manipulator was discussed in [Yurkevich and Shatalov (2000)].

12.4 Exercises

12.1 Consider the system (7.136) preceded by ZOH where $a = 3$. Design the discrete-time control law (12.25) to meet the following specifications: $\overline{\varepsilon}_r = 0$, $t_s^d \approx 2$ s, $\sigma^d \approx 5\%$. Run a computer simulation of the closed-loop system with zero initial conditions. Compare simulation results of the output response with the assignment for $r(t) = 1, \ \forall\, t > 0$.

12.2 Consider the system (7.137) preceded by ZOH where $a = 1$. Design the discrete-time control law (12.25) to meet the following specifications: $\overline{\varepsilon}_r = 0$, $t_s^d \approx 3$ s, $\sigma^d \approx 10\%$. Run a computer simulation of the closed-loop system.

12.3 Consider the system (7.138) preceded by ZOH. Design the discrete-time control law (12.25) to meet the following specifications: $\overline{\varepsilon}_r = 0$, $t_s^d \approx 2$ s, $\sigma^d \approx 0\%$. Run a computer simulation of the closed-loop system.

12.4 Consider the system (7.139) preceded by ZOH. Design the discrete-time control law (12.25) to meet the following specifications: $\overline{\varepsilon}_r = 0$, $t_s^d \approx 4$ s, $\sigma^d \approx 20\%$. Run a computer simulation of the closed-loop system.

12.5 Consider the system (7.140) preceded by ZOH. Design the discrete-time control law (12.25) to meet the following specifications: $\overline{\varepsilon}_r = 0$, $t_s^d \approx 3$ s, $\sigma^d \approx 10\%$. Run a computer simulation of the closed-loop system.

12.6 Consider the system (7.97) preceded by ZOH. Design the discrete-time control law (12.25) to meet the following specifications: $\overline{\varepsilon}_r = 0$, $t_s^d \approx 2$ s, $\sigma^d \approx 0\%$. Run a computer simulation of the closed-loop system.

12.7 Consider the system (7.144) preceded by ZOH. Design the discrete-time control law (12.25) to meet the following specifications: $\overline{\varepsilon}_r = 0$, $t_s^d \approx 6$ s, $\sigma^d \approx 0\%$. Run a computer simulation of the closed-loop system.

12.8 Consider the system (7.145) preceded by ZOH. Design the discrete-time control law (12.25) to meet the following specifications: $\bar{\varepsilon}_r = 0$, $t_s^d \approx 2$ s, $\sigma^d \approx 0\%$. Run a computer simulation of the closed-loop system.

12.9 Consider the system (7.146) preceded by ZOH. Design the discrete-time control law (12.25) to meet the following specifications: $\bar{\varepsilon}_r = 0$, $t_s^d \approx 3$ s, $\sigma^d \approx 10\%$. Run a computer simulation of the closed-loop system.

12.10 Consider the system (7.141) preceded by ZOH. Design the discrete-time control law (12.25) to meet the following specifications: $\bar{\varepsilon}_{r1} = 0$, $\bar{\varepsilon}_{r2} = 0$, $t_{s1}^d \approx 2$ s, $\sigma_1^d \approx 0\%$, $t_{s2}^d \approx 3$ s, $\sigma_2^d \approx 0\%$. Run a computer simulation of the closed-loop system.

12.11 Consider the system (7.142) preceded by ZOH. Design the discrete-time control law (12.25) to meet the following specifications: $\bar{\varepsilon}_{r1} = 0$, $\bar{\varepsilon}_{r2} = 0$, $t_{s1}^d \approx 3$ s, $\sigma_1^d \approx 0\%$, $t_{s2}^d \approx 1$ s, $\sigma_2^d \approx 0\%$. Run a computer simulation of the closed-loop system.

12.12 Consider the system (7.143) preceded by ZOH. Design the discrete-time control law (12.25) to meet the following specifications: $\bar{\varepsilon}_{r1} = 0$, $\bar{\varepsilon}_{r2} = 0$, $t_{s1}^d \approx 2$ s, $\sigma_1^d \approx 0\%$, $t_{s2}^d \approx 4$ s, $\sigma_2^d \approx 0\%$. Run a computer simulation of the closed-loop system.

12.13 Consider the system

$$\begin{aligned} \dot{x}_1 &= x_3, \quad \dot{x}_3 = x_1^2 + x_1 x_2 + u_1 - u_2, \\ \dot{x}_2 &= x_4, \quad \dot{x}_4 = x_1 + \sin(x_2) - u_1 - 2u_2, \\ y_1 &= x_1, \quad y_2 = x_2, \end{aligned}$$

which is preceded by ZOH. Check the invertibility and internal stability of this system and design the discrete-time control law (12.25) to meet the following specifications: $\bar{\varepsilon}_{r1} = 0$, $\bar{\varepsilon}_{r2} = 0$, $t_{s1}^d \approx 2$ s, $\sigma_1^d \approx 0\%$, $t_{s2}^d \approx 4$ s, $\sigma_2^d \approx 10\%$. Run a computer simulation of the closed-loop system.

Chapter 13

Control of distributed parameter systems

In this chapter the above design methodology is applied to systems governed by partial differential equations. The representation of the solution for initial and boundary value problems by Fourier series is the essential point, and leads to analysis of infinite-dimensional systems of differential equations. The chapter is concerned with the control problem for systems governed by parabolic equations, while the results can be extended for other types of partial differential equations. At the beginning, the system with infinite-dimensional control is considered; then the particularities of finite-dimensional control for distributed parameter systems are discussed.

13.1 One-dimensional heat system with distributed control

One-dimensional heat equation

Let us consider a heating process described by a one-dimensional parabolic equation given by

$$\frac{\partial x}{\partial t}(z,t) = \alpha^2 \frac{\partial^2 x}{\partial z^2}(z,t) + c(t)x(z,t) + w(z,t) + u(z,t), \tag{13.1}$$

where t is time, $t > 0$, z is the spatial variable, $0 < z < 1$, $x(z,0) = x^0(z)$ is the initial condition, $[\partial x(z,t)/\partial z]|_{z=0} = 0$ and $[\partial x(z,t)/\partial z]|_{z=1} = 0$ are the boundary conditions, $c(t)$ is an unknown varying parameter, $|c(t)| \le c_0 < \infty$, $w(z,t)$ is a distributed external disturbance unavailable for measurement, $u(z,t)$ is the distributed control, α^2 is a constant (we shall take $\alpha^2 = 1$). We assume also that for all functions $x(z,t)$, $x^0(z)$, $w(z,t)$, and

$u(z,t)$ the eigenfunction expansions

$$x(z,t) = \sum_{n=0}^{\infty} x_n(t)\varphi_n(z), \quad x^0(z) = \sum_{n=0}^{\infty} x_n^0 \varphi_n(z),$$

$$u(z,t) = \sum_{n=0}^{\infty} u_n(t)\varphi_n(z), \quad w(z,t) = \sum_{n=0}^{\infty} w_n(t)\varphi_n(z),$$

hold where $\varphi_n(z) = \varphi_n^0 \cos(\sqrt{\lambda_n}\, z)$ are eigenfunctions (also known as spatial modes or normal modes), $\lambda_n = n^2\pi^2$ are eigenvalues, $\varphi_0^0 = 1$, $\varphi_n^0 = \sqrt{2}\ \forall\, n = 1, 2, \ldots$.

In this case, the time functions $x_n(t)$ for $x(z,t)$ satisfy the equations [Wang (1972); Ray (1981)]

$$\begin{aligned} \dot{x}_n(t) &= \{c(t) - \lambda_n\} x_n(t) + w_n(t) + u_n(t), \\ x_n(0) &= x_n^0, \quad n = 0, 1, \ldots. \end{aligned} \tag{13.2}$$

Note that $x_n(t)$ is the weighting coefficient for the nth eigenfunction (mode) in the eigenfunction expansion of $x(z,t)$.

It is possible that some or all of the first n_0 equations are unstable due to variations of the parameter $c(t)$, where $n_0 = \mathrm{int}(\sqrt{c_0}/\pi)$. Here $\mathrm{int}(y)$ is the integer part of y.

Control problem statement

The control problem is to provide a desired spatial temperature distribution assigned by a function $x^d(z,t)$:

$$\lim_{t\to\infty} \sup_{0<z<1} \{x^d(z,t) - x(z,t)\} = 0, \tag{13.3}$$

where $x^d(z,t)$ is defined by the eigenfunction expansion

$$x^d(z,t) = \sum_{n=0}^{\infty} x_n^d(t)\varphi_n(z). \tag{13.4}$$

Moreover, the control transients $x(z,t) \to x^d(z,t)$ should have a desired behavior. These transients should not depend on the external disturbances $w(z,t)$ and varying parameter $c(t)$ of the system (13.1).

Insensitivity condition

Denote by $e_n(t) = x_n^d(t) - x_n(t)$ a realization error of the desired time function $x_n^d(t)$. Then the requirement (13.3) corresponds to

$$\lim_{t\to\infty} e_n(t) = 0, \qquad n = 0, 1, \ldots. \tag{13.5}$$

So, the desired behavior of the transients $x(z,t) \to x^d(z,t)$ can be provided if the process $x_n(t) \to x_n^d(t)$ satisfies the desired differential equation

$$\dot{x}_n(t) = F_n(x_n(t), x_n^d(t)) \tag{13.6}$$

for each time function $x_n(t)$. The parameters of (13.6) are selected in accordance with assigned transient performance indices. For instance, the linear differential equation in the form

$$\dot{x}_n(t) = T_n^{-1}[x_n^d(t) - x_n(t)] \tag{13.7}$$

is the most convenient in this case.

Denote $e_n^F = F_n - \dot{x}_n(t)$, where e_n^F is the realization error of the desired dynamics assigned by (13.6). As a result, the control problem (13.3) can be solved if

$$e_n^F = 0, \quad \forall\, n = 0, 1, \ldots. \tag{13.8}$$

This is the insensitivity condition of the transients in the system (13.1) with respect to the external disturbances $w(z,t)$ and varying parameter $c(t)$.

By (13.2) and (13.6) the expression (13.8) can be rewritten in the form

$$F_n(x_n(t), x_n^d(t)) + \{\lambda_n - c(t)\}x_n(t) - w_n(t) - u_n(t) = 0, \tag{13.9}$$
$$\forall\, n = 0, 1, \ldots$$

So, the discussed control problem has been reformulated as the requirement to provide the condition (13.9) or, in other words, to find a solution to (13.9) when its varying parameters are unknown.

The solution of (13.9) consists of the functions $u_n(t) = u_n^{LID}(t)$ defined by

$$u_n^{LID}(t) = F_n(x_n(t), x_n^d(t)) + \{\lambda_n - c(t)\}x_n(t) - w_n(t), \tag{13.10}$$
$$\forall\, n = 0, 1, \ldots,$$

where $u_n^{LID}(t)$ is called the linear inverse dynamics solution since the discussed control system (13.1) is linear. The control functions $u_n^{LID}(t)$ correspond to the above nonlinear inverse dynamics solution u^{NID} (see p. 193).

As a result, we see that the distributed control function

$$u^{LID}(z,t) = \sum_{n=0}^{\infty} u_n^{LID}(t)\varphi_n(z)$$

$$= \sum_{n=0}^{\infty} \left[F_n(x_n(t), x_n^d(t)) + \{\lambda_n - c(t)\}x_n(t) - w_n(t)\right] \varphi_n(z) \quad (13.11)$$

gives the desired behavior of transients $x(z,t) \to x^d(z,t)$.

Note that $u^{LID}(z,t)$ cannot be used in practice; otherwise, all parameters and external disturbances must be known and available for measurement in the system (13.1).

Realizability of desired behavior

Assume that the region of allowable values of the control function $u(z,t)$ is assigned by

$$|u(z,t)| \leq M_u < \infty, \qquad \forall\, t \geq 0 \text{ and } 0 < z < 1 \quad (13.12)$$

since in a real plant the control resource is always bounded. Here $M_u =$ const is such a bound.

Then, in accordance with (13.10), the solution of the discussed control problem exists and the desired behavior of transients $x(z,t) \to x^d(z,t)$ is realizable if

$$|u^{LID}(z,t)| \leq M_u < \infty, \quad \forall\, t \geq 0 \text{ and } 0 < z < 1. \quad (13.13)$$

From (13.7) and (13.13) we find that if the conditions

$$|x_n^0| \leq \frac{M_x^0}{n^4}, \quad \sup_{t\geq 0} |x_n^d(t)| \leq \frac{M_x^d}{n^4},$$

$$\sup_{t\geq 0} |w_n(t)| \leq \frac{M_w}{n^2}, \quad \forall\, n = 0, 1, \ldots, \quad (13.14)$$

are fulfilled, then the series (13.11) is absolutely convergent and a certain value M_u exists such that the requirement (13.13) is satisfied where

$$M_x^0 = \text{const}, \quad M_x^d = \text{const}, \quad M_w = \text{const}.$$

Control law

In order to ensure that $e_n^F = 0$ when the parameter $c(t)$ is varying and unknown and the distributed external disturbance $w(z,t)$ is unavailable for measurement, let us consider the control law for the equation of the time function $x_n(t)$ given by

$$\begin{aligned} &\mu_n^{q_n} u_n^{(q_n)} + d_{n,q_n-1}\mu_n^{q_n-1}u_n^{(q_n-1)} + \cdots + d_{n,1}\mu_n u_n^{(1)} \\ &+d_{n,0}u_n = k_n\{F_n(x_n, x_n^d) - x_n^{(1)}\}, \quad U_n(0) = U_n^0, \end{aligned} \tag{13.15}$$

where

$$\begin{aligned} &\mu_n > 0, \quad q_n \geq 1, \quad d_{n,j} > 0, \ \forall\, j = 1, \dots, q_n - 1, \quad d_{n,0} = 1 \text{ or } d_{n,0} = 0 \\ &U_n = \{u_n, u_n^{(1)}, \dots, u_n^{(q_n-1)}\}^T, \quad U_n \in \Omega_{U_n} \subset \mathbb{R}^{q_n}, \quad U_n(0) \in \Omega_{U_n}^0 \subset \Omega_{U_n}. \end{aligned}$$

The closed-loop system equations for the nth mode are

$$x_n^{(1)}(t) = \{c(t) - \lambda_n\}x_n(t) + w_n(t) + u_n(t), \ x_n(0) = x_n^0, \tag{13.16}$$

$$\begin{aligned} &\mu_n^{q_n} u_n^{(q_n)}(t) + d_{n,q_n-1}\mu_n^{q_n-1}u_n^{(q_n-1)}(t) + \cdots + d_{n,1}\mu_n u_n^{(1)}(t) \\ &\quad +d_{n,0}u_n(t) = k_n\{F_n(x_n, x_n^d) - x_n^{(1)}\}, \quad U_n(0) = U_n^0, \end{aligned} \tag{13.17}$$

where $n = 0, 1, \dots$.

Equations (13.16)–(13.17) are the singularly perturbed equations with small parameter μ. These are the same as (4.45)–(4.46), and the analysis is similar.

So, distributed controller design is reduced to controller design for each separate mode.

13.2 Heat system with finite-dimensional control

Modal form of heat system

Consider the distributed parameter system given by (13.1) with the distributed control $u(t,z)$ induced by the finite-dimensional control vector

$$\boldsymbol{u}(t) = \{u_1(t), u_2(t), \dots, u_p(t)\}^T,$$

where

$$u(t,z) = \sum_{i=1}^{p} s_i(z)u_i(t).$$

The behaviors of the time functions $x_n(t)$ for $x(t,z)$ are governed by the infinite-dimensional system of differential equations (the so-called modal form of the linear distributed system [Wang (1972); Ray (1981)]) given by

$$\dot{\boldsymbol{x}}_1(t) = \{c(t)I_p - \Lambda_1\}\boldsymbol{x}_1(t) + B_1\boldsymbol{u}(t) + \boldsymbol{w}_1(t),\ \ \boldsymbol{x}_1(0) = \boldsymbol{x}_1^0, \quad (13.18)$$

$$\dot{\boldsymbol{x}}_2(t) = \{c(t)I_\infty - \Lambda_2\}\boldsymbol{x}_2(t) + B_2\boldsymbol{u}(t) + \boldsymbol{w}_2(t),\ \ \boldsymbol{x}_2(0) = \boldsymbol{x}_2^0, \quad (13.19)$$

where

$$\begin{aligned} \boldsymbol{x}_1 &= \{x_0, x_1, \dots, x_{p-1}\}^T, \quad & \boldsymbol{x}_2 &= \{x_p, x_{p+1}, x_{p+2}, \dots\}^T, \\ \boldsymbol{w}_1 &= \{w_0, w_1, \dots, w_{p-1}\}^T, \quad & \boldsymbol{w}_2 &= \{w_p, w_{p+1}, w_{p+2}, \dots\}^T, \\ \Lambda_1 &= \operatorname{diag}\{\lambda_0, \dots, \lambda_{p-1}\}, \quad & \Lambda_2 &= \operatorname{diag}\{\lambda_p, \lambda_{p+1}, \lambda_{p+2}, \dots\}, \quad (13.20) \end{aligned}$$

$$B_1 = \begin{bmatrix} (s_1, \varphi_0\ \ ) & \cdots & (s_p, \varphi_0\ \ ) \\ (s_1, \varphi_1\ \ ) & \cdots & (s_p, \varphi_1\ \ ) \\ \vdots & \vdots & \vdots \\ (s_1, \varphi_{p-1}) & \cdots & (s_p, \varphi_{p-1}) \end{bmatrix}, B_2 = \begin{bmatrix} (s_1, \varphi_p\ \ ) & \cdots & (s_p, \varphi_p\ \ ) \\ (s_1, \varphi_{p+1}) & \cdots & (s_p, \varphi_{p+1}) \\ (s_1, \varphi_{p+2}) & \cdots & (s_p, \varphi_{p+2}) \\ \vdots & \vdots & \vdots \end{bmatrix}.$$

Here (s_i, φ_j) is defined as

$$(s_i, \varphi_j) = \int_0^1 s_i(z)\varphi_j(z)\,dz.$$

Assumption 13.1 The spatial distributions of the controls $u_i(t)$ defined by the functions $s_i(x)$ are chosen so that B_1 is invertible, i.e., $\det B_1 \neq 0$.

Control problem

Assume $y(t) = \boldsymbol{x}_1(t)$ is the output vector, available for measurement. Then the relative degree of the system (13.18)–(13.19) for each component of $y(t)$ is equal to 1.

The system (13.18)–(13.19) corresponds to the normal form of the state space representation given by (7.42)–(7.43) where, in contrast to (7.43), the internal subsystem (13.19) is the infinite-dimensional system of differential equations and the external subsystem (13.18) describes the behavior of the measurable output vector $y(t) = \boldsymbol{x}_1(t)$.

Consider the output regulation problem

$$\lim_{t\to\infty} \boldsymbol{e}(t) = 0, \quad (13.21)$$

where $\boldsymbol{e}(t) = \boldsymbol{x}_1^d(t) - \boldsymbol{x}_1(t)$ and the reference model of the desired behavior of $\boldsymbol{x}_1(t)$ is assigned by the stable differential equation

$$\dot{\boldsymbol{x}}_1(t) = F(\boldsymbol{x}_1(t), \boldsymbol{x}_1^d(t)) \tag{13.22}$$

and $\boldsymbol{x}_1^d = \boldsymbol{x}_1$ at the equilibrium of (13.22) for $\boldsymbol{x}_1^d = \text{const}$.

Denote by

$$\boldsymbol{e}^F = F(\boldsymbol{x}_1(t), \boldsymbol{x}_1^d(t)) - \dot{\boldsymbol{x}}_1(t) \tag{13.23}$$

the realization error of the desired dynamics (13.22).

As a result, the solution of the output regulation problem corresponds to the requirement

$$\boldsymbol{e}^F = 0, \tag{13.24}$$

where (13.24) is the insensitivity condition of the output transients in the system (13.18)–(13.19) with respect to the external disturbances $w(z,t)$ and varying parameter $c(t)$.

From (13.18),(13.22), (13.23), and (13.24) we can obtain the linear inverse dynamics solution given by

$$\boldsymbol{u}^{LID}(t) = \{B_1\}^{-1}[F(\boldsymbol{x}_1(t), \boldsymbol{x}_1^d(t)) - \{c(t)I_p - \Lambda_1\}\boldsymbol{x}_1(t) - \boldsymbol{w}_1(t)]. \tag{13.25}$$

Closed-loop system

In order to fulfill the requirement (13.24) let us consider the control law for the equation of the first p time functions $\boldsymbol{x}_1(t)$ given by

$$\begin{aligned} D_q\boldsymbol{\mu}^q\tilde{\boldsymbol{u}}^{(q)} + D_{q-1}\boldsymbol{\mu}^{q-1}\tilde{\boldsymbol{u}}^{(q-1)} + \cdots + D_1\boldsymbol{\mu}\tilde{\boldsymbol{u}}^{(1)} + D_0\tilde{\boldsymbol{u}} \\ = K_1\{F(\boldsymbol{x}_1, \boldsymbol{x}_1^d) - \boldsymbol{x}_1^{(1)}\}, \end{aligned} \tag{13.26}$$

where

$$\boldsymbol{u} = K_0\tilde{\boldsymbol{u}}. \tag{13.27}$$

The closed-loop system is

$$\dot{\boldsymbol{x}}_1 = \{c(t)I_p - \Lambda_1\}\boldsymbol{x}_1 + B_1K_0\tilde{\boldsymbol{u}} + \boldsymbol{w}_1,\ \boldsymbol{x}_1(0) = \boldsymbol{x}_1^0, \tag{13.28}$$

$$\dot{\boldsymbol{x}}_2 = \{c(t)I_\infty - \Lambda_2\}\boldsymbol{x}_2 + B_2K_0\tilde{\boldsymbol{u}} + \boldsymbol{w}_2,\ \boldsymbol{x}_2(0) = \boldsymbol{x}_2^0, \tag{13.29}$$

$$\begin{aligned} D_q\boldsymbol{\mu}^q\tilde{\boldsymbol{u}}^{(q)} + D_{q-1}\boldsymbol{\mu}^{q-1}\tilde{\boldsymbol{u}}^{(q-1)} + \cdots \\ + D_1\boldsymbol{\mu}\tilde{\boldsymbol{u}}^{(1)} + D_0\tilde{\boldsymbol{u}} = K_1\boldsymbol{e}^F,\ \tilde{U}(0) = \tilde{U}^0. \end{aligned} \tag{13.30}$$

By (13.22) and (13.23), and by substituting (13.28) into (13.30), we obtain

$$\dot{\boldsymbol{x}}_1 = \{c(t)I_p - \Lambda_1\}\boldsymbol{x}_1 + B_1K_0\tilde{\boldsymbol{u}} + \boldsymbol{w}_1, \ \boldsymbol{x}_1(0) = \boldsymbol{x}_1^0, \tag{13.31}$$

$$\dot{\boldsymbol{x}}_2 = \{c(t)I_\infty - \Lambda_2\}\boldsymbol{x}_2 + B_2K_0\tilde{\boldsymbol{u}} + \boldsymbol{w}_2, \ \boldsymbol{x}_2(0) = \boldsymbol{x}_2^0, \tag{13.32}$$

$$\begin{aligned} &D_q\boldsymbol{\mu}^q\tilde{\boldsymbol{u}}^{(q)} + D_{q-1}\boldsymbol{\mu}^{q-1}\tilde{\boldsymbol{u}}^{(q-1)} + \cdots + D_1\boldsymbol{\mu}\tilde{\boldsymbol{u}}^{(1)} + \Gamma\tilde{\boldsymbol{u}} \\ &\quad = K_1[F(\boldsymbol{x}_1, \boldsymbol{x}_1^d) - \{c(t)I_p - \Lambda_1\}\boldsymbol{x}_1 - \boldsymbol{w}_1], \ \tilde{U}(0) = \tilde{U}^0, \end{aligned} \tag{13.33}$$

where

$$\Gamma = D_0 + K_1B_1K_0.$$

Note that the internal subsystem (13.32) of the discussed system does not affect the behavior of the external subsystem (13.31); that is, (13.32) is an unobservable subsystem if $y(t) = \boldsymbol{x}_1(t)$ is the output variable. Therefore, instead of (13.31), (13.32), and (13.33), let us consider the closed-loop input-output system equations

$$\dot{\boldsymbol{x}}_1 = \{c(t)I_p - \Lambda_1\}\boldsymbol{x}_1 + B_1K_0\tilde{\boldsymbol{u}} + \boldsymbol{w}_1, \ \boldsymbol{x}_1(0) = \boldsymbol{x}_1^0, \tag{13.34}$$

$$\begin{aligned} &D_q\boldsymbol{\mu}^q\tilde{\boldsymbol{u}}^{(q)} + D_{q-1}\boldsymbol{\mu}^{q-1}\tilde{\boldsymbol{u}}^{(q-1)} + \cdots + D_1\boldsymbol{\mu}\tilde{\boldsymbol{u}}^{(1)} + \Gamma\tilde{\boldsymbol{u}} \\ &\quad = K_1[F(\boldsymbol{x}_1, \boldsymbol{x}_1^d) - \{c(t)I_p - \Lambda_1\}\boldsymbol{x}_1 - \boldsymbol{w}_1], \ \tilde{U}(0) = \tilde{U}^0, \end{aligned} \tag{13.35}$$

given that the asymptotic stability of the subsystem (13.32) is satisfied by reason of which subsystem (13.32) can be omitted.

The system (13.34), (13.35) are the singularly perturbed equations where fast and slow modes are induced as $\boldsymbol{\mu} \to 0$ and the fast-motion subsystem is given by

$$\begin{aligned} &D_q\boldsymbol{\mu}^q\tilde{\boldsymbol{u}}^{(q)} + D_{q-1}\boldsymbol{\mu}^{q-1}\tilde{\boldsymbol{u}}^{(q-1)} + \cdots + D_1\boldsymbol{\mu}\tilde{\boldsymbol{u}}^{(1)} + \Gamma\tilde{\boldsymbol{u}} \\ &\quad = K_1[F(\boldsymbol{x}_1, \boldsymbol{x}_1^d) - \{c(t)I_p - \Lambda_1\}\boldsymbol{x}_1 - \boldsymbol{w}_1], \ \tilde{U}(0) = \tilde{U}^0, \end{aligned} \tag{13.36}$$

where $\boldsymbol{x}_1$ is the frozen variable during the transients in (13.36).

Assume that the stability and sufficiently fast damping of the FMS transients in (13.36) are provided by selection of the control law parameters $D_i, K_1, K_0, \boldsymbol{\mu}$ and consider the (quasi-) steady state of the FMS (13.36). Then, by finding the limit $\boldsymbol{\mu} \to 0$ in (13.36), we obtain

$$\tilde{\boldsymbol{u}}(t) = \tilde{\boldsymbol{u}}^s(t),$$

where

$$\tilde{\boldsymbol{u}}^s = \Gamma^{-1}K_1[F(\boldsymbol{x}_1, \boldsymbol{x}_1^d) - \{c(t)I_p - \Lambda_1\}\boldsymbol{x}_1 - \boldsymbol{w}_1]. \tag{13.37}$$

Here $\tilde{\boldsymbol{u}}^s(t)$ is the control function in the closed-loop system which corresponds to the quasi-steady state of the FMS (13.36). From (13.25), (13.27), and (13.37) the expression

$$\begin{aligned}\tilde{\boldsymbol{u}}^s &= \tilde{\boldsymbol{u}}^{LID} + \{B_1K_0\}^{-1}K_1^{-1}D_0\{K_1^{-1}D_0 + B_1K_0\}^{-1} \\ &\quad\times[\{c(t)I_p - \Lambda_1\}\boldsymbol{x}_1 + \boldsymbol{w}_1 - F(\boldsymbol{x}_1, \boldsymbol{x}_1^d)] \qquad (13.38)\end{aligned}$$

results, where $\tilde{\boldsymbol{u}}^{LID}(t) = K_0^{-1}\boldsymbol{u}^{LID}(t)$ and $\boldsymbol{u}^{LID}(t)$ is given by (13.25).

Substitution of (13.37) into (13.34) yields the slow-motion subsystem of the first p time functions $x_n(t)$:

$$\begin{aligned}\dot{\boldsymbol{x}}_1 &= F(\boldsymbol{x}_1, \boldsymbol{x}_1^d) + K_1^{-1}D_0\{K_1^{-1}D_0 + B_1K_0\}^{-1} \\ &\quad\times[\{c(t)I_p - \Lambda_1\}\boldsymbol{x}_1 + \boldsymbol{w}_1 - F(\boldsymbol{x}_1, \boldsymbol{x}_1^d)]. \qquad (13.39)\end{aligned}$$

This is similar to (8.44).

13.3 Degenerated motions

Consider the closed-loop system (13.31)–(13.32) on the condition that the steady state of the FMS (13.36) is maintained, that is

$$\begin{aligned}\dot{\boldsymbol{x}}_1 &= F(\boldsymbol{x}_1, \boldsymbol{x}_1^d) + K_1^{-1}D_0\{K_1^{-1}D_0 + B_1K_0\}^{-1} \\ &\quad\times[\{c(t)I_p - \Lambda_1\}\boldsymbol{x}_1 + \boldsymbol{w}_1 - F(\boldsymbol{x}_1, \boldsymbol{x}_1^d)],\ \boldsymbol{x}_1(0) = \boldsymbol{x}_1^0, \qquad (13.40) \\ \dot{\boldsymbol{x}}_2 &= \{c(t)I_\infty - \Lambda_2\}\boldsymbol{x}_2 + \phi(\boldsymbol{x}_1, \boldsymbol{x}_1^d, \boldsymbol{w}_1, \boldsymbol{w}_2),\ \boldsymbol{x}_2(0) = \boldsymbol{x}_2^0. \qquad (13.41)\end{aligned}$$

Let $\boldsymbol{x}_1^d(t) = \text{const}$, $\forall\ t \geq 0$ and $D_0 = 0$. Assuming in the closed-loop system (13.40)–(13.41) that transient processes of $\boldsymbol{x}_1(t)$ have been finished, then from the subsystem (13.40) we obtain the algebraic equations $\boldsymbol{x}_1(t) = \boldsymbol{x}_1^d(t) = \text{const}$ and in this case dynamical processes take place for the internal variables $\boldsymbol{x}_2$ only. This system of the algebraic and differential equations

$$\begin{aligned}\boldsymbol{x}_1(t) &= \boldsymbol{x}_1^d(t) = \text{const},\ \forall\ t \geq 0, \qquad (13.42) \\ \dot{\boldsymbol{x}}_2 &= \{c(t)I_\infty - \Lambda_2\}\boldsymbol{x}_2 + \phi(\boldsymbol{x}_1^d, \boldsymbol{w}_1, \boldsymbol{w}_2),\ \boldsymbol{x}_2(0) = \boldsymbol{x}_2^0 \qquad (13.43)\end{aligned}$$

is called the degenerated system.

Denote by $\boldsymbol{x}_2^{deg}$ a solution of (13.43). Then we have spatially distributed degenerated motions of the form

$$x(z,t) = x^{deg}(z,t) = \sum_{n=0}^{p-1} x_n^d \varphi_n(z) + \sum_{n=p}^{\infty} x_n^{deg}(t)\varphi_n(z). \tag{13.44}$$

These determine an asymptotic accuracy of realization of the desired distribution (13.4). The second item in (13.44) gives a bounded contribution to the sum if the transients of $\boldsymbol{x}_2(t)$ are stable. Therefore, the dimension of the control vector $\boldsymbol{u}(t)$ should be chosen in accordance with the condition

$$p \geq \text{int}(\sqrt{c_0}/\pi),$$

which guarantees degenerated system stability.

13.4 Estimation of modes

We have assumed that the vector $y(t) = \boldsymbol{x}_1(t)$ is measurable. In practice, the vector $\boldsymbol{x}_1(t)$ is usually estimated indirectly by means of a measurable finite-dimensional vector

$$\boldsymbol{y} = \{y_1, y_2, \ldots, y_m\}^T, \quad m \geq p,$$

where $y_j(t) = x(z_j, t)$. Then, in accordance with [Wang (1972)], instead of $\boldsymbol{x}_1(t)$ the estimate of $\boldsymbol{x}_1(t)$ given by

$$\tilde{\boldsymbol{x}}_1(t) = [\Phi^T\Phi]^{-1}\Phi^T\boldsymbol{y}(t) \tag{13.45}$$

can be used to implement the control law (13.26) where

$$\Phi = \begin{bmatrix} \varphi_0(z_1) & \varphi_1(z_1) & \cdots & \varphi_{p-1}(z_1) \\ \varphi_0(z_2) & \varphi_1(z_2) & \cdots & \varphi_{p-1}(z_2) \\ \vdots & \vdots & \vdots & \vdots \\ \varphi_0(z_m) & \varphi_1(z_m) & \cdots & \varphi_{p-1}(z_m) \end{bmatrix}.$$

The relationship between $\tilde{\boldsymbol{x}}_1(t)$ and $\boldsymbol{x}_1(t)$ is given by

$$\tilde{\boldsymbol{x}}_1(t) = \boldsymbol{x}_1(t) + \Delta\boldsymbol{x}_1(t),$$

where $\Delta\boldsymbol{x}_1(t)$ is the estimation error defined by

$$\Delta\boldsymbol{x}_1(t) = [\Phi^T\Phi]^{-1}\Phi^T\Delta\boldsymbol{y}(t)$$

and

$$\Delta \boldsymbol{y} = \{\Delta y_1, \Delta y_2, \ldots, \Delta y_m\}^T, \qquad \Delta \boldsymbol{y}_j(t) = \sum_{k=p}^{\infty} x_k(t)\varphi_k(z_j).$$

Note that in the closed-loop system the influence of $\Delta \boldsymbol{x}_1(t)$ is similar to that of sensor noise. In accordance with (13.45), we should choose z_j so that

$$\text{rank } \Phi = p.$$

13.5 Notes

Fourier analysis is widely used for analysis and design of control systems with distributed parameters based on the conventional methods of control system theory [Wang (1972); Ray (1981); Smagina *et al.* (2002)]. In this chapter the results presented in [Yurkevich (1992a); Yurkevich (1992a)] are extended in the context of control system design with the highest derivative in feedback for systems governed by partial differential equations based on the design procedures developed in the preceding chapters.

The main advantage of the approach is that the desired transients and control accuracy for the first p controlled modes are guaranteed despite unknown external disturbances and varying parameters of the system.

Finally, we should note that the approach may be used for nonlinear time-varying distributed parameter systems. In this case it is necessary to use Ritz-Galerkin type ideas.

13.6 Exercises

13.1 Prove that $\varphi_n(z)$ are the eigenfunctions of the system (13.1), where

$$[\partial x(z,t)/\partial z]|_{z=0} = 0, \qquad [\partial x(z,t)/\partial z]|_{z=1} = 0$$

are the boundary conditions.

13.2 Determine the eigenfunctions $\varphi_n(z)$ of the system (13.1) where

$$x(z,t)|_{z=0} = 0, \qquad x(z,t)|_{z=1} = 0$$

are the boundary conditions.

13.3 Consider a heating process described by the one-dimensional parabolic equation (13.1) with $|c(t)| \leq 25$. Determine the minimum number of modes to be controlled in order to guarantee closed-loop system stability.

13.4 Design the discrete-time controller for the nth mode of the system given by (13.1) based on the pseudo-continuous approach. Derive the relationship between sampling period T_s, phase margin of the FMS, and other controller parameters where: (a) $q_n = 1$; (b) $q_n = 2$.

13.5 Design the discrete-time controller for the time function $x_0(t)$ of the system given by (13.1) based on the pseudo-continuous approach to meet the following specifications: $\bar{\varepsilon}_{r0} = 0$; $t^d_{s0} \approx 3$ s; $\sigma^d_0 \approx 0\%$; $q_0 = 1$. Determine the sampling period T_s such that the phase margin of the FMS will meet the requirement $\varphi(\tau) \geq 0.2$ rad. Compare simulation results of the step output response of the closed-loop control system for the 0th mode with the assignment.

13.6 Design the discrete-time controller for the time function $x_1(t)$ of the system given by (13.1) based on the pseudo-continuous approach to meet the following specifications: $\bar{\varepsilon}_{r1} = 0$; $t^d_{s1} \approx 1$ s; $\sigma^d_1 \approx 0\%$; $q_1 = 1$. Determine the sampling period T_s such that the phase margin of the FMS will meet the requirement $\varphi(\tau) \geq 0.3$ rad. Compare simulation results of the step output response of the closed-loop control system for the 1st mode with the assignment.

13.7 Design the discrete-time controller (12.48) for the time function $x_n(t)$ of the system given by (13.1). Derive the expression of the velocity error due to disturbance $w_n(t)$ where: (a) $q_n = 1$; (b) $q_n = 2$. Check the result by computer simulation of the closed-loop system.

Appendix A

Proofs

A.1 Proof of expression (8.29)

From (8.28) we get

$$\begin{aligned} u^s &= [D_0 + KG]^{-1}K\{F - f\} \\ &= u^{NID} - u^{NID} + [D_0 + KG]^{-1}K\{F - f\}. \end{aligned}$$

By taking into account (8.18), we obtain

$$\begin{aligned} u^s &= u^{NID} + \{G^{-1} - [D_0 + KG]^{-1}K\}\{f - F\} \\ &= u^{NID} + \{[D_0 + KG]^{-1}[D_0 + KG]G^{-1} - [D_0 + KG]^{-1}K\}\{f - F\} \\ &= u^{NID} + [D_0 + KG]^{-1}\{[D_0 + KG]G^{-1} - K\}\{f - F\} \\ &= u^{NID} + [D_0 + KG]^{-1}D_0G^{-1}\{f - F\}. \end{aligned}$$

A.2 Proof of expression (8.42)

By taking $\boldsymbol{\mu} \rightarrow 0$ in (8.41), we obtain

$$\begin{aligned} \tilde{u}^s &\doteq [D_0 + K_1GK_0]^{-1}K_1\{F - f\} \\ &= \tilde{u}^{NID} - \tilde{u}^{NID} + [D_0 + K_1GK_0]^{-1}K_1\{F - f\}. \end{aligned}$$

In accordance with (8.43), we get

$$\begin{aligned}\tilde{u}^s &= \tilde{u}^{NID} + \{[GK_0]^{-1} - [D_0 + K_1GK_0]^{-1}K_1\}\{f - F\}\\ &= \tilde{u}^{NID} + \{[GK_0]^{-1} - [K_1K_1^{-1}D_0 + K_1GK_0]^{-1}K_1\}\{f - F\}\\ &= \tilde{u}^{NID} + \{[GK_0]^{-1} - [K_1^{-1}D_0 + GK_0]^{-1}\}\{f - F\}\\ &= \tilde{u}^{NID} + \{[GK_0]^{-1}[K_1^{-1}D_0 + GK_0][K_1^{-1}D_0 + GK_0]^{-1}\\ &\quad -[K_1^{-1}D_0 + GK_0]^{-1}\}\{f - F\}\\ &= \tilde{u}^{NID} + [GK_0]^{-1}K_1^{-1}D_0[K_1^{-1}D_0 + GK_0]^{-1}\{f - F\}.\end{aligned}$$

A.3 Proof of expression (8.65)

By taking into account (8.62), (8.63), and (8.64), we obtain

$$\begin{aligned}\mathbf{e}^F &= F(Y, R) - y_*\\ &= \mathcal{T}^{-1}e^s - 0.\end{aligned}$$

From (8.57) we get

$$\mathcal{T}^{-1}e^s = K_1^{-1}D_0[K_1^{-1}D_0 + G^*K_0]^{-1}\{\mathcal{T}^{-1}e^s - H_s^*\}.$$

The above expression may be rewritten in the following form:

$$\begin{aligned}&\{I - K_1^{-1}D_0[K_1^{-1}D_0 + G^*K_0]^{-1}\}\mathcal{T}^{-1}e^s\\ &= -K_1^{-1}D_0[K_1^{-1}D_0 + G^*K_0]^{-1}H_s^*.\end{aligned}$$

Hence,

$$\begin{aligned}&\{[K_1^{-1}D_0 + G^*K_0][K_1^{-1}D_0 + G^*K_0]^{-1}\\ &-K_1^{-1}D_0[K_1^{-1}D_0 + G^*K_0]^{-1}\}\mathcal{T}^{-1}e^s\\ &= -K_1^{-1}D_0[K_1^{-1}D_0 + G^*K_0]^{-1}H_s^*.\end{aligned}$$

Consequently,

$$G^*K_0[K_1^{-1}D_0 + G^*K_0]^{-1}\mathcal{T}^{-1}e^s = -K_1^{-1}D_0[K_1^{-1}D_0 + G^*K_0]^{-1}H_s^*.$$

Therefore,

$$e^s = -\mathcal{T}\,[K_1^{-1}D_0 + G^*K_0][G^*K_0]^{-1}K_1^{-1}D_0[K_1^{-1}D_0 + G^*K_0]^{-1}H_s^*.$$

A.4 Proof of expression (11.37)

(i) From (11.34) and (11.38), we get

$$\begin{aligned}\sum_{j=1}^{n} \tilde{\beta}_j &= \sum_{j=1}^{n} (d_j - \lambda_0 b_j) \\ &= \Delta d_s - \Delta\lambda_0 \sum_{j=1}^{n} (b_j^0 + \Delta b_j) - \lambda_0^0 \sum_{j=1}^{n} \Delta b_j \,. \end{aligned} \tag{A.1}$$

(ii) Denote

$$\chi = \Delta d_s \left[1 - \sum_{j=1}^{n} \tilde{\beta}_j \right]^{-1} . \tag{A.2}$$

From (11.27), (11.28) and (11.29), we get

$$\lambda_0 \left[\sum_{j=1}^{n} b_j \right] \left[1 - \sum_{j=1}^{n} \tilde{\beta}_j \right]^{-1} = 1 + \chi. \tag{A.3}$$

(iii) By substituting (11.22) and (11.36) into (11.32), we obtain

$$\begin{aligned} y_k &= \sum_{j=1}^{n} [a_j y_{k-j} + \hat{b}_j w_{k-j}] \\ &\quad + \lambda_0 \left[\sum_{j=1}^{n} b_j \right] \left[1 - \sum_{j=1}^{n} \tilde{\beta}_j \right]^{-1} \sum_{j=1}^{n} [(a_j^d - a_j) y_{k-j} + b_j^d r_{k-j} - \hat{b}_j w_{k-j}]. \end{aligned}$$

The above expression can be rewritten in the form:

$$\begin{aligned} y_k &= \sum_{j=1}^{n} [a_j y_{k-j} + \hat{b}_j w_{k-j}] + (1+\chi) \sum_{j=1}^{n} [(a_j^d - a_j) y_{k-j} + b_j^d r_{k-j} - \hat{b}_j w_{k-j}] \\ &= \sum_{j=1}^{n} a_j^d y_{k-j} + \sum_{j=1}^{n} b_j^d r_{k-j} + \chi \sum_{j=1}^{n} \{[a_j^d - a_j] y_{k-j} + b_j^d r_{k-j} - \hat{b}_j w_{k-j}\}. \end{aligned}$$

By (A.1) and (A.2) we have that the above expression is identical to (11.37).

A.5 Proof of expressions (11.40)–(11.41)

From (11.37) and (11.39), we obtain

$$\left[1 - \sum_{j=1}^{n} a_j^d\right] y^s = \sum_{j=1}^{n} b_j^d r + \chi \sum_{j=1}^{n} \{[a_j^d - a_j] y^s + b_j^d r - \hat{b}_j w^s\}.$$

By taking into account (11.42) and (A.2), the above expression can be rewritten in the form:

$$\phi_1 y^s = \left[1 + \Delta d_s - \sum_{j=1}^{n} \tilde{\beta}_j\right] \left[\sum_{j=1}^{n} b_j^d\right] r - \Delta d_s \left[\sum_{j=1}^{n} \hat{b}_j\right] w^s.$$

Hence,

$$\begin{aligned} e^s &= r - y^s \\ &= \left\{ 1 - \frac{\left[1 + \Delta d_s - \sum_{j=1}^{n} \tilde{\beta}_j\right] \sum_{j=1}^{n} b_j^d}{\phi_1} \right\} r - \frac{\Delta d_s \sum_{j=1}^{n} \hat{b}_j}{\phi_1} w^s. \end{aligned}$$

By taking into account (11.8) and (11.42), we get

$$e^s = -\Delta d_s \frac{1 - \sum_{j=1}^{n} a_j}{\phi_1} r - \Delta d_s \frac{\sum_{j=1}^{n} \hat{b}_j}{\phi_1} w^s.$$

A.6 Proof of expression (11.47)

(i) From (11.43) we obtain

$$1 - \sum_{j=1}^{n} \tilde{\beta}_j = \lambda_0 \sum_{j=1}^{n} b_j. \tag{A.4}$$

(ii) Let us consider a stationary behavior of control variable

$$u_k = u_0 + u^v T_s k, \quad k = 0, 1, \ldots \tag{A.5}$$

given that the conditions (11.43), (11.44), and (11.45) are satisfied. Hence, we have

$$\sum_{j=1}^{n} \tilde{\beta}_j u_{k-j} = u_0 \sum_{j=1}^{n} \tilde{\beta}_j + u^v T_s k \sum_{j=1}^{n} \tilde{\beta}_j - u^v T_s \sum_{j=1}^{n} j\tilde{\beta}_j, \tag{A.6}$$

$$\sum_{j=1}^{n} \hat{b}_j w_{k-j} = w^v T_s k \sum_{j=1}^{n} \hat{b}_j - w^v T_s \sum_{j=1}^{n} j\hat{b}_j. \tag{A.7}$$

From closed-loop system equations (11.32)–(11.33), by taking into account (11.48)–(11.49) and (A.6)–(A.7), we obtain

$$\begin{aligned}\left[1 - \sum_{j=1}^{n} a_j\right] y^v &= u_0 \sum_{j=1}^{n} b_j + u^v T_s k \sum_{j=1}^{n} b_j - u^v T_s \sum_{j=1}^{n} j b_j \\ &\quad + w^v T_s k \sum_{j=1}^{n} \hat{b}_j - w^v T_s \sum_{j=1}^{n} j\hat{b}_j,\end{aligned} \tag{A.8}$$

$$\begin{aligned}u_0 + u^v T_s k &= u_0 \sum_{j=1}^{n} \tilde{\beta}_j + u^v T_s k \sum_{j=1}^{n} \tilde{\beta}_j - u^v T_s \sum_{j=1}^{n} j\tilde{\beta}_j \\ &+ \lambda_0 \left\{ y^v \sum_{j=1}^{n} [a_j^d - a_j] + r \sum_{j=1}^{n} b_j^d - w^v T_s k \sum_{j=1}^{n} \hat{b}_j + w^v T_s \sum_{j=1}^{n} j\hat{b}_j \right\}.\end{aligned} \tag{A.9}$$

Since (A.8) and (A.9) are satisfied for all k, the equations (A.8) and (A.9) can be decomposed into

$$0 = u^v T_s k \sum_{j=1}^{n} b_j + w^v T_s k \sum_{j=1}^{n} \hat{b}_j, \tag{A.10}$$

$$u^v T_s k = u^v T_s k \sum_{j=1}^{n} \tilde{\beta}_j - \lambda_0 w^v T_s k \sum_{j=1}^{n} \hat{b}_j, \tag{A.11}$$

and

$$\left[1 - \sum_{j=1}^{n} a_j\right] y^v = u_0 \sum_{j=1}^{n} b_j - u^v T_s \sum_{j=1}^{n} j b_j - w^v T_s \sum_{j=1}^{n} j\hat{b}_j, \tag{A.12}$$

$$\begin{aligned}u_0 &= u_0 \sum_{j=1}^{n} \tilde{\beta}_j - u^v T_s \sum_{j=1}^{n} j\tilde{\beta}_j \\ &+ \lambda_0 \left\{ y^v \sum_{j=1}^{n} [a_j^d - a_j] + r \sum_{j=1}^{n} b_j^d + w^v T_s \sum_{j=1}^{n} j\hat{b}_j \right\}.\end{aligned} \tag{A.13}$$

From (A.10) we get

$$u^v = -\left[\sum_{j=1}^{n} \hat{b}_j\right]\left[\sum_{j=1}^{n} b_j\right]^{-1} w^v. \tag{A.14}$$

From (11.43) and (A.14) it follows that (A.11) holds for all k.
(iii) Let us derive u_0 from (A.13) and substitute u_0 into (A.12). As a result, by taking into account (11.8) and (A.4), we obtain

$$\left[1-\sum_{j=1}^{n} a_j^d\right] y^v - r\sum_{j=1}^{n} b_j^d = T_s\left[\sum_{j=1}^{n} \hat{b}_j\right]\left\{\frac{\sum_{j=1}^{n} j\tilde{\beta}_j}{1-\sum_{j=1}^{n}\tilde{\beta}_j} + \frac{\sum_{j=1}^{n} jb_j}{1-\sum_{j=1}^{n} b_j}\right\} w^v.$$

Finally, by (11.8) and (11.46), we get

$$e_w^v = -T_s\left[\sum_{j=1}^{n} \hat{b}_j\right]\left[1-\sum_{j=1}^{n} a_j^d\right]^{-1}\left\{\frac{\sum_{j=1}^{n} j\tilde{\beta}_j}{1-\sum_{j=1}^{n}\tilde{\beta}_j} + \frac{\sum_{j=1}^{n} jb_j}{1-\sum_{j=1}^{n} b_j}\right\} w^v.$$

A.7 Proof of expression (11.51)

(i) Let us consider a stationary behavior of output and control variables

$$y_k = y_0 + y^v T_s k, \quad k = 0, 1, \ldots, \tag{A.15}$$

$$u_k = u_0 + u^v T_s k, \quad k = 0, 1, \ldots, \tag{A.16}$$

given that the conditions (11.43), (11.48), and (11.49) are satisfied.

From (11.50), (A.15)–(A.16), and due to assumption that $e_r^v = \text{const}$, we obtain

$$y^v = r^v \quad \text{and} \quad e_r^v = -y_0. \tag{A.17}$$

(ii) From closed-loop system equations (11.32)–(11.33), by taking into account (A.15)–(A.16), (A.17), and (11.39), we obtain

$$[y_0 + r^v T_s k]\left[1 - \sum_{j=1}^{n} a_j\right] = -r^v T_s \sum_{j=1}^{n} j a_j + [u_0 + u^v T_s k]\sum_{j=1}^{n} b_j$$
$$-u^v T_s \sum_{j=1}^{n} j b_j + w^s \sum_{j=1}^{n} \hat{b}_j, \qquad (A.18)$$

$$[u_0 + u^v T_s k]\left[1 - \sum_{j=1}^{n} \tilde{\beta}_j\right] = -u^v T_s \sum_{j=1}^{n} j\tilde{\beta}_j$$
$$+\lambda_0 \left\{ [y_0 + r^v T_s k]\sum_{j=1}^{n}[a_j^d - a_j] - r^v T_s \sum_{j=1}^{n} j[a_j^d - a_j]\right.$$
$$\left. + r^v T_s k \sum_{j=1}^{n} b_j^d - r^v T_s \sum_{j=1}^{n} j b_j^d - w^s \sum_{j=1}^{n} \hat{b}_j \right\}. \qquad (A.19)$$

Since (A.18) and (A.19) are satisfied for all k, the equations (A.18) and (A.19) can be decomposed into

$$r^v T_s k \left[1 - \sum_{j=1}^{n} a_j\right] = u^v T_s k \sum_{j=1}^{n} b_j, \qquad (A.20)$$

$$u^v T_s k \left[1 - \sum_{j=1}^{n} \tilde{\beta}_j\right] = \lambda_0 \left\{ r^v T_s k \sum_{j=1}^{n}[a_j^d - a_j] + r^v T_s k \sum_{j=1}^{n} b_j^d \right\}, \qquad (A.21)$$

and

$$y_0 \left[1 - \sum_{j=1}^{n} a_j\right] = -r^v T_s \sum_{j=1}^{n} j a_j + u_0 \sum_{j=1}^{n} b_j - u^v T_s \sum_{j=1}^{n} j b_j + w^s \sum_{j=1}^{n} \hat{b}_j, \qquad (A.22)$$

$$u_0 \left[1 - \sum_{j=1}^{n} \tilde{\beta}_j\right] = -u^v T_s \sum_{j=1}^{n} j\tilde{\beta}_j + \lambda_0 \left\{ y_0 \sum_{j=1}^{n}[a_j^d - a_j]\right.$$
$$\left. - r^v T_s \sum_{j=1}^{n} j[a_j^d - a_j] - r^v T_s \sum_{j=1}^{n} j b_j^d - w^s \sum_{j=1}^{n} \hat{b}_j \right\}. \qquad (A.23)$$

From (A.20) we get

$$u^v = \left[1 - \sum_{j=1}^{n} a_j\right]\left[\sum_{j=1}^{n} b_j\right]^{-1} r^v. \qquad (A.24)$$

From (11.43) and (A.24) it follows that (A.21) holds for all k.
(iii) Let us derive u_0 from (A.23). By substituting u_0 into (A.22) and by taking into account (11.8) and (A.4), we obtain

$$y_0\left[1-\sum_{j=1}^{n} a_j^d\right]=-r^v T_s\left\{\sum_{j=1}^{n} j[a_j^d+b_j^d]+\left[1-\sum_{j=1}^{n} a_j\right]\left[\frac{\sum_{j=1}^{n} j\tilde{\beta}_j}{1-\sum_{j=1}^{n}\tilde{\beta}_j}+\frac{\sum_{j=1}^{n} jb_j}{1-\sum_{j=1}^{n} b_j}\right]\right\}.$$

Finally, since $y_0=-e_r^v$, we have

$$e_r^v=T_s\left[1-\sum_{j=1}^{n} a_j^d\right]^{-1}\left\{\sum_{j=1}^{n} j[a_j^d+b_j^d]+\left[1-\sum_{j=1}^{n} a_j\right]\left[\frac{\sum_{j=1}^{n} j\tilde{\beta}_j}{1-\sum_{j=1}^{n}\tilde{\beta}_j}+\frac{\sum_{j=1}^{n} jb_j}{1-\sum_{j=1}^{n} b_j}\right]\right\}r^v.$$

A.8 Proof of expression (12.56)

From (11.38), (12.52), (12.5), (12.54)–(12.55), and by taking into account that $g=$ const, we obtain

$$\sum_{j=1}^{n}\hat{b}_j=T_s^n\sum_{j=1}^{n}\frac{\epsilon_{n,j}}{n!}=T_s^n, \tag{A.25}$$

$$\sum_{j=1}^{n}b_j=T_s^n g\sum_{j=1}^{n}\frac{\epsilon_{n,j}}{n!}=T_s^n g, \tag{A.26}$$

$$\sum_{j=1}^{n}\tilde{\beta}_j=\sum_{j=1}^{n}\Delta d_j=\Delta d_s. \tag{A.27}$$

By substituting (A.25)–(A.27) into (11.41), we obtain

$$e_w^s=T_s^n\Delta d_s\left\{\left[1-\sum_{j=1}^{n} a_j^d\right][1-\Delta d_s]-\Delta d_s\sum_{j=1}^{n}(a_j^d-a_{n,j})\right\}^{-1} w^s. \tag{A.28}$$

Relation (12.52) is valid given that $T_s\to 0$; hence, from (A.28) we obtain

$$\lim_{T_s\to 0}\frac{e_w^s}{T_s^n}=\Delta d_s\left\{\left[1-\sum_{j=1}^{n} a_j^d\right][1-\Delta d_s]-\Delta d_s\sum_{j=1}^{n}(a_j^d-a_{n,j})\right\}^{-1} w^s.$$

A.9 Proof of expression (12.57)

(i) Assume that

$$\Delta d_s = 0.$$

From (A.26) and (A.27), we get

$$\phi_2 = \sum_{j=1}^{n} j \Delta d_j + \sum_{j=1}^{n} j \frac{\epsilon_{n,j}}{n!}. \tag{A.29}$$

From (A.25), (A.29), and (11.47), we obtain

$$e_w^v = -T_s^{n+1} \left[1 - \sum_{j=1}^{n} a_j^d \right]^{-1} \left[\sum_{j=1}^{n} j \Delta d_j + \sum_{j=1}^{n} j \frac{\epsilon_{n,j}}{n!} \right] w^v. \tag{A.30}$$

Relation (12.52) is valid given that $T_s \to 0$; hence, from (A.30) we obtain

$$\lim_{T_s \to 0} \frac{e_w^s}{T_s^{n+1}} = - \left[1 - \sum_{j=1}^{n} a_j^d \right]^{-1} \left[\sum_{j=1}^{n} j \Delta d_j + \sum_{j=1}^{n} j \frac{\epsilon_{n,j}}{n!} \right] w^v. \tag{A.31}$$

(ii) From (12.6), we get

$$\begin{aligned} \frac{d}{dz} \left[z^n - \frac{\mathcal{E}_n(z)}{n!} \right] \Bigg|_{z=1} &= \left[n z^{n-1} - \frac{1}{n!} \frac{d\mathcal{E}_n(z)}{dz} \right] \Bigg|_{z=1} \\ &= \frac{n+1}{2}. \end{aligned} \tag{A.32}$$

(iii)

$$\frac{d}{dz}\left[z^n - \frac{\mathcal{E}_n(z)}{n!}\right]\bigg|_{z=1} = \{nz^{n-1} - [n!]^{-1}\left[\epsilon_{n,1}(n-1)z^{n-1}\right.$$

$$\left.+\epsilon_{n,2}(n-2)z^{n-2} + \cdots + \epsilon_{n,n-1}(n-(n-1))z^{n-n}\right]\}\big|_{z=1}$$

$$= n - \frac{1}{n!}\left[\epsilon_{n,1}(n-1) + \epsilon_{n,2}(n-2) + \cdots + \epsilon_{n,n-1}(n-(n-1))\right]$$

$$= n - \frac{n}{n!}\left[1 - 1 + \sum_{j=1}^{n-1}\epsilon_{n,j}\right] + \frac{1}{n!}\left[n - n + \sum_{j=1}^{n-1} j\epsilon_{n,j}\right]$$

$$= n - \frac{n}{n!}\left[-1 + \sum_{j=1}^{n}\epsilon_{n,j}\right] + \frac{1}{n!}\left[-n + \sum_{j=1}^{n} j\epsilon_{n,j}\right]$$

$$= n - \frac{n}{n!}\left[-1 + n!\right] + \frac{1}{n!}\left[-n + \sum_{j=1}^{n} j\epsilon_{n,j}\right]$$

$$= \frac{1}{n!}\sum_{j=1}^{n} j\epsilon_{n,j}. \tag{A.33}$$

(iv) Finally, from (A.31), (A.32) and (A.33), we get (12.57).

Appendix B

Notation system

$A_{FMS}(z)$	characteristic polynomial of discrete-time FMS
$D_{FMS}(s)$	characteristic polynomial of continuous-time FMS
$\deg A(s)$	degree of polynomial $A(s)$
$\det B$	determinant of matrix B
$\dim M$	dimension of manifold M
e	error of the reference input realization where $e = r - y$
e^F	error of the desired dynamics realization
e^s	steady-state error of the reference input realization
e_w^v	velocity error due to ramp disturbance $w(t)$
$\overline{e}_w^v$	relative velocity error due to ramp disturbance $w(t)$
e_r^v	velocity error in the presence of ramp input $r(t)$
$\overline{e}_r^v$	relative velocity error due to ramp reference input $r(t)$
I_m	identity matrix, where $I_m \in \mathbb{R}^{m \times m}$
$\mathrm{Im}\ s_i$	imaginary part of s_i
k_0	high gain
$\lg(\omega)$	base 10 logarithm of the elements of ω, i.e., $\lg(\omega) = \log_{10}(\omega)$
$L(\omega)$	Bode amplitude diagram
n	degree of differential equation of plant model
n_s	high-frequency sensor noise
q	degree of differential or difference equation of controller
p	operator of differentiation, $p = d(\cdot)/dt$
r	reference input
$\mathbb{R}^m$	vector space of the mth dimension
$\mathrm{Re}\, s_i$	real part of s_i
$\mathrm{Re}\, \lambda_i(A)$	real part of eigenvalue λ_i of A
t	time
t_p	peak time of the output variable $y(t)$

t_s	settling time of the output variable $y(t)$
T	time constant
T_s	sampling period
u	input (control variable)
w	external disturbance or varying parameter
$G_{yr}(s)$	transfer function, $G_{yr}(s) = y(s)/r(s)$
X	state vector of the plant model
X^T	transposition of the vector X
x^s	steady state of $x(t)$
y	output variable of the system
y^s	final value of the output variable $y(t)$, i.e., $y^s = \lim_{t\to\infty} y(t)$
$y(t_p)$	peak value of the output variable $y(t)$
y_k	output of discrete-time system where $y_k = y(t)\|_{t=kT_0}$
α	relative degree (invertibility index)
$\overline{\varepsilon}_F$	relative error of desired dynamics realization
$\overline{\varepsilon}_r$	relative error of reference input realization
$\nabla_u V$	partial derivative of V with respect to u, i.e., $\nabla_u V = \partial V/\partial u$
η	degree of time-scale separation between fast and slow modes
μ	small parameter
ζ_{FMS}	damping ratio of FMS
σ	percentage maximum overshoot of step response
τ	time delay or time constant
φ_i	phase margin of ith FMS on Nyquist diagram
ω	angular frequency
ω_c	crossover frequency
ω^d	damped frequency
Ω_u	bounded set of allowable values of control variable
Ω_X	bounded subset of vector space of $\mathbb{R}^n$
$1(t)$	unit step function
$\|x\|_2$	Euclidean norm of an element $x = \{x_1, x_2, \ldots, x_n,\}^T$
$r(t) \stackrel{t}{\equiv} 0$	$r(t)$ identically equals zero for all $t \in [0, \infty)$
FMS	fast-motion subsystem
LTI	linear time-invariant system
SMS	slow-motion subsystem
ZOH	zero-order hold
■	end of the proof

Bibliography

Åström, K.J., Hagander, P. and Sternby, J. (1984). Zeros of sampled systems, *Automatica*, **20**, pp. 31–38.

Åström, K.J. and Wittenmark, B. (1997). *Computer-controlled systems: theory and design*, 3rd ed., Prentice-Hall information and system sciences series, Upper Saddle River, N.J. : Prentice Hall, 557 p.

Atassi, A.N. and Khalil, H. (1999). A separation principle for the stabilization of a class of nonlinear systems, *IEEE Trans. Automat. Contr.*, **AC-44**, 9, pp. 1672–1687.

Atherton, D.P. (1981). *Stability of nonlinear systems*, (Series: Electronic and electrical engineering research studies. Control theory and applications studies series. 1), Chichester, New York : Research Studies Press, 231 p.

Batenko, A. P. (1977). *Final state control of driven plants*, Moscow: Sov. radio, 256 p. (in Russian)

Barbot, J.P., Monaco, S. and Normand-Cyrot, D. (1999). Discrete-time approximated linearization of SISO systems under output feedback, *IEEE Trans. Automat. Contr.*, **AC-44**, 9, pp. 1729–1733.

Barbot, J.P. Pantalos, N., Monaco, S. and Normand-Cyrot, D. (1994). On the control of regular ϵ perturbed nonlinear systems, *Int. J. of Control*, **59**, 5, pp. 1255–1279.

Bartolini, G., Ferrara, A., Usai, E. and Utkin, V.I. (2000). On multi-input chattering-free second-order sliding mode control, *IEEE Trans. Automat. Contr.*, **AC-45**, 9, pp. 1711–1717.

Błachuta M.J. (1997). On zeros of sampled systems. *Proc. of American Control Conf.* Albuquerque. NM, pp. 3205–3210.

Błachuta, M.J. (1999a). On zeros of pulse transfer functions, *IEEE Trans. Automat. Contr.*, **AC-44**, 6, pp. 1229–1234.

Błachuta, M.J. (1999b). *Contributions to the theory of discrete-time control for continuous-time systems*, Scientific Bulletins of the Silesian Technical University, s. Automatyka, Vol. 127, 190 p.

Błachuta, M.J., Yurkevich, V.D. and Wojciechowski, K. (1992). Aircraft control based on localization method, *Preprints of IFAC Workshop on Motion Control for Intelligent Automation*, Perugia, Italy, **2**, pp. 19-24.

Błachuta, M.J., Yurkevich, V.D. and Wojciechowski, K. (1995). Design of aircraft 3D motion control using dynamic contraction method, *Proc. of IFAC-Workshop "Motion Control"*, Munich, pp. 323–330.

Błachuta, M.J., Yurkevich, V.D. and Wojciechowski, K. (1997). Design of analog and digital aircraft flight controllers based on dynamic contraction method, *Proc. of the 1997 AIAA Guidance, Navigation and Control Conf.*, New Orleans, Louisiana, part 3, pp. 1719–1729.

Błachuta, M.J., Yurkevich, V.D. and Wojciechowski, K. (1999). Robust quasi NID aircraft 3D flight control under sensor noise, *Int. J. Kybernetika*, **35**, 5, pp. 637–650.

Błachuta M.J., Shatalov B.V., Yurkevich V.D., and Wojciechowski K. (2000). Design of digital controller for aircraft longitudinal motion, *Preprints of the 11th IFAC Workshop on Control Applications of Optimization*, St.Petersburg, Russia, **1**, pp. 76–81.

Bosgra, O.K. and Kwakernaak, H. (2000). *Design methods for control systems*, Course for the Dutch Graduate Network oı. Systems and Control, Thech. Un. of Delfa and Un. of Twente, Winter Term, 2000-2001.

Boychuk, L. M. (1966). An inverse method of the structural synthesis of automatic control nonlinear systems, *Automation*, Kiev: Naukova Dumka, No.6, pp. 7–10. (in Russian).

Brockett, R. W. (1965). Poles, zeros, and feedback: state space interpretation, *IEEE Trans. Automat. Contr.*, **AC-10**, 2, pp. 129–135.

Brockett, R. W. and Mesarovic, M.D. (1965). The reproducibility of multivariable systems, *J. of Math. Analysis and Appl.*, **11**, pp. 548–563.

Brogan, W.L. (1991). *Modern control theory*, 3rd ed., Prentice Hall, 653 p.

Brunovský, P. (1970). A classification of linear controllable systems, *Kybernetika*, 3. pp. 173–188.

Bullinger, E. and Allgöwer, F. (1997). An adaptive high-gain observer for nonlinear systems, *Proc. 36th IEEE Conf. on Decision and Control*, San Diego, California, pp. 4348–4353.

Byrnes, C. I., Castro, R. and Isidori, A. (1988). Linear model matching with prescribed tracking error and internal stability for nonlinear systems, *Proc. of 8th Int. Conf. on Analysis and Optimization of Systems,* France, pp. 249–258.

Byrnes, C. I. and Isidori, A. (1991). Asymptotic stabilization of minimum phase nonlinear systems, *IEEE Trans. on Automat. Contr.*, **AC-36**, 10, pp. 1122–1137.

Canudas de Wit, C. and Slotine, J.-J. E. (1991). Sliding observers in robot manipulators, *Automatica*, **27**, 5, pp. 859–864.

Chen, Chi-Tsong. (1993). *Analog and digital control system design: transfer-function, state-space, and algebraic methods*, Fort Worth, Toronto : Saunders College Pub., 600 p.

Dabroom, A.M., and Khalil, H.K. (2001). Output feedback sampled-data control of nonlinear systems using high-gain observers, *IEEE Trans. on Automat. Contr.*, **AC-46**, 11, pp. 1712–1725.

Davis, J.H. and Hirschorn, R.M. (1988). Tracking control of a flexible robot link,

IEEE Trans. Automat. Contr., **AC-33**, 3, pp. 238–248.

Davison, E.J. (1976). The robust decentralized control of a general servomechanism problem, *IEEE Trans. Automat. Contr.*, **AC-21**, 1, pp. 14–24.

DeCarlo, R.A., Zak, S.H. and Matthews, G.P. (1988). *Variable structure control of nonlinear multivariable systems: a tutotial*, *Proc. of the IEEE*, **76**, 3, pp. 212–232.

Djemai, M., Barbot, J.P. and Khalil, H.K. (1999). Digital multirate control for a class of nonlinear singularly perturbed systems, *Int. J. Control,* **72,** 10, pp. 851–865.

Dorato, P. (1969). On the inverse of linear dynamical systems, *IEEE Trans. Systems Science and Cybernetics*, **SSC-5**, 1, pp. 43–48.

Dorf, R.C. and Bishop, R.H. (2001). *Modern control systems*, 9th ed., Upper Saddle River, NJ: Prentice Hall, 831 p.

Drazenovic, B. (1969). The invariance conditions in variable structure systems, *Automatica*, **5**, 3, pp. 287–295.

El'sgol'tz, L.E. and Norkin, S.B. (1973). *Introduction to the theory and application of differential equations with deviating arguments*, Series Mathematics in science and engineering. Vol. 105, New York: Academic Press, 357 p.

Emelyanov, S.V. (1963). On peculiarities of variable structure control systems with discontinuous switching functions, *Dokl. Akad. Nauk SSSR,* **153**, pp. 776–778.

Emelyanov S.V., Korovin, S.K. and Levant, A. (1993). Higher-order sliding modes in control systems, *Differential equations*, **29**, 11, pp. 1627–1647.

Emre, E. and Hüseyin, Ö. (1974). Invertibility criteria for linear multivariable systems, *IEEE Trans. Automat. Contr.*, **AC-19**, 5, pp. 609–610.

Escobar, G., Ortega, R., Sira-Ramirez, H., Vilain, J.-P. and Zein, I. (1999). An experimental comparison of several nonlinear controllers for power converters, *Control Systems Magazine, IEEE*, **19**, 1, pp. 66–82.

Esfandiary, F., and Khalil, H. (1992). Output feedback stabilization of fully linearizable systems. *Int. J. Control,* **56,** pp. 1007–1037.

Fehrmann, R., Mutschkin, W. and Neumann, R. (1989). Entwurf und praktische Erprobung eines digitalen *PL*-Reglers, *Messen Steuern Regeln*, Berlin: VEB Verlag Technik, **32**, 2, pp. 70–72. (in German).

Filippov, A.F. (1964). *Differential equations with discontinuous right-hand side*, American Mathematical Society Translations, American Mathematical Society, Providence, RI, Series 2, Vol. 2, pp. 199–231.

Filippov, A.F. (1988). *Differential equations with discontinuous right- hand side*, Kluwer Academic Publishers, Dordrecht, The Netherland.

Fomin V.N., Fradkov, A.L., and Yakoubovich, V.A. (1981). *Adaptive control of dynamical systems*, Moscow: Nauka, 448 p. (in Russian).

Fradkov, A.L., Miroshnik, I.V. and Nikiforov, V.O. (1999). *Nonlinear and adaptive control of complex systems*, (Series: Mathematics and its applications, Vol. 491), Kluwer Academic Publishers, 510 p.

Franklin, G.F., Powell, J.D. and Emami-Naeini A. (2002). *Feedback control of dynamic systems*, 4th ed., Prentice Hall, 910 p.

Francis, B.A. (1977). The linear multivariable regulator problem, *SIAM J. Contr.*

and Optim., **15**, pp. 486–505.

Fridman, L. and Levant, A. (1996). Higher order sliding modes as the natural phenomena of control theory, *In: Robust Control Variable Structure and Lyapunov Techniques*, Garafalo F. and Glielmo G. (Eds.), (Lecture Notes in Control and Information Science, Springer Verlag, London, Vol. 217), pp. 107–133.

Fridman, L.M., Shustin, E.I. and Fridman, E.M. (1996). Steady modes and sliding modes in the relay control systems with time delay, *Proc. of 35th IEEE Conf. on Decision and Control*, **4**, pp. 4601–4606.

Gauthier, J.P., Hammouri, H. and Kupka, I. (1991). Observers for nonlinear systems, *Proc. of 30th IEEE Conf. on Decision and Control*, **2**, pp. 1483–1489.

Gerashchenko E. I. and Gerashchenko S.M. (1975). *A method of motion separation and optimization of non-linear systems*, Moscow: Nauka, 296 p. (in Russian)

Gouaisbaut, F., Perruquetti, W. and Richard, J.P. (1999). A sliding mode control for linear systems with input and state delays, *Proc. of 38th IEEE Conf. on Decision and Control*, **4**, pp. 4234–4239.

Graham, D. and Lathrop, R.C. (1953). The synthesis of optimum response: Criteria and standard forms, *Trans. of the AIEE,* **72**, Part 2 - Applications and Industry, pp. 273–288.

Hammouri, H. and Marchand, N. (2000). High gain observer for a class of implicit systems, *Proc. of 39th IEEE Conf. on Decision and Control*, **1**, pp. 804–808.

Han, J.D., Wang, Y.C., Tan, D.L. and Xu, W.L. (2000). Acceleration feedback control for direct-drive motor system, *Proc. of IEEE/RSJ Int. Conf. on Intelligent Robots and Systems, (IROS 2000)*, **2**, pp. 1068–1074.

Hauser, J., Sastry, S. and Meyer, G. (1992). Nonlinear control design for slightly non-minimum phase systems: Application to V/STOL aircraft, *Automatica*, **28**, pp. 665–679.

Hernandez, J. and Barbot, J.P. (1996). Sliding observer-based feedback control for flexible joints manipulator, *Automatica*, **32**, 9, pp. 1243–1254.

Hirschorn, R.M. (1979a). Invertible of multivariable nonlinear control systems, *IEEE Trans. Automat. Contr.*, **AC-24**, 6. pp. 855–865.

Hirschorn, R.M. (1979b). Invertibility of nonlinear control systems, *SIAM J. Contr. and Optim.*, **17**, 2. pp. 289–297.

Hirschorn, R.M. and Aranda-Bricaire E. (1998). Global approximate output tracking for nonlinear systems, *IEEE Trans. Automat. Contr.*, **AC-43**, 10. pp. 1389–1398.

Hoppensteadt, F.C. (1966). Singular perturbations on the infinite time interval, *Trans. of the American Mathematical Society*, **123**, pp. 521–535.

Hori, Y. (1988). Novel servo system design method based on the acceleration control, *Proc. of 14th Annual Conf. (IECON '88)*, Industrial Electronics Society, **2**, pp. 447–452.

Huang J. and Rugh W.J. (1990). On a nonlinear multivariable servomechanism problem, *Automatica*, **26**, pp. 963–972.

Ioannou, P.A. and Kokotović, P.V. (1983). *Adaptive systems with reduced models,*

(Lecture Notes in Control and Information Sciences, Vol. 47), New-York: Springer-Verlag, 162 p.

Isidori, A. (1995). *Nonlinear control systems*, 3rd ed., New York: Springer-Verlag, 549 p.

Isidori, A. and Byrnes, C.I. (1990). Output regulation of nonlinear systems, *IEEE Trans. Automat. Contr.*, **AC-35**, 2, pp. 131–140.

Isidori, A., Sastry, S.S., Kokotović, P.V. and Byrnes, C.I. (1992). Singularly perturbed zero dynamics of nonlinear systems, *IEEE Trans. Automat. Contr.*, **AC-43**, 10, pp. 1625–1631.

Itkis, U. (1976). Control systems of variable structure, Toronto; New York: Wiley, 214 p.

Jury, E.I. (1958). *Sampled data control systems*, New York: Wiley, 453 p.

Khalil, H.K. (1987). Output feedback control of linear two-time-scale systems, *IEEE Trans. Automat. Contr.*, **AC-32**, 9, pp. 784–792.

Khalil, H.K. (1994). Robust servomechanism output feedback controllers for feedback linearizable systems, *Automatica*, **30**, 10, pp. 1587–1599.

Khalil, H.K. (2000). Universal integral controllers for minimum-phase nonlinear systems, *IEEE Trans. Automat. Contr.*, **AC-45**, 3, pp. 490–494.

Khalil, H.K. (2002). *Nonlinear Systems*, 3rd ed., Upper Saddle River, N.J. : Prentice Hall, 750 p.

Kharitonov, V.L. and Torres Munoz, J.A. (2002). Recent results on the robust stability of multivariate polynomials, *IEEE Trans. on Circuits and Systems I: Fundamental Theory and Applications*, **49**, 6, pp. 715–724.

Klimushchev, A.I. and Krasovskii, N.N. (1962). Uniform asymtotic stability of systems of differential equations with a small parameter in the derivative terms, *J. Appl. Math. Mech.*, **25**, pp. 1011–1025.

Kokotović, P.V. (1984). Applications of singular perturbation techniques to control problems, *SIAM Review*, **26**, 4, pp. 501–550.

Kokotović, P.V. and Khalil, H.K. (1986). *Singular perturbations in systems and control*, IEEE Press, 494 p.

Kokotović, P.V., Khalil, H.K. and O'Reilly, J. (1986). *Singular perturbation methods in control: analysis and design*, Academic Press, 371 p.

Kokotović, P.V., O'Malley, R.E. and Sannuti, P. (1976). *Singular perturbations and order reduction in control – an overview*, *Automatica*, **12**, pp. 123–132.

Kolmanovskii, V.B. and Nosov, V.R. (1986). *Stability of functional differential equations*, (Series: Mathematics in science and engineering, Vol. 180), London, Orlando : Academic Press, 217 p.

Kotnik, P.T., Yurkovich, S. and Özgüner, Ü. (1988). Acceleration feedback control for a flexible manipulator arm, *Proc. of IEEE Int. Conf. on Robotics and Automation*, **1**, pp. 322–323.

Kotta, Ü. (1995). *Inversion method in the discrete-time nonlinear control systems synthesis problem*, (Lecture Notes in Control and Information Sciences, Vol. 205), New-York: Springer-Verlag, 152 p.

Krasovskii N.N. (1963). *Stability of motion: applications of Lyapunov's second method to differential systems and equations with delay*, Stanford, Calif., Stanford University Press, 188 p.

Krutko, P.D. (1988). The principle of acceleration control in automated system design problems, *Sov. J. Comput. Syst. Sci.*, (USA), **26**, 4, pp. 47–57.

Krutko, P.D. (1991). Optimization of control systems with respect to local functionals characterizing the energy of motion, *Sov. Phys.-Dokl.*, (USA), **36**, 9, pp. 623–625.

Krutko, P.D. (1995). Optimization of multidimensional dynamic systems using the criterion of minimum acceleration energy, *Sov. J. Comput. Syst. Sci.*, (USA), **33**, 4, pp. 27–42.

Krutko, P.D. (1996). Control of motion in Eulerian systems: synthesis of algorithms using the method of inverse problems of dynamics, *Sov. J. Comput. Syst. Sci.*, (USA), **34**, 2, pp. 1–21.

Kuo, B.C. and Golnaraghi, F. (2003). *Automatic control systems*, 8th ed., New York: John Wiley & Sons, Inc., 609 p.

Landau, I.D. (1994). Robust digital control of systems with time delay (the Smith predictor revisited), *Proc. 33th IEEE Conf. on Decision and Control*, **1**, pp. 865–869.

Levant, A. (1998). Robust exact real-time differentiation, *Automatica*, **34**, 3, pp. 379–384.

Levy, S. and Sivan, R. (1966). On the stability of a zero-output systems, *IEEE Trans. Automat. Contr.*, **AC-11**, 2, pp. 315–316.

Lindorff, D.P. (1965). *Theory of sampled-data control systems*, New York: John Wiley & Sons, Inc., 305 p.

Litkouhi, B. and Khalil, H. (1985). Multirate and composite control of two-time-scale discrete-time systems, *IEEE Trans. Automat. Contr.*, **AC-30**, 7, pp. 645–651.

Luenberger, D.G. (1967). Canonical forms for linear multivariable systems, *IEEE Trans. Automat. Contr.*, **AC-12**, 3, pp. 290–293.

Lun, J.Y.S., Walker, M.W. and Paul, R.P.C. (1980). Resolved acceleration control of mechanical manipulator, *IEEE Trans. Automat. Contr.*, **AC-25**, 3, pp. 468–474.

Luo, G. and Saridis, G. (1985). *L-Q* design of PID controllers for robot arms, *IEEE Journal of Robotics and Automation*, **RA-1**, 3, pp. 152–159.

Mahmoud, N.A. and Khalil, H. (1996). Asymptotic regulation of minimum-phase nonlinear systems using output feedback, *IEEE Trans. Automat. Contr.*, **AC-41**, 10, pp. 1402–1412.

Marino R. (1985). High-gain feedback in non-linear systems, *Int. J. of Control*, **42**, 6. pp. 1369–1385.

Marconi, L. (1998). Asymptotic tracking and transient improvement in output regulation of nonlinear uncertain systems, *Proc. 37th IEEE Conf. on Decision and Control*, Tampa, Florida, USA, pp. 3099–3104.

Meerov, M. V. (1965). *Structural synthesis of high-accuracy automatic control systems*, (Pergamon Press international series of monographs on automation and automatic control, Vol. 6), Oxford, New York : Pergamon Press, 341 p.

Mees, A.I. and Bergen, A. (1975). Describing functions revisited, *IEEE Trans. Automat. Contr.*, **AC-20**, 4, pp. 473–478.

Mickens, R.E. (1981). *An introduction to nonlinear oscillations*, New York : Cambridge University Press, 224 p.

Mutschkin W.S. (1988). Direkte digitale regelung nach dem prinzip der lokalisation, *Messen Steuern Regeln,* (Berlin: VEB Verlag Technik), **31**, 7, pp. 315–317. (in German).

Naidu D.S. (1988). *Singular perturbation methodology in control systems*, IEE control engineering series, Vol. 34, London, 287 p.

Naidu, D.S. and Rao, A.K. (1985). *Singular perturbation analysis of discrete control systems*, Berlin, New York: Springer-Verlag, 195 p.

Narendra, K.S. and Oleng, N.O. (2002). Exact output tracking in decentralized adaptive control systems, *IEEE Trans. Automat. Contr.*, **AC-47**, 2, pp. 390–395.

Neimark, Ju.I. (1947). *D*-subdivision and spaces of quasi-polynomials, *Prikl. Mat. Meh.*, **13**, 4, pp. 349–380.

Nijmeijer, H. and Schaft, A.J. van der. (1990). *Nonlinear dynamical control systems,* Springer-Verlag, 467 p.

Nyquist, H. (1932). Regeneration theory, *Bell Systems Tech. J.*, Jan., pp. 126–147.

Ogata, K. (1994). *Discrete-time control systems*, 2nd ed., Upper Saddle River, NJ: Prentice Hall, 745 p.

O'Reilly, J. (1983). *Observers for linear systems*, (Series: Mathematics in Science and Engineering. Vol. 170), London: Academic Press, 246 p.

Palmor, Z.J. (1996). Time-delay compensation: Smith predictor and its modifications, *In: The control handbook, Ed. by W.S. Levine.*, Boca Raton, FL: CRC Press, pp. 224–237.

Paltov, I.P. (1975). *Transients performance and compensator design for nonlinear control systems*, (Uniform title: Kachestvo protsessov i sintez korrektiruyushchikh ustroystv v nelineynykh avtomaticheskikh sistemakh), Nauka, Moscow, 368 p. (in Russian).

Patel, R.V. and Toda, M. (1980). Quantitative measures of robustness for multivariable systems, *Joint Automatic Control Conference*, TP8-A, 1980.

Perruquetti, W. and Barbot, J.P (Eds.) (2002). *Sliding mode control in engineering*, Marcel Dekker, New York, 409 p.

Petrov, B.N. and Krutko, P.D. (1980). Inverse problems of the dynamics of controlled systems, Nonlinear models, *Eng. Cybern.*, (USA), **18**, 5, pp. 102–108.

Piazzi, A. and Marro, G. (1996). Robust stability using interval analysis, *Int. J. of Systems Science*, **27**, 12, pp. 1381–1390.

Popov, E.P. and Krutko, P.D. (1979). Symmetry in automatic control systems, *Sov. Phys.-Dokl.*, (USA), **24**, 2, pp. 91–92.

Porter, W.A. (1970). Diagonalization and inverses for nonlinear systems, *Int. J. of Control*, **11**, 1, pp. 67–76.

Ray, W. H. (1981). *Advanced Process Control*, McGraw-Hill chemical engineering series, New York : McGraw-Hill, 376 p.

Readman, M. and Belanger, P.R. (1991). Acceleration feedback control for flexible joint robots. *Proc. of 30-th IEEE Conf. on Decision and Control*, pp. 1385–1390.

Riedle, B.D. and Kokotović, P.V. (1985). Stability analysis of adaptive systems with unmodelled dynamics, *Int. J. of Control*, **41**, 2. pp. 389–402.

Rosenbrock, H.H. (1970). *State-space and multivariable theory*, London: Nelson, 257 p.

Rosenbrock, H.H. (1973). The zeros of a systems, *Int. J. of Control*, **18**, 2. pp. 297–299.

Sain, M.K. and Massey, J.L. (1969). Invertibility of linear time-invariant dynamical systems, *IEEE Trans. Automat. Contr.*, **AC-14**, 2, pp. 141–149.

Saksena, V., O'Reilly, J. and Kokotović, P.V. (1984). Singular perturbations and time-scale methods in control theory: survey 1976-1983, *Automatica*, **20**, 3, pp. 273–293.

Sanders, S.R. (1993). On limit cycles and the describing function method in periodically switched circuits, *IEEE Trans. Circuits and Systems I: Fundamental Theory and Applications*, **40**, 9, pp. 564–572.

Santilli, R.M., (1978). *The inverse problem in Newtonian mechanics*, New York : Springer-Verlag, 266 p.

Sastry, S.S. (1999). *Nonlinear systems: analysis, stability and control*, (Series: Interdisciplinary applied mathematics: Vol. 10), Springer, 667 p.

Sastry, S.S. Hauser, J. and Kokotović, P.V. (1989). Zero dynamics of regularly perturbed systems may be singularly perturbed, *Systems and Control Letters*, **14**, pp. 299–314.

Schmidt P.B. and Lorenz R.D. (1990). Design principles and implementation of acceleration feedback to improve performance of DC drives, *Conf. Record of IEEE IAS Annual Mtg.*, pp. 422–427.

Schneider, D.M. (1988). Control of processes with time delay, *IEEE Trans. on Industry Applications*, **24**, 2, pp. 186–191.

Siljak, D.D. (1991). *Decentralized control of complex systems*, (Series: Mathematics in science and engineering, Vol. 184), Boston : Academic Press, 527 p.

Silverman, L.M. (1969). Inversion of multivariable linear systems, *IEEE Trans. Automat. Contr.*, **AC-14**, 3, pp. 270–276.

Singh, S. N. (1980). Decoupling of invertible non-linear systems with state feedback and precompensation, *IEEE Trans. Automat. Contr.*, **AC-25**, 6, pp. 1237–1239.

Sira-Ramirez, H. (1989). Nonlinear variable structure systems in sliding mode: the general case, *IEEE Trans. Automat. Contr.*, **AC-34**, 11, pp. 1186–1188.

Sira-Ramirez, H. (2002). Sliding modes, passivity, and flatness, *In: Sliding Mode Control in Engineering*, Perruquetti, W. and Barbot, J.P (Eds.), New York: Marcel Dekker, pp. 163–189.

Slotine, J.-J. E. and Li, W. (1991). *Applied nonlinear control*, Prentice Hall, 461 p.

Smagina, E.M., Nekhamkina, O. and Sheintuch, M. (2002). Stabilization of fronts in a reaction-diffusion system: Application of Gershgorin theorem, *Ind. Eng. Chem. Res.*, **41**, 8, pp. 2023–2032.

Smith, O.J.M. (1957). Close control of loops with dead time, *Chemical Engineering Progress*, **34**, 5, pp. 217–219.

Sobolev, S.L. (1977). On the roots of Euler polynomials, *Soviet Math. Dokl.*, 1977, **18**, pp. 935–938.

Spong, M.W. and Vidyasagar, M. (1989). *Robot dynamics and control*, New York: Wiley, 336 p.

Studenny, J. and Belanger, P.R. (1984). Robot manipulator control by acceleration feedback, *Proc. of 23-th IEEE Conf. on Decision and Control*, pp. 1070-1072.

Studenny, J. and Belanger, P.R. (1986). Robot manipulator control by acceleration feedback: stability, design and performance issues. *Proc. of 25-th IEEE Conf. on Decision and Control*, pp. 80–85.

Studenny, J., Belanger, P.R. and Daneshmend L.K. (1991). A digital implementation of the acceleration feedback control law on a PUMA 560 manipulator. *Proc. of 30-th IEEE Conf. on Decision and Control*, pp. 2639–2648.

Sussmann, H.J. and Kokotović, P.V. (1991). The peaking phenomenon and the global stabilization of nonlinear systems, *IEEE Trans. Automat. Contr.*, **AC-36**, 4, pp. 424–440.

Suvorov, A. (1991). Synthesis of control systems based on the method of localization of given fast oscillating process, *Proc. of the Int. Workshop on Control System Synthesis: Theory and Application*, (27 May - 1 June, 1991), Novosibirsk, USSR, pp. 93–99.

Tang, Y., Tomizuka, M., Guerrero, G. and Montemayor, G. (2000). Decentralized robust control of mechanical systems, *IEEE Trans. Automat. Contr.*, **AC-45**, 4, pp. 771–776.

Taylor, J. H. (2000). *Describing Functions*, In Supplement 1 of the Electrical and Electronics Engineering Encyclopedia, New York: John Wiley & Sons, Inc., pp. 77–98.

Teel, A.R., Nešić, D. and Kokotović, P.V. (1998). A note on input-to-state stability of sampled-data nonlinear systems. *Proc. of 37th IEEE Conf. on Decision and Control*, Tampa, Florida, USA, pp. 2473–2478.

Teel, A. and Praly, L. (1995). Tools for simiglobal stabilization by partial state and output feedback. *SIAM J. Control Optim.*, **33**, pp. 1443–1488.

Tikhonov, A.N. (1948). On the dependence of the solutions of differential equations on a small parameter, *Mathematical Sb.*, Moscow, **22**, pp. 193–204 (in Russian).

Tikhonov, A.N. (1952). Systems of differential equations containing a small parameter multiplying the derivative, *Mathematical Sb.*, Moscow, **31**, 3, pp. 575–586 (in Russian).

Tornambè, A. (1992). High-gain observers for nonlinear systems, *Int. J. Systems Science*, **23**, 6, pp. 1475–1489.

Tsypkin, Ya.Z. (1964). *Sampling systems theory and its application*, (Int. series of monographs on electronics and instrumentation, Vol. 22, Vol. 23), New York: Macmillan, 742 p.

Tsypkin, Ya.Z. (1971). *Adaptation and Learning in Automatic Systems*, (Series: Mathematics in science and engineering, Vol. 73), New York: Academic Press, 291 p.

Tsypkin, Y. Z. and Polyak, B. T. (1991). Frequency domain criteria for L_p-robust stability of continuous linear systems, *IEEE Trans. Automat. Contr.*, **AC-36**, 12, pp. 1464–1469.

Tudoroiu, N., Yurkevich, V. and Khorasani, K. (2003a). Discrete-time control system design for a reactive ion etching (RIE) system, *Proc. of Int. Conf.*

"Physics and Control", 20-22 Aug., 2003, St.-Petersburg, Russia.

Tudoroiu, N., Khorasani, K. and Yurkevich, V. (2003b). Robust two-time scale control system design for reactive ion etching system, *Proc. of European Control Conf. (ECC 2003)*, 1-4 September 2003, University of Cambridge, Cambridge, UK.

Tustin, A. (1947). A method of analyzing the behaviour of linear systems in terms of time series, *JIEE*, London, **94**, part IIA, pp. 130–142.

Utkin, V.I. (1977). Variable structure systems with sliding modes, *IEEE Trans. Automat. Contr.*, **AC-22**, 2, pp. 212–222.

Utkin, V.I. (1992). *Sliding Modes in Control and Optimization*, Springer-Verlag, 286 p.

Utkin, V.I. (1994). Sliding mode control in discrete-time and difference systems, In: *Variable Structure and Lyapunov Control, Ed. by A.S.I Zinober*, Springer-Verlag.

Utkin, V.I. and Vostrikov, A.S. (1978). Control systems with decoupling motions, *Prepr. of 7th IFAC World Congress*, Finland, **2**, pp. 967–973.

Vasileva, A.B. (1963). Asymptotic behavior of solutions to certain problems involving nonlinear differential equations containing a small parameter multiplying the highest derivatives, *Russian Math. Surveys*, **18**, 3, pp. 13–81.

Vostrikov, A.S. (1977a). On the synthesis of control units of dynamic systems, *Systems Science.* Wrocław: Techn. Univ., **3**, 2, pp. 195–205.

Vostrikov, A.S. (1977b). *Optimal and adaptive systems*, Novosib. Electr. Instit., Novosibirsk, 65 p. (in Russian).

Vostrikov, A.S. (1979). *Control of dynamic systems*, Novosib. Electr. Instit., Novosibirsk, 112 p. (in Russian).

Vostrikov, A.S. (1988). Localisation principle in problem of synthesising automatic control systems, *Izv. Vyssh. Uchebn. Zaved. Priborostr.*, **31**, 2, pp. 42–49 (in Russian).

Vostrikov, A.S. (1990). *Synthesis of nonlinear systems by means of localisation method*, Novosibirsk State University, Novosibirsk, 120 p. (in Russian).

Vostrikov, A.S. and Sarycheva, O.M. (1982). Synthesis of control systems with the derivative of a Lyapunov function in the control law, *Sov. Autom. Control*, (USA), **14**, 4, pp. 34–40.

Vostrikov, A.S., Utkin, V.I. and Frantsuzova, G.A. (1982). Systems with state vector derivative in the control, *Automation & Remote Control*, (USA), **43**, 3, part 1, pp. 283–286.

Vostrikov, A.S. and Yurkevich, V.D. (1991). Decoupling of multi-channel nonlinear time-varying systems by derivative feedback, *Systems Science*, Wroclaw, **17**, 4, pp. 21–33.

Vostrikov, A.S. and Yurkevich, V.D. (1993a). Design of multi-channel systems with the velocity vector in the control law, *Automation & Remote Control*, (USA), **54**, 2, part 1, pp. 214–226.

Vostrikov, A.S. and Yurkevich, V.D. (1993b). Design of control systems by means of localisation method, *Prepr. of 12-th IFAC World Congress*, **8**, pp. 47–50.

Vukić, Z., Kuljača, L., Donlagić, D. and Tešnjak, S. (2003). *Nonlinear control systems*, New York : Marcel Dekker, Inc., 394 p.

Wang, P.K.C. (1972). Modal feedback stabilization of a linear distributed systems, *IEEE Trans. Automat. Contr.*, **AC-17**, 4, pp. 552–555.

Wang, S.H. and Davison, E.J. (1973). A new invertibility criterion for linear multibariable systems, *IEEE Trans. Automat. Contr.*, **AC-18**, 5, pp. 538–539.

Weller, S.R., Moran, W., Ninness, B. and Pollington, A.D. (1997). Sampling zeros and Euler-Frobenius polynomials, *Proc. of 36th IEEE Conf. on Decision and Control*, San Diego, CA, pp. 1471–1476.

Wolovich, W.A. (1994). *Automatic control systems: Basic analysis and design*, Saunders College Publishing, Harcourt Brace College Publishers, 450 p.

Young, K.D., Kokotović, P.V. and Utkin, V.I. (1977). A singular pertubation analysis of high-gain feedback systems, *IEEE Trans. Automat. Contr.*, **AC-22**, 3, pp. 931–938.

Young, K.D. and Özgüner, Ü. (1999). *Variable structure systems, sliding mode, and nonlinear control*, (Series: Lecture notes in control and information science, Vol. 247), London, New York: Springer, 386 p.

Young, K.D., Utkin, V.I. and Özgüner, Ü. (1999). A control engineer's guide to sliding mode control, *IEEE Trans. Contr. Syst. Tech.*, **7**, 3, pp. 328–342.

Yurkevich, V.D. (1992a). Control of distributed parameter systems: localisation method, *IEE Proc., part-D*, **139**, 2, pp. 141–146.

Yurkevich, V.D. (1992b). Control of parabolic systems on the basis of the localisation method, *IEE Proc., part-D*, **139**, 6, pp. 511–516.

Yurkevich, V.D. (1993a). Sampled control system synthesis by means of dynamic contraction method, *Prepr. of 2nd Int. IFAC Workshop on Nonsmooth and Discontinuous Problems of Control and Optimization*, Chelyabinsk, Russia, pp. 157–158.

Yurkevich, V.D. (1993b). Control of uncertain systems: dynamic compaction method, *Proc. of 9th Int. Conf. on Systems Engineering*, University of Nevada, Las Vegas, pp. 636–640.

Yurkevich, V.D. (1994). Design of aircraft longitudinal motion control using dynamic compaction method, *Proc. of 3rd Int. Workshop on Advanced Motion Control*, University at Berkeley, pp. 1029–1038.

Yurkevich, V.D. (1995a). A new approach to design of control systems under uncertainty: dynamic contraction method, *Prepr. of 3rd IFAC Symposium on Nonlinear Control Systems Design*, Tahoe City, California, **2**, pp. 443–448.

Yurkevich, V.D. (1995b). Decoupling of uncertain continuous systems: dynamic contraction method, *Proc. of 34th IEEE Conf. on Decision & Control*, New Orleans, Louisiana, **1**, pp. 196–201.

Yurkevich, V.D. (1996). Synthesis of discrete control systems by the method of dynamic contraction, *J. Comput. Syst. Sci. Int.*, (USA), **34**, 2, pp. 155–165.

Yurkevich, V.D. (1997). Recent results of dynamic contraction method (survey), *Proc. of IFAC Workshop on Singular Solutions and Perturbations in Control Systems*, Pereslavl-Zalessky, Russia, pp. 133–138.

Yurkevich, V.D. (1999a). Robust two-time-scale discrete-time system design, *Prepr. 14th IFAC World Congress*, Beijing, China, **G**, pp. 343–348.

Yurkevich, V.D. (1999b). Design of robust sampled-data control systems for MIMO uncertain nonlinear time-varying plants, *Proc. of 6th St.-Petersburg Symposium on Adaptive Systems Theory*, St.-Petersburg, Russia, **1**, pp. 206–209.

Yurkevich, V.D. (2000a). *Design of two-time-scale nonlinear time-varying control systems*, (Series: Analysis and Design of Nonlinear Systems, Ed. by G.A.Leonov and A.L.Fradkov), St.-Petersburg: Nauka, 287 p. (in Russian).

Yurkevich, V.D. (2000b). Universal digital controller for MIMO continuous plants with uncertainty, *UKACC Int. Conf. Control-2000*, Cambridge, UK.

Yurkevich, V.D. (2002). Design of digital MIMO control systems with two-time-scale motions, *Prepr. 15th IFAC World Congress*, Barcelona, Spain.

Yurkevich, V.D., Błachuta, M.J. and Wojciechowski, K. (1991). Stabilization system for the aircraft longitudinal motion, (Uniform title: Synteza ukladu stabilizacji podluxnego ruchu samolotu), *Archiwum Automatyki i Robotyki*, Warzawa, **36**, 3-4, pp. 517–535. (in Polish).

Yurkevich, V.D., Błachuta, M.J. and Wojciechowski, K. (1997). Design of digital controllers for MIMO non-linear time-varying systems based on dynamic contraction method, *Proc. of European Control Conf. (ECC'97)*, Brussels, Belgium.

Yurkevich, V.D., Błachuta, M.J. and Wojciechowski, K. (1998). Robust digital controller design for MIMO non-linear time-varying systems, *Prepr. 4th IFAC Symp. Nonlinear Control Systems Design (NOLCOS'98)*, Enschede, The Netherlands, **3**, pp. 854–859.

Yurkevich, V.D., Błachuta, M.J. and Wojciechowski, K. (2000). Design of the universal digital flight controller for an aircraft with not exactly known parameters, *Proc. of 3rd IFAC Symposium on Robust Control Design (ROCOND-2000)*, Prague, Czech Republic.

Yurkevich V.D. and Shatalov B.V. (2001). Tracking digital controller for robot manipulator, *Preprints of 5-th IFAC Symposium on Nonlinear Control Systems (NOLCOS'01)*, St.Petersburg, Russia, **2**, pp. 526–531.

Zadeh, L.A. and Desoer, C.A. (1963). *Linear system theory: the state space approach*, New York: McGraw-Hill, 628 p.

Zhong-Ping Jiang, Repperger, D.W. and Hill, D.J. (2001). Decentralized nonlinear output-feedback stabilization with disturbance attenuation, *IEEE Trans. Automat. Contr.*, **AC-46**, 10, pp. 1623–1629.

Index